AF361755

Light Weight Alloys

Light Weight Alloys

Processing, Properties and Their Applications

Special Issue Editors

Maurizio Vedani
Riccardo Casati

MDPI • Basel • Beijing • Wuhan • Barcelona • Belgrade • Manchester • Tokyo • Cluj • Tianjin

Special Issue Editors

Maurizio Vedani
Politecnico di Milano
Italy

Riccardo Casati
Politecnico di Milano
Italy

Editorial Office
MDPI
St. Alban-Anlage 66
4052 Basel, Switzerland

This is a reprint of articles from the Special Issue published online in the open access journal *Metals* (ISSN 2075-4701) (available at: https://www.mdpi.com/journal/metals/special_issues/light_weight_alloys).

For citation purposes, cite each article independently as indicated on the article page online and as indicated below:

LastName, A.A.; LastName, B.B.; LastName, C.C. Article Title. *Journal Name* **Year**, *Article Number*, Page Range.

ISBN 978-3-03928-919-6 (Pbk)
ISBN 978-3-03928-920-2 (PDF)

Cover image courtesy of Maurizio Vedani.

Contents

About the Special Issue Editors

Maurizio Vedani is Full Professor of Metallurgy at Politecnico di Milano and Chief of the Material Section of the Department of Mechanical Engineering. His main research focus is on the microstructure and mechanical behavior of metals and metallic alloys during manufacturing, with special attention to light metals such as aluminium alloys, metal matrix composites, titanium, and magnesium alloys. He has authored or co-authored more than 230 scientific papers. In recent years, he has specifically researched ultrafine grained metal alloys produced by various severe plastic deformation methods, including degradable magnesium and zinc alloys for biomedical applications and novel alloys for additive manufacturing.

Riccardo Casati is Associate Professor of Metallurgy at the Department of Mechanical Engineering of Politecnico di Milano. His research deals with the evolution of microstructure and mechanical behavior of metallic materials during manufacturing and service. In recent years, his scientific activity has mainly focused on the design of alloys and composites produced by powder metallurgy and additive manufacturing processes. He is co-author of more than hundred scientific papers published in scientific journals and conference proceedings.

Preface to "Light Weight Alloys"

There is growing interest in light metallic alloys for a wide number of applications owing to their processing efficiency and long service life. The reduction of greenhouse gas emissions, sustainability, and recyclability are central topics in discussions of environmental issues and climate change. Reducing the overall weight of transport systems on land, and in the sea and sky, represents one of the first and most important challenges to addressing these goals. It is expected that, driven by increasingly strict environmental legislation, the broad concept of "lightweight design" will gain increased attention in the near future by the transportation sector and for the mass production of goods in general.

Lightweight design means "the art and science of making things—parts, products or structures—as light as possible, within constraints". Constraints are set by the expected functions, shape and size limitations, loading history, and service environment. As a result, lightweighting implies the optimization of shapes in terms of putting material volumes only where they are needed, the accurate design of structures to prevent failures with full awareness on safety margins, and the use of light, high-performance materials.

The topic of light materials occupies the lion's share in this perspective and light metal alloys based on aluminum, magnesium, and titanium are important classes of engineering materials for this purpose. Aluminum, the most popular of the above products, has been widely applied to the transportation, construction, and packaging industries, with a global annual production of more than 63 million tons of primary aluminum in 2017. Magnesium alloys have receiving renewed and increasing interest in last decades for new applications in the automotive, aerospace, and ICT sector (computers, communication, and consumer electronic products). Titanium alloys are mostly used in the strategic aerospace industry, with important applications in the chemical and biomedical fields.

A deep knowledge of the physical metallurgy and the processability of theses light alloys is considered of great importance in order to fully exploit their potential and make the production of light structural parts more efficient.

The collection of papers published here addresses a wide range of topics including aluminum, magnesium, and titanium alloys, and the common characterizing of the performance of the alloys after their manufacturing by conventional and innovative processing routes.

The editors would like to acknowledge the valuable contributions of all the authors who made the publication of this issue on light alloy metallurgy possible.

Maurizio Vedani, Riccardo Casati
Special Issue Editors

Article

Influence of Temperature-Dependent Properties of Aluminum Alloy on Evolution of Plastic Strain and Residual Stress during Quenching Process

Yuxun Zhang [1,2], Youping Yi [1,2], Shiquan Huang [1,2,*] and Hailin He [1]

[1] State Key Laboratory of High Performance Complex Manufacturing, Central South University, Changsha 410083, China; zhangyuxun198@163.com (Y.Z.); yyp@csu.edu.cn (Y.Y.); hehailin0621@126.com (H.H.)

[2] School of Mechanical and Electrical Engineering, Central South University, Changsha 410083, China

* Correspondence: huangsqcsu@sina.com; Tel.: +86-135-4898-1584

Received: 31 March 2017; Accepted: 15 June 2017; Published: 21 June 2017

Abstract: To lessen quenching residual stresses in aluminum alloy components, theory analysis, quenching experiments, and numerical simulation were applied to investigate the influence of temperature-dependent material properties on the evolution of plastic strain and stress in the forged 2A14 aluminum alloy components during quenching process. The results show that the thermal expansion coefficients, yield strengths, and elastic moduli played key roles in determining the magnitude of plastic strains. To produce a certain plastic strain, the temperature difference increased with decreasing temperature. It means that the cooling rates at high temperatures play an important role in determining residual stresses. Only reducing the cooling rate at low temperatures does not reduce residual stresses. An optimized quenching process can minimize the residual stresses and guarantee superior mechanical properties. In the quenching process, the cooling rates were low at temperatures above 450 °C and were high at temperatures below 400 °C.

Keywords: aluminum alloy; quenching process; material property; cooling rate; plastic strain; residual stress

---

1. Introduction

Heat-treatable aluminum alloys are widely used to fabricate forged components used in aerospace and aircraft industry for weight reduction. Solution quenching and aging treatments are applied to the aluminum alloy components to obtain high mechanical properties [1]. For this purpose, fast cooling rates are required to avoid or limit precipitation during the quenching process [2]. However, high cooling rates result in serious inhomogeneous deformations and lead to high residual stresses [3], which deteriorates the mechanical properties and dimensional stability [4,5], and also have important impact on fatigue properties [6–8]. Therefore, it is important to study how to control the residual stresses.

Residual stresses can be relieved by plastic deformation. For example, Koç et al. [9] found that compression and stretching processes could reduce the residual stress of 7050 forged blocks by more than 90%. However, this technique cannot be used for complicated cross-section components [10]. Many researchers employed vibration methods to release the residual stresses [11,12] and studied their mechanisms [13]. However, this technique is confined to large components since the high amplitude of vibration deforms the thinner or smaller components [5]. Heat treatment methods are also applied to relieve the residual stresses. Dong et al. [10] successfully lowered the residual stress with uphill quenching and thermal-cold cycling processes. Sun et al. [14] relieved the residual stress by repeated heating of the samples at high heating rates and subsequent artificial aging treatment.

Although the technologies described above are effective in relieving the residual stresses, the need of special tools and procedures increases their costs. Therefore, an economical approach is to minimize the residual stresses during quenching. Residual stresses decrease with reducing cooling rates. Our previous work showed that, as compared to the mechanical properties, the residual stress decreased faster with decreasing cooling rates [15]. Dong et al. [10] balanced the residual stress and mechanical properties of the thin aluminum alloy plates with warm water at 80 °C. The mechanical performances are mainly determined by the cooling rates in the quenching sensitivity temperature range. The quenching sensitivity can be studied by time-temperature- transformation/properties (TTT/TTP) curves. For example, by using of TTP curves, Li et al. [16] showed that the mechanical performances of the 6063 aluminum alloy are determined by the cooling rates during the quenching sensitivity temperature range from 410–300 °C. Based on the results, they proposed a step quenching method to balance the mechanical properties and residual stresses. Its cooling rates are high in the quenching sensitivity temperature range to guarantee mechanical performances, and they are low in the others ranges to reduce residual stresses. Our previous work [15] showed that such a step quenching method could balance the residual stress and mechanical performances. However, the step quenching technology is mainly designed based on the characteristics of the quench-induced phase transformation of the material. It is best to utilize the characteristics of quenching residual stress evolution and quench-induced phase transformation during quenching to design the quenching technology. Nallathambi et al. [17] studied the influence of thermal, metallurgical, and mechanical properties on the final distortion and residual stresses during quenching. However, the effect of material properties on the evolution of residual stress at different temperatures is still unclear. This information is needed to guide the designing of quenching technology.

This work investigated the effect of temperature-dependent material properties of forged 2A14 aluminum alloy on the evolution of stress and plastic strain during quenching process using a constructed model and numerical simulation methods. Moreover, the effect of cooling rates on residual stresses was studied by quenching the samples with different quenching technologies. Table 1 listed all the nomenclatures used in this paper.

Table 1. Nomenclature.

T	Temperature
ΔT	temperature difference between the temperature of the unit and reference temperature 0 °C
T_g	temperature difference between the two units of the model
α	thermal expansion coefficient
E	elastic modulus
E_p	plastic modulus
$\sigma_{0.2}$	yield strength
i	number of the unit
ε^T	thermal expansion strain
ε	total strain
ε^e	elastic strain
ε^p	plastic strain
σ	thermal stress
PEEQ	equivalent plastic strain
PE22	plastic strain in y direction
S22	stress in y direction

2. Methods

2.1. Theoretical Analysis

Residual stresses are determined by inhomogeneous plastic strains. They usually increase with increasing magnitude of plastic strain. During quenching, inhomogeneous temperature distribution causes inhomogeneous thermal expansion strain (ε^T), which causes the thermal stresses and strains to maintain the force and shape balance. When thermal stress exceeds the yield strength, plastic strain occurs. Material properties change with temperatures. The yield strengths and elastic moduli of 2A14 aluminum alloy decrease with increasing temperatures. This implies that the characteristics of evolution of thermal stresses and plastic strains may be different at different temperatures during quenching. In this work, a model with two units was proposed to investigate these characteristics (Figure 1). As shown in Figure 1a, the temperatures of the two units are initially the same, leading to the same thermal expansions. The thermal expansion was obtained using Equation (1), where 0 °C was taken as the reference temperature. As quenching proceeded, a temperature difference (T_g) between the two units appeared, as shown in Figure 1b; it resulted in strains to balance the unequal thermal expansions. To simplify the analysis, we presumed that the final heights of the two units are the same (Equation (2)). Moreover, the two units have the same magnitude of thermal stresses. The strain comprises elastic strain and plastic strain, which are described by Equation (4) in reference [18]. In this model, we presumed that the thermal stress always exceeds the yield strength (Equation (3)). Combining Equations (1)–(4), the plastic strain of unit i is obtained from Equation (5). Equations (6)–(8) proposed three variants, A, B, and C, to simplify the formation of Equation (5). Equation (9) was obtained by combining Equations (5)–(8), and was used to investigate the influence of temperature-dependent material properties on plastic strains at different temperatures, but with the same temperature difference. As shown in Table 2, the plastic moduli were obtained from a previous work [19]. The other material properties are described in Section 2.3.

$$\varepsilon^T = \alpha \Delta T \tag{1}$$

$$\alpha_i T_i - \varepsilon_i = \alpha_{i+1} T_{i+1} + \varepsilon_{i+1}; \ \sigma_i = \sigma_{i+1} \tag{2}$$

$$\varepsilon_i = \varepsilon_i^e + \varepsilon_i^p \tag{3}$$

$$\varepsilon_i^e = \frac{\sigma_i}{E_i}; \ \varepsilon_i^p = \frac{(\sigma_i - \sigma_{0.2,i})}{E_{p,i}} \tag{4}$$

$$\varepsilon_i^p = \frac{(\alpha_i T_i - \alpha_{i+1} T_{i+1}) + \left(\dfrac{\sigma_{0.2,i+1}}{E_{p,i+1}} + \dfrac{\sigma_{0.2,i}}{E_{p,i}} \right)}{\left(\dfrac{1}{E_{p,i+1}} + \dfrac{1}{E_{p,i}} + \dfrac{1}{E_{i+1}} + \dfrac{1}{E_i} \right) E_{p,i}} - \frac{\sigma_{0.2,i}}{E_{p,i}} \tag{5}$$

$$A = (\alpha_i T_i - \alpha_{i+1} T_{i+1}) \tag{6}$$

$$B = \left(\frac{\sigma_{0.2,i+1}}{E_{p,i+1}} + \frac{\sigma_{0.2,i}}{E_{p,i}} \right) \tag{7}$$

$$C = \left(\frac{1}{E_{p,i+1}} + \frac{1}{E_{p,i}} + \frac{1}{E_{i+1}} + \frac{1}{E_i} \right) \tag{8}$$

$$\varepsilon_i^p = \frac{A + B}{C E_{p,i}} - \frac{\sigma_{0.2,i}}{E_{p,i}} \tag{9}$$

where α, E, E_p, and $\sigma_{0.2}$ are the thermal expansion coefficient, elastic modulus, plastic modulus, and yield strength, respectively. T is the temperature of the unit. ΔT is the temperature difference between T and reference 0 °C. ε^T, ε, ε^e, ε^p, and σ are the thermal expansion, total strain, elastic strain, plastic strain, and thermal stress, respectively. The subscript i is the number of the unit.

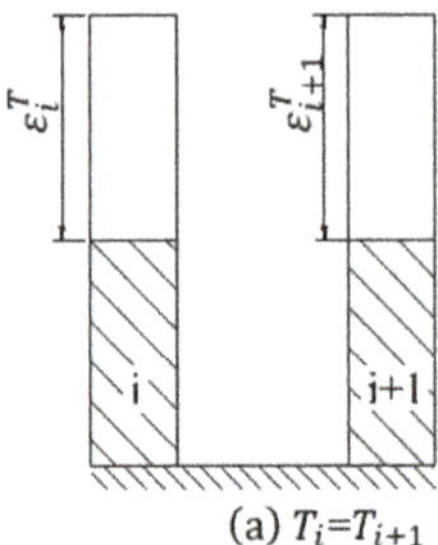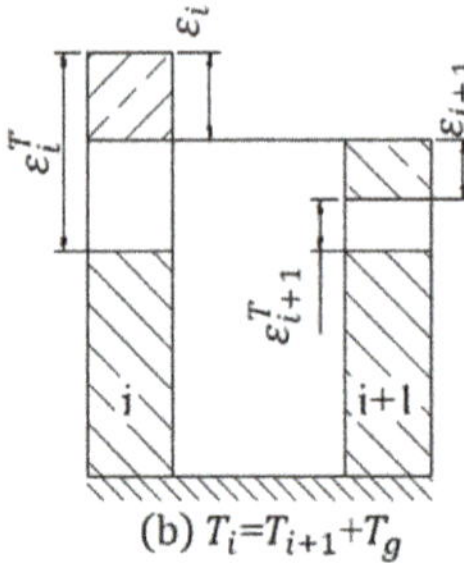

(a) $T_i=T_{i+1}$ (b) $T_i=T_{i+1}+T_g$

Figure 1. Two-unit model: (**a**) uniform temperatures at the beginning of quenching; and (**b**) temperature difference appeared during quenching.

Table 2. Plastic moduli at different temperatures.

Temperature (°C)	30	130	230	330	430	530
E_p (MPa)	9000	5000	2000	1000	400	200

2.2. Quenching Experiment: Material and Heat Treatment

Quenching residual stresses are affected by products' material properties, structure and inhomogeneous temperature distribution during quenching. The temperature distribution is determined by the cooling rates; increasing cooling rate increases the temperature gradient. The influence of cooling rates on residual stress was investigated by quenching the forged 2A14 aluminum alloy samples with different quenching technologies. The samples were machined from a commercial forged component 78 mm in thickness; their chemical compositions are listed in Table 3. As shown in Figure 2a, the size of the samples is 110 mm × 100 mm × 70 mm, and its thickness direction is along the *z*-axis. Their *x*, *y*, and *z* directions are along the long transverse direction, short transverse direction, and thickness direction of the component, respectively. Similar to our previous work [15], the two end surfaces (110 mm × 100 mm) were taken as the main heat-dispersing surfaces, while the other four surfaces were encapsulated with about 20-mm-thick asbestos. As shown in Table 4, the samples were quenched immediately after the solution treatment at 500 °C for 4 h. Sample A1 was quenched with room temperature water (about 20 °C). Sample A2 was first quenched in room temperature water until the temperature at the central point P0 decreased to about 410 °C, and then it was cooled in room temperature air. The transfer time was smaller than 1 s. Sample A3 was quenched by a step quenching technology using a spray-quenching device designed by us, as shown in Figure 3. Detailed information of the device is present in our submitted patent (CN 201710016344.5). As shown in Figure 2a, the temperatures at points P0 and P1 during quenching were monitored by using Φ1 mm naked type K thermocouples deeply embedded into the samples. The cooling history of P0 was used to stand for the cooling rates of the samples with different quenching technologies. The temperature difference between points P0 and P1 was used to estimate the inhomogeneous temperature distribution of the samples.

Table 3. Chemical composition of the studied material (wt %).

Cu	Mg	Mn	Si	Fe	Ni	Zn	Ti	Al
4.28	0.6	0.81	0.94	0.15	0.003	0.01	0.04	Balance

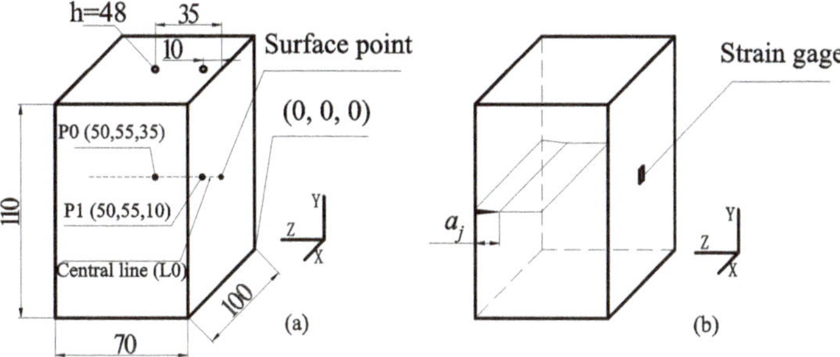

Figure 2. Schematic of quenching of a sample: (**a**) sizes and measured points (the unit used in this figure is "mm"); and (**b**) cutting plane of slitting method.

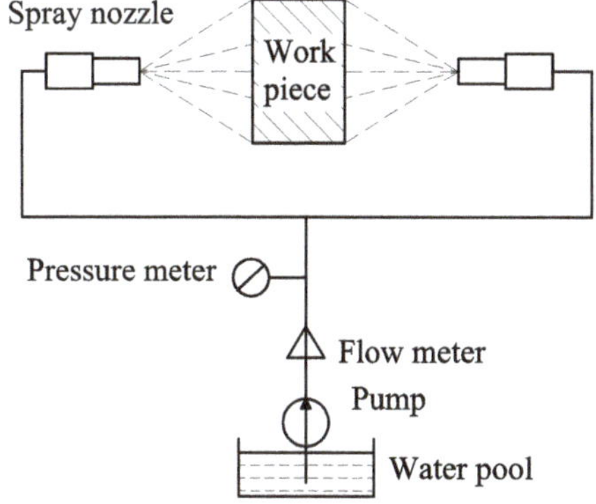

Figure 3. Schematic of spray quenching experiment.

Table 4. Heat treatments applied to 2A14 aluminum alloy samples.

Samples	Solution Treatment	Quenching Technology	Ageing Treatment
A1	500 °C/4 h	20 °C	165 °C/9 h
A2	500 °C/4 h	Step 1: 20 °C Step 2: air	165 °C/9 h
A3	500 °C/4 h	Spray quenching	165 °C/9 h

Refering to References [10,20], the residual stresses of the as-quenched samples were measured by the slitting method. As shown in Figure 2b, a wire-electrode cutting machine was used to cut incrementally along the cutting plane. With the incremental increase in the depth (a_j) of the cutting plane, the residual stresses of the blocks were released. The strains (ε_y) in the y direction were measured by using foil strain gauges (BX120-5AA) with 5 mm gauge lengths, which were connected in 1/4 bridging mode. The central line of the gauge was at the mid-length of the sample. The strains were the functions of cutting depths (a_j). We presumed that the residual stress $\sigma_y(z)$ along the cutting plane is the function of z. As in Equation (10), it can be described by the polynomial P_i (70 − z) and the undetermined coefficient A_i. At the same time, as in Equation (11), the measured stains (ε_y) at different depths (a_j) can be described by the undetermined coefficient A_i and compliance function C_i (a_j). C_i (a_j) is the strain at the measured point with the depth (a_j) of the cutting plane increasing incrementally, when P_i (70 − z) stress was applied along the z direction of the sample. The numerical simulation method was used to calculate C_i (a_j). As shown in Equation (12), the least square method is used to obtain A_i using the measured strains and calculated strains given in Equation (11). In this paper, the residual stress (Equation (10)) was not calculated, and, instead, the measured strains were used to estimate the residual stress of the samples, which underwent different quenching treatments.

$$\sigma_y(z) = \sum_{i=1}^{n} A_i P_i(70 - z) = PA \tag{10}$$

$$\varepsilon_y(a_j) = \sum_{i=1}^{n} A_i C_i(a_j) = CA \tag{11}$$

$$\frac{\partial}{\partial A_i} \sum_{j=1}^{m} \left[\varepsilon(a_j) - \sum_{k=1}^{n} A_k C_k(a_j, P_k) \right]^2 = 0 \ i = 1, 2, \ldots, n \tag{12}$$

The tensile test samples in x direction were machined from the mid-thickness part of the aged samples, and were used to evaluate the mechanical properties. Three test specimens were taken from every sample with dimensions of 2×8 mm^2 and a gauge length 30 mm, utilizing a 25 mm gauge length extensometer according to GB/T 1685-2013. The samples were tested at a strain rate of 0.0011 s^{-1}.

2.3. Numerical Simulation

The numerical simulation software ABAQUS standard with coupled temperature–displacement analysis was used to further study the influence of material properties and cooling rates at different temperatures on plastic strain and residual stress during quenching. The size of the model was the same as that of the samples used for the heat treatment experiments. Only the two end surfaces (110 mm $\times$ 100 mm) exchanged heat with water during quenching. Due to the symmetry, only an eighth of the sample was used to reduce calculating time, as shown in Figure 4. The displacement in the normal direction of the three symmetry planes was restricted. Similar to the experiments, only end surface of this model exchanged heat with the environment. The element type used in this model is an 8-node thermally coupled brick, trilinear displacement and temperature element (C3D8T), and the number of the elements was 6160. The incremental time in the analysis is chosen automatically by the computer program and the solution technique is full Newton method.

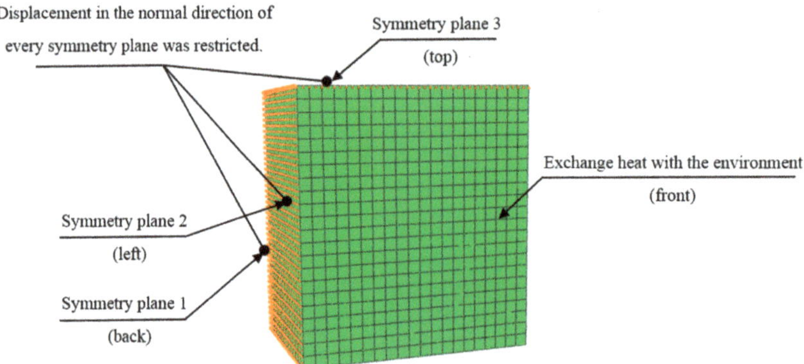

Figure 4. Numerical simulation model.

The density of this material is set constant at a value of 2800 kg·m^{-3}. Figure 5 shows the yield stress at different plastic strains used in the simulation model, where the yield points at different temperatures are the small value of the curves. As shown in Table 5, the thermal expansion coefficients, elasticity moduli, conductivities, and specific heat capacities were obtained from the literature [21,22]. The convective heat transfer coefficient of aluminum/air was set constant at 0.2 kW·m^{-2}·s^{-1}.

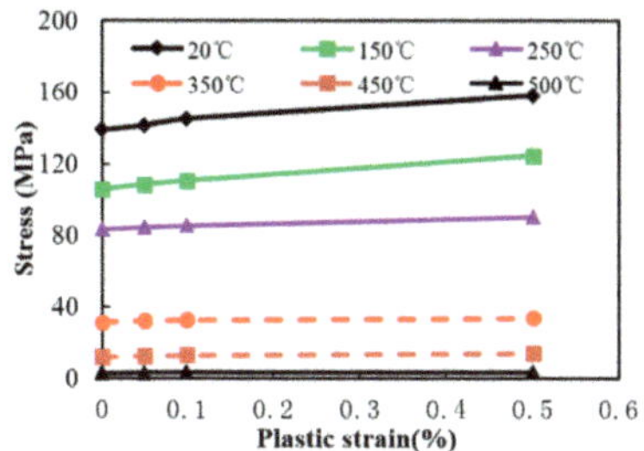

Figure 5. Yield strengths at different temperatures.

Table 5. Thermal properties.

Temperature (°C)	20	100	200	300	400	500
Conductivities ($W{\cdot}m^{-1}{\cdot}K^{-1}$)	114.3	122.3	130.8	145.1	124.5	122.7
Specific heat capacities ($J{\cdot}kg^{-1}{\cdot}K^{-1}$)	809	860	897	922	872	985
Elasticity moduli (GPa)	81.5	66.2	49.3	31.0	25.3	–
Thermal expansion coefficients (10^{-5})	2.08	2.19	2.61	2.70	2.68	2.73

The temperature-dependent material properties changed with temperature, and affected the evolution of thermal stress and plastic strain at different temperatures during the quenching process. The thermal expansion coefficient, elastic modulus, and yield strength play key roles in the evolution of thermal stress and plastic strain. Numerical simulation was used to analyze the effect by comparing the properties of samples M0–M3 with different material properties, as shown in Table 6. Sample M0 used the measured material properties of the studied material, as shown in Table 5 and Figure 5. For samples M1–M3, only one kind of measured material properties was adjusted and they are the thermal expansion coefficient, elastic modulus, and yield strength, respectively. The material properties were adjusted by taking the value of the corresponding material properties at 20 °C as reference values, and multiplying the reference values with the multiplier factors at different temperatures (as shown in Figure 6) to obtain the corresponding values at different temperatures. The Mises stress and equivalent plastic strain (PEEQ) along the central line (L0) were used to estimate the level of residual stresses and plastic strains of the samples.

Table 6. Simulation scheme.

Samples Material Properties	M0	M1	M2	M3
Thermal expansion coefficients	–	adjusting	–	–
Elasticity moduli	–	–	adjusting	–
Yield strengths	–	–	–	adjusting

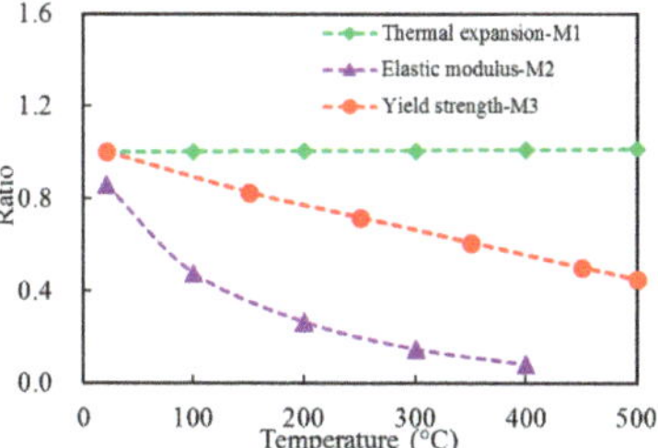

Figure 6. Multiplier factors of thermal expansion coefficient, elastic modulus, and yield strength at different temperatures.

ABAQUS was also used to study the influence of cooling rates on the residual stresses by simulating the quenching process. Prior to this, the heat transfer coefficients of aluminum/water were obtained by using the Deform 2D inverse heat transfer module. The measured temperature vs. time curves at point P0 were used. The Mises stresses, principal stresses in the y direction (S22), equivalent plastic strain (PEEQ), and plastic strains in the y direction (PE22) along the central line (L0) were used to estimate the plastic strains and stresses of the samples.

3. Results and Analysis

3.1. Influence of Material Properties at Different Temperatures on the Evolution of Stress and Strain

Material properties change with temperatures. Thus, the evolution of stress and plastic strain are different at different temperatures. This section will investigate the evolution of plastic strain at different temperatures during quenching, and study how the changing of material properties affect the evolution of residual plastic strain and stress after quenching treatment.

First, the two-unit model (Figure 1) was used to investigate the plastic strains at different temperatures with the same temperature differences (T_g). Replace the coefficients of Equation (9) with the material properties and set T_g as 50 °C, 100 °C, and 150 °C, respectively. The plastic strains (ε_i^p) at different temperatures of unit i were obtained. As shown in Figure 7, the plastic strain increases with increasing temperature difference and temperature. The results indicate that to produce a certain plastic strain, the temperature difference should be increased with decreasing temperature. Consequently, the temperature differences at high temperatures should be reduced to minimize the residual stresses.

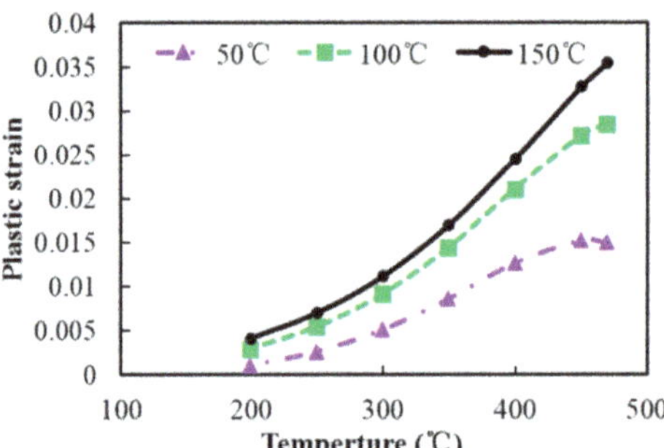

Figure 7. Plastic strains of unit i of the model at different temperatures with a certain temperature difference (T_g).

Figure 8 shows the simulation results for the samples (M0–M3) with different material properties. Equivalent plastic strain (PEEQ) and Mises stresses along the central line (L0) represent the plastic strains and residual stresses of the samples, respectively. As shown in Figure 8b,c, the plastic strains and residual stresses of the samples M1–M3 are lower than those of the sample M0. Figure 8a compares the adjusting material properties used for samples M1–M3 with the ones for sample M0. Like the treatments in Section 2.3, the values of every kind of measured material properties at 20 °C were considered to be reference values, and the material properties at different temperatures were divided by the corresponding reference value. Compared with the material properties used for sample M0, the thermal expansion coefficients used for sample M1 are smaller, the elastic moduli used for sample M2 are smaller, and the yield strengths for sample M3 are bigger. Reducing thermal expansion coefficients reduced thermal expansion and thermal stresses during quenching, which resulted in the decrease of plastic strains. Reducing the elastic modulus resulted in an increase in the allowable elastic strain at a certain thermal strain, and this led to the reduction in the plastic strain during quenching.

The increase in the yield strength reduced the plastic strain at a certain thermal stress. Therefore, the plastic strains and residual stresses of samples M1–M3 are smaller than the ones of sample M0.

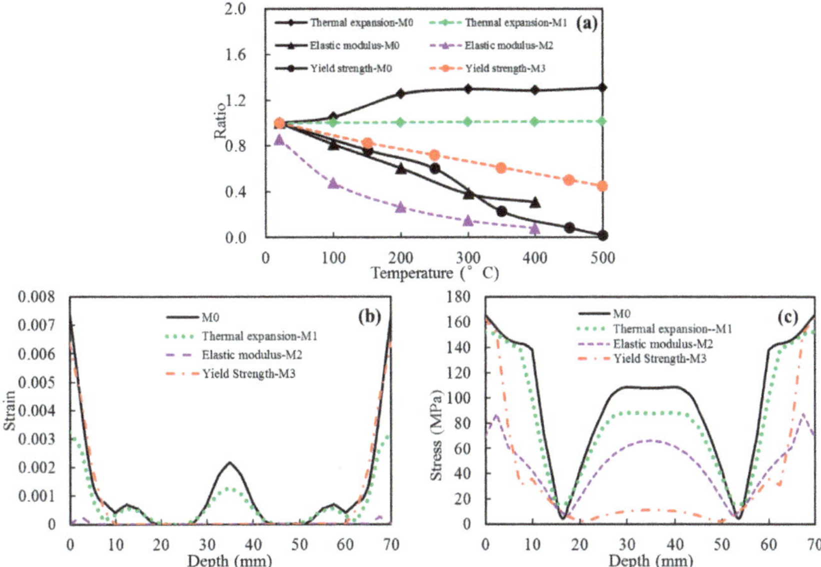

Figure 8. Residual equivalent plastic strain (PEEQ) and Mises stresses along the central line L0 of the sample (110 mm × 100 mm × 70 mm) in Figure 2a using different thermal expansion coefficients, elastic moduli, and yield strengths: (**a**) normalized measured and adjusting material properties; (**b**) PFEQ; and (**c**) Mises stresses.

3.2. Influence of Cooling Rates at Different Temperatures on Residual Stress

The section above indicates that to produce a certain plastic strain, the magnitude of temperature difference should be increased with decreasing temperatures. For real components, the residual stresses are determined by inhomogeneous plastic deformations. The evolution of plastic strain and stress is affected by the temperature distributions in the components during quenching. The temperature difference increased with increasing cooling rate. At the same time, the cooling rates determine mechanical properties, and this relationship has been studied using time-temperature- properties (TTP) curves in our previous work [15]. To optimize the quenching process, this section will investigated the influence of cooling rates at different temperatures on the evolution of plastic strain and final residual stress by quenching the samples with different quenching technologies.

Figure 9 shows the cooling rates at P0, temperature differences between points P0 and P1, strains measured by the slitting method, and tensile properties of the samples. The results show that reducing the cooling rates in a low temperature range does not reduce residual stresses. However, an optimized step quenching process can minimize the residual stresses and result in good mechanical properties. As shown in Figure 9a,b, the cooling rates and temperature differences between P0 and P1 of Samples A1 and A2 are almost the same in the temperature range from 500 to 420 °C, and the cooling rates and temperature differences of Sample A2 are much lower than those of Sample A1 at temperatures below 400 °C. However, the residual stresses of the two samples are almost the same, as shown in Figure 9c. This means that changing the cooling rates at low temperatures does not change the residual stresses for the Samples A1 and A2. In the case of Sample A3, the cooling rates are lower at temperatures above 300 °C, especially above 450 °C, than those of Sample A1. The cooling rates are higher than those of the sample A1 at temperatures below about 300 °C. The temperature differences of Sample A3 are lower at temperatures above 400 °C and slightly higher at temperatures below 350 °C. However, the residual stresses of Sample A3 are much lower than those of Sample A1 due to the lower cooling rates at high temperatures. This is because plastic strain occurred more easily at high temperatures, according to the results and analysis presented in Section 3.1. Moreover, as shown in Figure 9d, the tensile properties of sample A3 are close to those of Sample A1, because the cooling rates are high in the quenching sensitivity range from 300 to 400 °C [15]. This means that residual stresses and mechanical performances can be balanced by employing an optimized quenching technology with low cooling rates in high temperature range and high cooling rates in other temperature ranges.

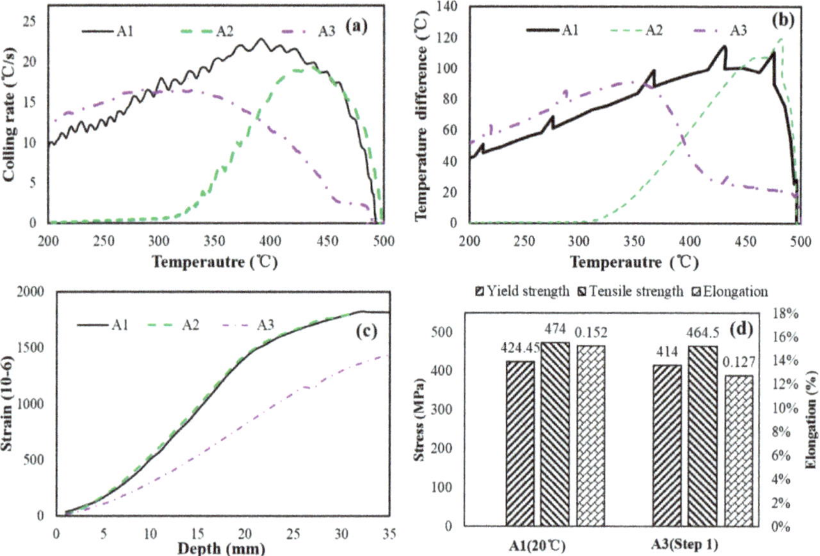

Figure 9. Results of Samples A1–A3 with different quenching technology: (**a**) cooling rates at P0 during the quenching process; (**b**) temperature differences between points P0 and P1 during the quenching process; (**c**) strains measured by the slitting method after quenching treatment; and (**d**) tensile properties after aging treatments.

The numerical simulation method was used to further study the influence of cooling rates on the evolution of residual stress. Figure 10 shows the evolution of the plastic strain at P0, plastic strains and residual stresses along L0. As shown in Figure 10a, the plastic strains (plastic strains in the y direction (PE22) and equivalent plastic strain (PEEQ)) at P0 of Sample A1 and A2 vs. temperature curves are almost the same. The plastic strains reached the magnitudes at about 490 °C and remained unchanged thereafter; even the cooling rates of Sample A2 are much lower below 400 °C. According to the results and analysis presented in Section 3.1, to produce a certain plastic strain, the temperature difference should be increased with decreasing temperatures. This explains that reducing the cooling rates at low temperature range does not reduce the plastic strains at P0. Consequently, as shown in Figure 10b,c, the plastic strains (PE22 and PEEQ) and residual stress (Mises stress and principal stresses in the y direction (S22)) along L0 of the two samples are almost the same. Moreover, in the case of Sample A3, the step quenching technology produces lower plastic strains and residual stresses along L0 due to the low cooling rates at high temperatures. The results coincide with the experimental results in the above paragraph.

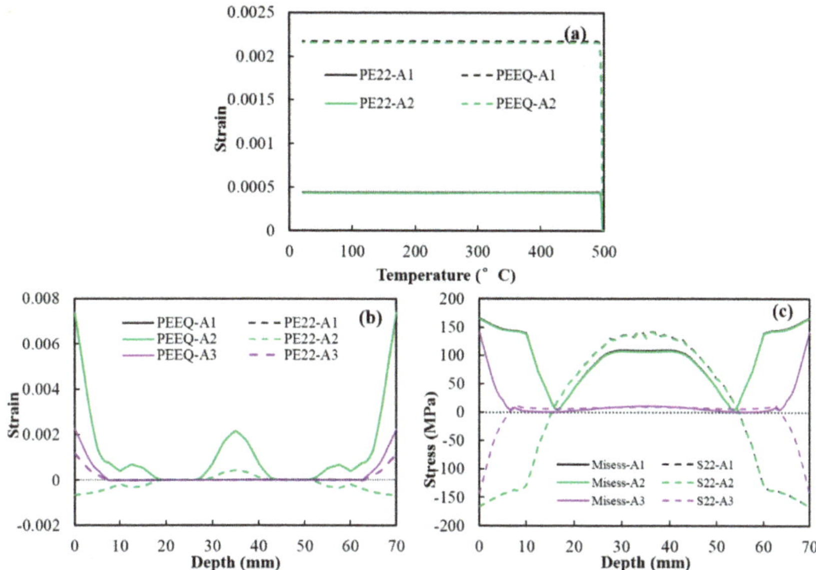

Figure 10. Simulation results of the quenching samples: (**a**) evolution of the plastic strain (plastic strains in the y direction (PE22) and equivalent plastic strain (PEEQ)) at point P0; (**b**) plastic strains (PE22 and PEEQ) along the central line L0; and (**c**) residual stresses (Mises stress and principal stresses in the y direction (S22)) along L0.

4. Discussion

The influence of temperature-dependent material properties on the evolution of plastic strain at different temperatures during the quenching process was investigated, and the influence of cooling rates at different temperatures on the evolution of plastic strain and residual stress for forged 2A14 aluminum alloy components during the quenching process was also analyzed.

During quenching process, the thermal expansions at high-temperature locations than at low-temperature locations. The difference in the thermal expansions resulted in thermal strains and stresses to balance the inhomogeneous thermal expansions. This usually produced inhomogeneous plastic strains, which resulted in residuals stress after the quenching treatment.

Equation (9) shows that the thermal expansion coefficients, elastic moduli, and yield strengths play key roles in determining the magnitude of plastic strains. Figure 8 shows that compared with the sample M0, decreasing the thermal expansion coefficients and increasing yield strengths at high temperatures decreases the residual plastic strains and stresses along the central line L0 of the sample M1 and M3, respectively; further, decreasing the elastic moduli decreases the residual plastic strains and stresses along line L0 of the sample M2. This can be explained as follows: During the quenching process, decreasing thermal expansion coefficient decreased the thermal expansion and thermal stress, which resulted in a decrease in the plastic strain. Further, increasing yield strength reduced the plastic strain at a certain thermal stress; reducing the elastic modulus resulted in the rise of allowable elastic strain at a certain thermal strain, and then led to the reduction in the plastic strain. The reduction in the plastic strains during quenching decreased residual plastic strains and stresses of the components.

The paragraph above implies that increasing thermal expansion coefficient and decreasing the yield strength causes an increase in the plastic strain. However, decreasing elastic modulus decreases the plastic strain. For the studied material, the thermal expansion coefficient increase with the increasing temperature; the elastic modulus and yield strength decrease with increasing temperature. Using the model in Section 2.1, the evolution of the plastic strain at different temperatures during quenching process was studied. Figure 7 shows that the plastic strain increases with increasing temperature at the same temperature difference between the two units; the plastic strain increases with increasing temperature difference at the same temperatures. This means that the influence of the thermal expansion coefficients and yield strengths changing with temperatures on plastic strains are more serious than the effect of the changes in the elastic moduli. Thus, the plastic strains increase with temperature.

Quenching residual stress of aluminum alloy components are caused by the inhomogeneous temperature distribution. Decreasing the temperature gradients decreases the plastic strain during the quenching process, resulting in lower residual stresses. Decreasing the cooling rates decreases the temperature gradient. Many papers reported that reducing the cooling rate can minimize residual stresses [10,15,23]. However, the above paragraphs in this section imply that for this material component, the residual plastic strains and stresses are mainly determined by the cooling rates in a high-temperature zone. Figure 9 shows that compared with the sample A1, reducing the cooling rates at low temperatures below 400 °C does not reduce the measured strains of the sample A2, as indicated by the slitting method results; in contrast, reducing the cooling rates at high temperatures above about 450 °C reduces the strains of sample A3 sharply, as per measurement results of the slitting method. The strains are in proportion with the residual stress of the samples. Figure 10 shows the residual strains and stresses simulation results along the central line L0 of the samples A1–A3, and they show a similar trend as the quenching experiment results. These results confirmed that the cooling rate at the high-temperature zone insignificantly affects the residual stress of this studied material component.

Time-temperature-properties (TTP) curves of this studied material [15] show that mechanical performances are mainly determined by cooling rates in the quenching sensitivity temperature range. According to this conclusion and the results presented above, an optimized step quenching technology was proposed to balance the residual stresses and mechanical properties. By quenching sample A3 with this cooling path, the residual stresses were reduced significantly and the tensile properties changed slightly, compared with sample A1 quenched with water at 20 °C, as shown in Figures 9 and 10. In this quenching treatment, the cooling rates were low at high temperatures above 450 °C to minimize the residual stresses, and they increased in the other temperature ranges, resulting in good mechanical properties.

5. Conclusions

After analyzing and summarizing the results above, several main conclusions were inferred.

Plastic strains increase with the temperature when the temperature difference remains unchanged.

The cooling rates at high temperatures play a key role in determining the magnitude of residual stresses. Only reducing the cooling rates at low temperatures cannot reduce the residual stress and plastic strains.

Residual stresses and mechanical properties can be balanced with an optimized quenching technology. The cooling rates of the sample are low at high temperatures to reduce residual stress and are high at the other temperatures to improve mechanical properties.

Acknowledgments: This work was financially supported by State Key Laboratory of High Performance Complex Manufacturing (zzyjkt2014-02) and the fund of Jiangsu Province for the transformation of scientific and technological achievements (BA2015075).

Author Contributions: Yuxun Zhang, Shiquan Huang and Youping Yi conceived and designed the model, numerical simulation and experiments; Yuxun Zhang performed the experiments; Yuxun Zhang, Shiquan Huang and Hailin He analyzed the data; Youping, Yi contributed reagents/materials/analysis tools; Yuxun Zhang wrote the paper; Youping Yi, Shiquan Huang and Hailin He helped modify the manuscript.

References

1. Xiao, Y.; Xie, S.; Liu, J.; Wang, T. *Practical Handbook of Aluminum Technology*; Metallurgical Industry Press: Beijing, China, 2005.
2. Li, S.; Huang, Z.; Chen, W.; Liu, Z.; Qi, W. Quench sensitivity of 6351 aluminum alloy. *Trans. Nonferr. Met. Soc.* **2013**, *23*, 46–52. [CrossRef]
3. Robinson, J.S.; Tanner, D.A.; Petegem, S.V.; Evans, A. Influence of quenching and aging on residual stress in Al–Zn–Mg–Cu alloy 7449. *Mater. Sci. Technol.* **2013**, *28*, 420–430. [CrossRef]
4. Dolan, G.P.; Robinson, J.S. Residual stress reduction in 7175-T73, 6061-T6 and 2017A-T4 aluminium alloys using quench factor analysis. *J. Mater. Process. Technol.* **2004**, *153*, 346–351. [CrossRef]
5. Shalvandi, M.; Hojjat, Y.; Abdullah, A.; Asadi, H. Influence of ultrasonic stress relief on stainless steel 316 specimens: A comparison with thermal stress relief. *Mater. Des.* **2013**, *46*, 713–723. [CrossRef]
6. Citarella, R.; Carlone, P.; Lepore, M.; Sepe, R. Hybrid technique to assess the fatigue performance of multiple cracked FSW joints. *Eng. Fract. Mech.* **2016**, *162*, 38–50. [CrossRef]
7. Lepore, M.A.; Citarella, R.; Carlone, P.; Sepe, R. DBEM crack propagation in friction stir welded aluminum joints. *Adv. Eng. Softw.* **2015**, *101*, 50–59.
8. Citarella, R.G.; Cricrì, G.; Lepore, M.; Perrella, M. Assessment of Crack Growth from a Cold Worked Hole by Coupled FEM-DBEM Approach. *Key Eng. Mater.* **2013**, *577–578*, 669–672. [CrossRef]
9. Koç, M.; Culp, J.; Altan, T. Prediction of residual stresses in quenched aluminum blocks and their reduction through cold working processes. *J. Mater. Process. Technol.* **2006**, *174*, 342–354. [CrossRef]
10. Dong, Y.B.; Shao, W.Z.; Jiang, J.T.; Zhang, B.Y.; Zhen, L. Minimization of Residual Stress in an Al-Cu Alloy Forged Plate by Different Heat Treatments. *J. Mater. Eng. Perform.* **2015**, *24*, 2256–2265. [CrossRef]
11. Jiang, G.; He, W.; Zheng, J. Mechanism and experimental research on high frequency vibratory stress relief. *J. Zhejiang Univ. Eng. Sci.* **2009**, *43*, 1269–1272.
12. McGoldrick, R.T.; Saunders, H.E. Some experiments in stress-relieving castings and welded structures by vibration. *J. Am. Soc. Nav. Eng.* **1943**, *55*, 589–609. [CrossRef]
13. Yang, Y.P. Understanding of Vibration Stress Relief with Computation Modeling. *J. Mater. Eng. Perform.* **2008**, *18*, 856–862. [CrossRef]
14. Sun, Y.; Jiang, F.; Zhang, H.; Su, J.; Yuan, W. Residual stress relief in Al–Zn–Mg–Cu alloy by a new multistage interrupted artificial aging treatment. *Mater. Des.* **2016**, *92*, 281–287. [CrossRef]
15. Zhang, Y.; Yi, Y.; Huang, S.; Dong, F. Influence of quenching cooling rate on residual stress and tensile properties of 2A14 aluminum alloy forgings. *Mater. Sci. Eng. A* **2016**, *674*, 658–665. [CrossRef]
16. Li, H.Y.; Zeng, C.T.; Han, M.S.; Liu, J.J.; Lu, X.C. Time-temperature-property curves for quench sensitivity of 6063 aluminum alloy. *Trans. Nonferr. Met. Soc.* **2013**, *23*, 38–45. [CrossRef]
17. Nallathambi, A.K.; Kaymak, Y.; Specht, E.; Bertram, A. Sensitivity of material properties on distortion and residual stresses during metal quenching processes. *J. Mater. Process. Technol.* **2010**, *210*, 204–211. [CrossRef]
18. Chen, M. *Elasticity and Plasticity*, 1st ed.; China Science Publishing: Beijing, China, 2007; p. 421.
19. Dong, X.; Zheng, T.; Yang, W.; Xiao, J. Coupled Thermal-mechanical Analysis of the Heat-treatment Process of an Aluminum Alloy Forging. *Hot Work. Technol.* **2002**, *13*, 17–18.
20. Nervi, S.; Szabó, B.A. On the estimation of residual stresses by the crack compliance method. *Comput. Method Appl. Mach. Eng.* **2007**, *196*, 3577–3584. [CrossRef]
21. Liu, W.; Guo, L.; Ma, Y.; Wang, J.; Liu, D. Microstructure and flow behavior of 2A14 aluminum alloy during hot deformation. *Chin. J. Nonferr. Met.* **2013**, *23*, 2091–2097.
22. Zhao, Y.; Lin, S.; He, Z.; Wu, L. Numerical simulation of 2014 aluminium alloy friction stir welding process. *Chin. J. Mech. Eng.* **2006**, *42*, 92–97. [CrossRef]
23. Tanner, D.A.; Robinson, J.S. Reducing residual stress in 2014 aluminium alloy die forgings. *Mater. Des.* **2008**, *29*, 1489–1496. [CrossRef]

 *metals*

Article

Characteristics of the Dynamic Recrystallization Behavior of Ti-45Al-8.5Nb-0.2W-0.2B-0.3Y Alloy during High Temperature Deformation

Lin Xiang, Bin Tang *, Xiangyi Xue, Hongchao Kou and Jinshan Li

State Key Laboratory of Solidification Processing, Northwestern Polytechnical University,
Xi'an 710072, Shaanxi, China; xlin0731@163.com (L.X.); xuexy@nwpu.edu.cn (X.X.);
hchkou@nwpu.edu.cn (H.K.); ljsh@nwpu.edu.cn (J.L.)
* Correspondence: toby@nwpu.edu.cn; Tel.: +86-29-8846-0294

Received: 7 June 2017; Accepted: 4 July 2017; Published: 8 July 2017

Abstract: The dynamic recrystallization (DRX) behavior of Ti-45Al-8.5Nb-0.2W-0.2B-0.3Y (at %) alloy has been investigated through hot compression tests. The tests were executed at a temperature range of 1000–1200 °C and a strain rate range of 0.001–1 s^{-1} under a true strain of 0.9. It was found that the α_2 phase which is produced during heat treatment is reduced during hot compression due to thermo-mechanical coupling. The value of the activation energy is 506.38 KJ/mol. With the increase in deformation temperature and the decrease in strain rate, DRX is more likely to occur, as a result of sufficient time and energy for the DRX process. Furthermore, the volume fraction of high angle grain boundaries increases to 89.01% at a temperature of 1200 °C and the strain rate of 0.001 s^{-1}, meaning completely dynamic recrystallization. In addition, DRX is related to the formation of twin boundaries. The volume fraction of twin boundaries rises to 16.93% at the same condition of completely dynamic recrystallization.

Keywords: titanium aluminides; hot compression; dynamic recrystallization; microstructure

1. Introduction

TiAl-based alloys are considered a promising candidate for high-temperature structural applications in the aerospace and automotive industries [1–4] because of their outstanding properties such as low density, high specific yield strength, high specific stiffness, good oxidation resistance and creep properties. However, natural brittleness, a key factor, limits the application of conventional TiAl-based alloys. In order to improve the hot workability of TiAl-based alloys, many studies have been performed over the last two decades [5,6]. It has been found that the addition of Nb can improve the workability of ductile β/B2 phase at elevated temperatures [7,8], and Nb and W tend to enrich in β/B2 phase, which contributes to the stability of β/B2 phase [9].

It is commonly agreed that service properties are significantly connected with microstructures. It is also well-known that hot deformation is an effective method to optimize the microstructure, to further improve these properties. Dynamic recovery, phase transformations and DRX will occur during deformation at high temperature, resulting in the variation of size, morphology, fraction and distribution of phases [4,10–14]. Moreover, microstructure evolution is affected by hot deformation parameters, especially by deformation temperature, strain rate and degree of deformation [15,16]. The deformation behaviors of TiAl-based alloys of which the initial microstructure consists of the fully lamella or nearly lamella have been studied [17], indicating that the deformation mechanism of γ phase is mainly DRX and that of α phase is dynamic recovery during deformation at 1275 °C. When strain reaches critical levels, the nuclei of DRX grains are generated, which will grow with increasing deformation degree [18]. The DRX mechanism of a TiAl-based alloy in a single α-hcp phase region has also been investigated [19]. However, the phase transformation and DRX behavior of TiAl-based alloys with two β + γ phases needs to be further characterized in detail.

The present paper studies the phase transformation and DRX behavior of a TNB (high Nb containing TiAl-based alloy) alloy with β + γ two phases. For this purpose, the hot compression tests were conducted at different deformation temperatures and strain rates. The analysis of hot deformation behavior will be executed based on X-ray diffraction, microstructure and EBSD.

2. Materials and Methods

High-purity Ti, Al, Nb, W, B and Y ($\geq$99.9%) were used as the raw materials. An ingot of TNB alloy with a nominal composition of Ti-45Al-8.5Nb-0.2W-0.2B-0.3Y (at %) was fabricated using vacuum arc remelting (VAR melting) method, and the practical composition is listed in Table 1. Subsequently, the ingot was HIPed (hot isostatic pressed) at 1280 °C for 4 h under a pressure of 140 MPa. The dimension of the ingot after HIP is Φ 220 mm × 190 mm. In order to investigate hot deformation behavior, uniaxial compression tests were performed on a gleeble-3500 thermo simulation machine. Cylindrical samples for compression tests with 8 mm in diameter and 12 mm in height were cut from the HIPed ingot, which deformed at temperatures ranging from 1000 °C to 1200 °C and strain rates from 0.001 s^{-1} to 1 s^{-1} under a true strain 0.9. To reduce the friction between two end sides of cylindrical sample and pressure heads of thermo simulation machine, graphite lubricant was adopted. Prior to deformation, samples were heated to target temperatures at a rate of 10 K/s, and then kept 300 s for uniformity of temperature. After the hot compression test, samples were quenched immediately to retain the deformed microstructure at elevated temperature.

Table 1. Chemical composition of present TiAl alloy.

Element	Al	Nb	W	B	Y	Ti
at %	44.43	8.40	0.21	0.19	0.28	Bal.

The initial microstructure after hot isostatic pressing is shown in Figure 1a. Phase composition is determined by XRD (in Figure 1b), which mainly consists of black γ phase, white β phase and gray α₂ phase. The lamellar colony is composed of γ lath and a small amount of α₂ lath, which is surrounded by γ and β phase with the irregular morphologies. Some bright bulk particles are located in the lamellar colony, while others along the colony boundary were demonstrated to be impurities with B and Y elements [13], which are beneficial to decrease the size of the colony [20,21].

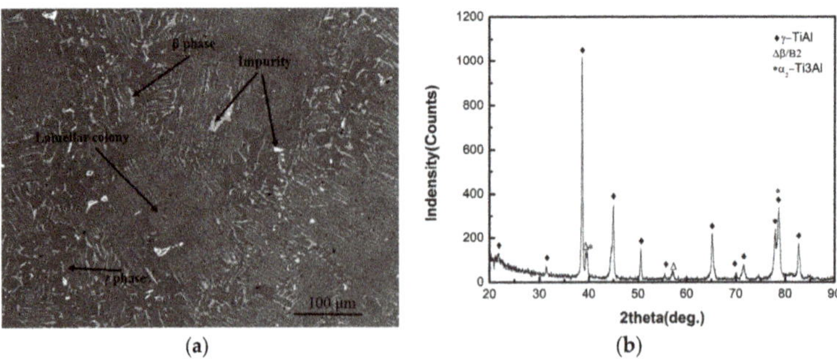

Figure 1. (**a**) The initial microstructure; and (**b**) X-ray diffraction pattern of HIPed alloy.

The microstructures were characterized via scanning electron microscopy (TESCAN, Brno city, Czech Republic) in the back scattered mode (SEM-BSE) and electron backscattered diffraction (EBSD). Dynamic recrystallization grains were investigated by EBSD which performed on TESCAN Field

Emission Scanning Electron. The specimens for EBSD were prepared by Vibrational Polish for 8 h. X-ray diffraction were adopted to analyze the phases of alloy, which was carried out using Cu K_α radiation with 2θ range from $20°$ to $90°$, and the voltage, current and scanning step were 40 kV, 30 mA and 0.03, respectively.

3. Results and Discussion

3.1. Analysis of Flow Curves

3.1.1. Characteristics of Flow Curves

Figure 2 shows flow stress curves of the present TiAl alloy deformed at different deformation temperatures and strain rates. It is obvious that flow stress is sensitive to deformation temperatures and strain rates. With the increase of deformation temperature and the decrease of strain rate, flow stress decreases dramatically. With the increase of strain, flow stress reaches the peak value at the initial deformation stage, followed by flow softening. The softening stage of the present TiAl alloy persists for long time before reaching the steady-state regime. It is a typical characteristic of TiAl alloy that the deformation is dominated by DRX [22]. The critical strains of specimens deformed at different conditions are shown in Table 2. All values of critical strain are below 0.2. Such a low value indicates that the onset of DRX is very easy. With the decrease of strain rate, the value of critical strain gradually decreases, which indicates that DRX easily occurs at a low strain rate.

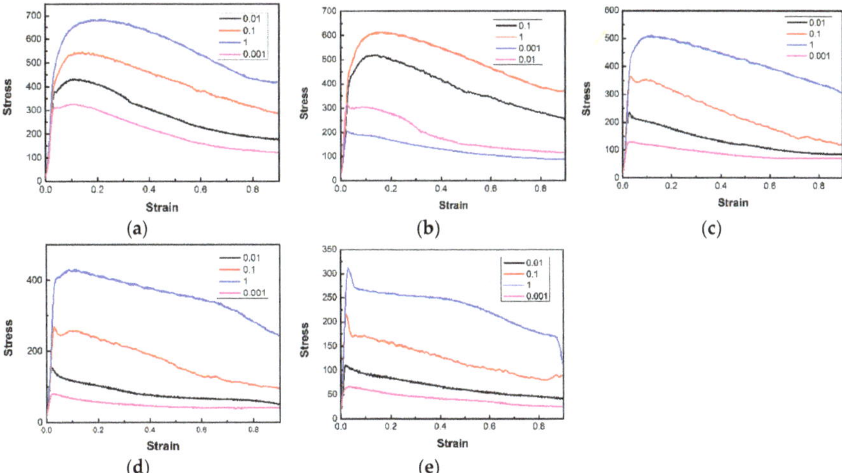

Figure 2. Flow curves of specimens deformed at (**a**) 1000 °C; (**b**) 1050 °C; (**c**) 1100 °C; (**d**) 1150 °C; (**e**) 1200 °C.

Table 2. The critical strain of the present alloy deformed at different conditions.

Parameters	1000 °C	1050 °C	1100 °C	1150 °C	1200 °C
$0.001\ \text{s}^{-1}$	0.05594	0.01272	0.01219	0.0136	0.00804
$0.01\ \text{s}^{-1}$	0.05747	0.01574	0.01651	0.01621	0.01039
$0.1\ \text{s}^{-1}$	0.06517	0.07555	0.01843	0.02057	0.01471
$1\ \text{s}^{-1}$	0.15034	0.12692	0.0619	0.08271	0.01828

3.1.2. Activation Energy Q

Flow behavior of TiAl alloy can be well-described by the Arrhenius equation. The flow stress and deformation parameters can be expressed as

$$Z = \dot{\varepsilon}\exp(\frac{Q}{RT}) = A[\sinh(\alpha\sigma)]^n,\tag{1}$$

For the different stress regime, Equation (1) can express it as follows:

$$\dot{\varepsilon} = B\sigma^{n'}\,(\alpha\sigma < 0.8),\tag{2}$$

$$\dot{\varepsilon} = B'\exp(\beta\sigma)\,(\alpha\sigma > 1.2),\tag{3}$$

where Q is activation energy, n and n' are stress exponents, α, A, B, and B' are material constants, and R is the gas constant. The constant α can be calculated by Equation (4):

$$\alpha = \beta/n',\tag{4}$$

Taking the logarithm of both sides of Equations (2) and (3), and linear curve fitting (Figure 3a,b), constants β and n' can be worked out. It is important to note that the flow stress is taken at the peak strain in Figure 3. Then, constant α of the present alloy can be determined, and the value is 3.4×10^{-3}.

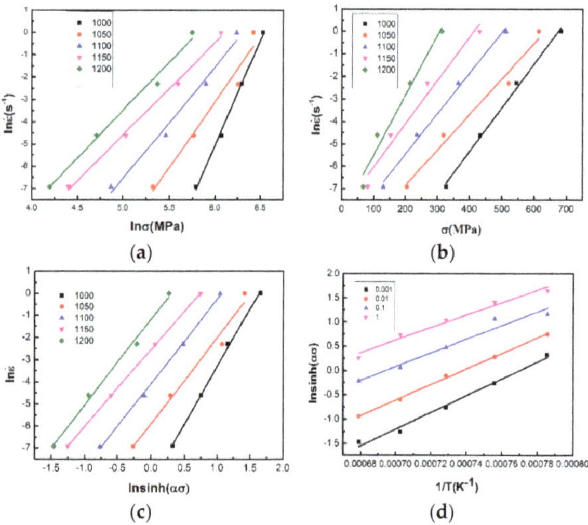

Figure 3. Plot of (**a**) $\ln\dot{\varepsilon} - \ln\sigma$; (**b**) $\ln\dot{\varepsilon} - \sigma$; (**c**) $\ln\dot{\varepsilon} - \ln[\sinh(\alpha\sigma)]$; (**d**) $\ln[\sinh(\alpha\sigma)] - 1/T$. It is important to note that the flow stress is taken at the peak strain.

The activation energy Q can be calculated by Equation (5):

$$Q = R\left\{\frac{\partial\ln\dot{\varepsilon}}{\partial\ln[\sinh(\alpha\sigma)]}\right\}_T\left\{\frac{\partial\ln[\sinh(\alpha\sigma)]}{\partial(1/T)}\right\}_\varepsilon,\tag{5}$$

Taking logarithm of both sides of Equation (1), the slope of $\ln\dot{\varepsilon} - \ln[\sinh(\alpha\sigma)]$ and $\ln[\sinh(\alpha\sigma)] - 1/T$ curves can be obtained, as shown in Figure 3c,d. Thus, the value of activation energy Q is 506.38 KJ/mol, which is higher than activation energy for Ti self-diffusion (260 KJ/mol), Al self-diffusion (390 KJ/mol) and Ti-Al inter-diffusion (295 KJ/mol) in single γ-TiAl alloys [23–25].

Such a high value of activation energy Q indicates that DRX appears in deformation and dominates the softening stage.

3.2. Analysis of Microstructure

Figure 4a is the X-ray diffraction pattern of specimens deformed at the strain rate of 0.001 s^{-1} and heat treatment for 1200 s. According to the binary phase diagram of TiAl alloy with 8 (at %) Nb published in [26], α_2 phase will be produced during heat treatment in the adopted temperatures. There are obvious peaks of α_2 phase in heat treatment specimens. In Figure 4b,c, γ phases, β/B2 phases and α_2 phases are represented by blue, red and yellow phases, respectively. Figure 4b is the phase map showing the phase composition of specimen with solution at 1200 °C for 1200 s. The yellow α_2 phase can be observed clearly. Reversely, it is found that the peaks of α_2 phase disappears in the compressed samples, which means that no α_2 phase exists. Figure 4c, the phase map, also demonstrates that no yellow α_2 phase exists. The result indicates that α_2 phase, which is produced during heat treatment, is reduced during hot compression at the temperatures range of 1000–1200 °C. TiAl-based alloy is a low stacking fault energy alloy, and for the present alloy, stacking fault energy further decreases on account of high Nb conditions [3]. Hence the onset of DRX happens more easily during hot deformation. The strain energy releases because of DRX, leading to a decrease in driving forces for producing α_2 phase.

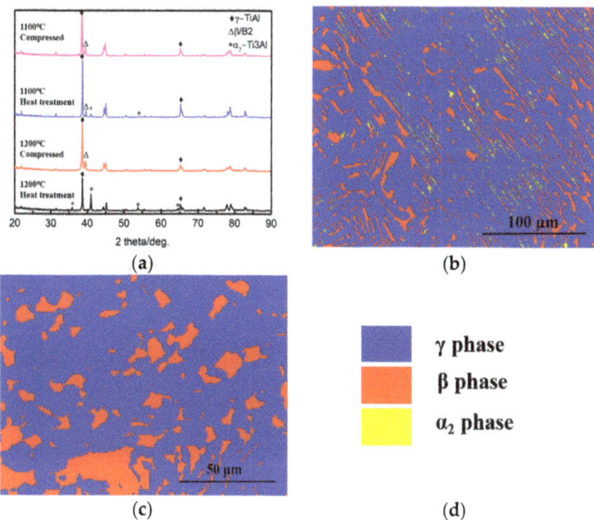

Figure 4. (**a**) X-ray diffraction patterns of specimens deformed at strain rate of 0.001 s^{-1} and heat treatment for 1200 s; phase map corresponds to (**b**) heat treatment at 1200 °C for 1200 s; and (**c**) deformed at 1200 °C.

Microstructures of deformed specimens at the strain rate of 0.01 s^{-1} and different temperatures ranging from 1000 °C to 1200 °C are shown in Figure 5. The β phase uniformly distributes in γ matrix. Elongated β phase, perpendicular to the direction of compression, is distributed in γ grain boundaries at the deformation temperature of 1000 °C. With the increase of temperature, a mass of elongated β phases are broken and refined (in Figure 5b,c). At a much higher deformation temperature, as shown in Figure 5d,e, the morphology of β phase gradually tends to be equiaxed and spherical. In addition, a vast number of γ DRX grains are observed at all chosen temperatures. However, DRX process proceeds less effectively at a low temperature, which leads to the existence of primary coarse grains. On the other hand, the β phase bends and distributes in γ grain boundary at high strain rate, depicted in

Figure 6a. Moreover, no DRX grains are observed. Instead, DRX grains are investigated obviously at low strain rate (Figure 6b). It is a common view that the short deformation time is caused by the high strain rate, which results in insufficient time for recrystallization.

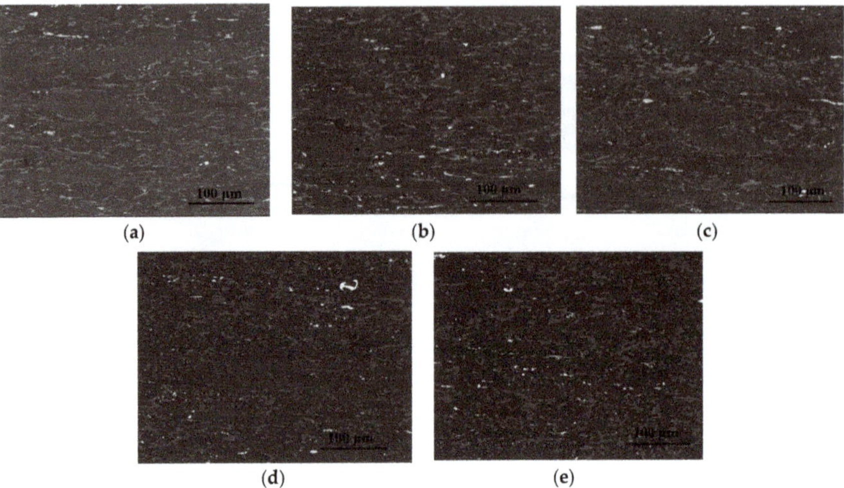

(a)   (b)   (c)

(d)   (e)

Figure 5. Microstructures of specimens isothermally deformed at the strain rate of 0.01 s^{-1} and different temperatures (**a**) 1000 °C; (**b**) 1050 °C; (**c**) 1100 °C; (**d**) 1150 °C; (**e**) 1200 °C.

(a)   (b)

Figure 6. Microstructures of specimens isothermally deformed at the temperature of 1200 °C and different strain rates of (**a**) 1 s^{-1}; (**b**) 0.001 s^{-1}.

Figure 7, the IPF maps and grain boundary maps overlapped with phases, shows the characteristics of grains and grain boundaries in phases at different conditions. In the EBSD map images, γ phases, β/B2 phases and α_2 phases are represented by blue, red and yellow phases, respectively. The black lines represent grain boundaries with a misorientation angle between 2° and 15°, while the bright blue lines are used for a misorientation angle of more than 15°. Twin boundaries (TBs) with $57° \leq \theta \leq 63°$ are expressed by the pink lines. Generally, two types of grain boundaries, low angle grain boundaries (LAGBs) and high angle grain boundaries (HAGBs), are defined based on misorientation angles. The LAGBs are the typical feature of substructure forming during hot deformation, while HAGBs usually are caused by DRX and grains grow up via merging grains [27]. As is well known, grain boundaries with misorientation angle between 2° and 15° are LAGBs and those with orientation angle more than 15° are HAGBs.

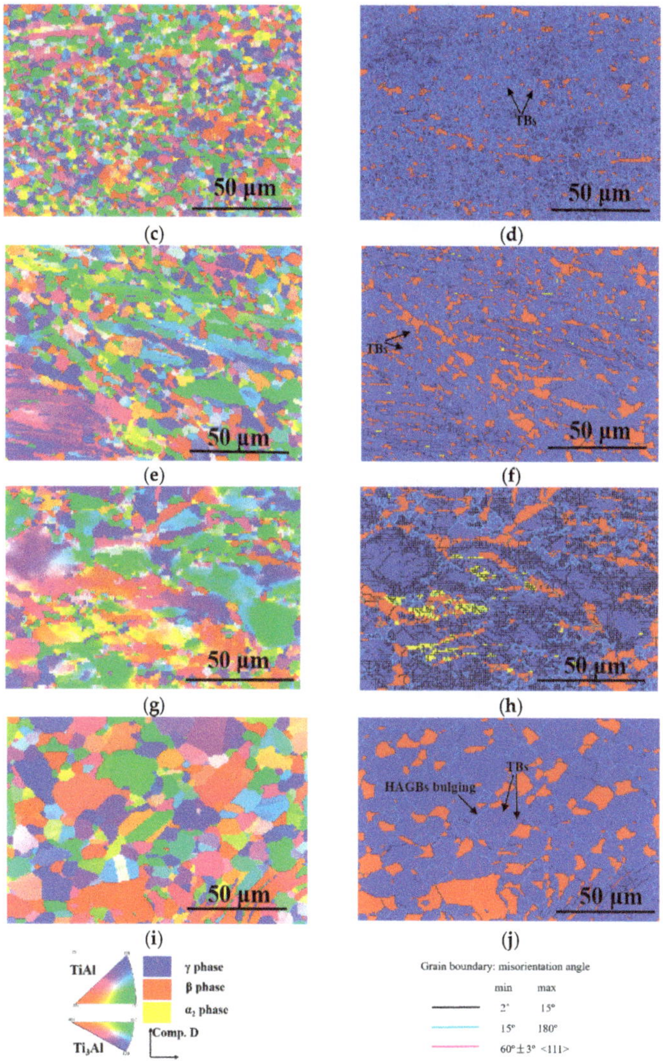

Figure 7. IPF maps and grain boundary maps overlapped with phases at different processing parameters: (**a,b**) 1000 °C, 0.01 s^{-1}; (**c,d**) 1100 °C, 0.01 s^{-1}; (**e,f**) 1200 °C, 0.01 s^{-1}; (**g,h**) 1200 °C, 0.1 s^{-1}; (**i,j**) 1200 °C, 0.001 s^{-1}.

In Figure 7a, a high number of fine and equiaxed γ grains are observed, whereas coarse γ grains still exist. Figure 7b shows the grain boundary map overlapped with phases which corresponds to Figure 7a. A large number of LAGBs can be investigated in coarse grains, which means coarse γ grains are deformed primary grains. It can also be found that grain boundaries of fine and equiaxed grains in Figure 7a are HAGBs, proving that those grains are DRX grains. In general, DRX was divided into 3 types: (i) discontinuous dynamic recrystallization which comprises grain nucleation and grain growth, is usually observed in low stacking fault energy metals; (ii) continuous dynamic recrystallization which has been observed in high stacking fault energy metals; (iii) geometric dynamic recrystallization generated by the fragmentation of the initial grains [28]. High Nb containing TiAl-based alloy is low stacking fault energy alloy. Thus, the type of DRX is mainly discontinuous dynamic recrystallization in TiAl-based alloy. As deformation temperature increases to 1100 °C (Figure 7c,d), the number of γ coarse grains with low angle decreases, and DRX grains are larger than those at the temperature of 1000 °C. Figure 8 shows the distribution of misorientation angle at different conditions. As shown in Figure 8a,b, the fraction of HAGBs increases from 78.34% to 82.24%. Because high temperature can provide DRX process with more active energy, which is beneficial to nucleation and growth of grains. Further raising deformation temperature to 1200 °C, a tiny number of γ coarse grains with low angle can be observed (in Figure 7e,f), i.e., the extent of DRX is extremely high, but residual lamella still exists. The grain boundaries of β grains are mainly HAGBs. In a word, as the deformation temperature arising, it is helpful for DRX process. It is worth noting that most of twin boundaries locate in DRX grains. Moreover, the area fraction of TBs is 2.67% at the deformation temperature of 1000 °C, while it increases to 17.87% with the deformation temperature rising to 1200 °C.

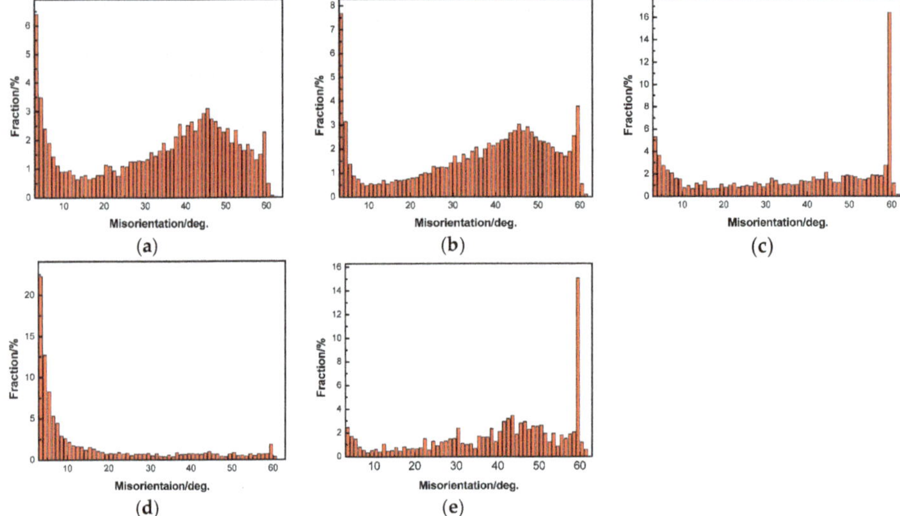

Figure 8. Misorientation angle distribution of deformed samples at different conditions: (**a**) 1000 °C, 0.01 s^{-1}; (**b**) 1100 °C, 0.01 s^{-1}; (**c**) 1200 °C, 0.01 s^{-1}; (**d**) 1200 °C, 0.1 s^{-1}; (**e**) 1200 °C, 0.001 s^{-1}.

As the strain rate increases to 0.1 s^{-1}, a few DRX grains can be found in Figure 7g. The fraction of LAGBs and HAGBs are 66.92% and 33.08%, respectively. Discontinuous dynamic recrystallization includes nucleation and growth processes of grains. The previous study has demonstrated that the nucleation process completes in a flash, while the growth process is overwhelmingly slow [22]. Deformation time is inadequate for DRX at high strain rate, and high temperature can activate a portion of slip system which benefits the formation of substructures. Therefore, the volume of LAGBs increased dramatically. Meanwhile, yellow α_2 phase is observed in Figure 7h. Because driving force for producing α_2 phase is not consumed completely by DRX. On the contrary, deformation time for DRX is guaranteed at the strain rate of 0.001 s^{-1}, and completely DRX occurs during hot compression. As a result, the volume fraction of LAGBs drops dramatically to 10.99%, and that of HAGBs increases to 89.01%. It is difficult to find TBs surrounded by HAGBs in Figure 7h, and the volume of TBs is only about 1.69%, depicted in Figure 8d. Reversely, a large number of TBs can be found at low strain rate as shown in Figure 7j, and the volume fraction of TBs increases to 16.93%.

Interestingly, the majority of DRX grains are twin-related to the parent grains. The opposite grain boundaries of DRX grains are HAGBs and bulged. This phenomenon corresponds to previous studies in some alloys such as copper [29], and this twin-related nucleation mechanism is regarded as a typical feature of discontinuous DRX. Grain-boundary migrations are activated by high strain energy, followed by twin nucleation. The twins always generate within the parent grains, and opposite grain boundaries with misorientation angle more than 15° migrate towards the neighboring grains. Moreover, the volume fraction of TBs increases with the enhancement of DRX extent, which also indicates that the formation of TBs is linked with discontinuous DRX.

4. Conclusions

In this study, uniaxial hot compression tests of Ti-45Al-8.5Nb-0.2W-0.2B-0.3Y alloy were carried out, and the results were analyzed using SEM, X-ray diffraction and EBSD techniques. Some conclusions drawn in this paper were as follows:

(1) Flow stress is sensitive to the temperature and strain rate. With an increase in temperature and a decrease in strain rate, flow stress decreases significantly. The critical strain is blow 0.2, and the value of activation energy Q is 506.38 KJ/mol.

(2) α_2 phase, produced during heat treatment, is reduced by thermo-mechanical coupling during hot compression. The yellow α_2 phase can be observed clearly in a phase map of the heating treatment specimen. However, no yellow α_2 phase exists in deformed specimen.

(3) With an increase in temperature and a decrease in strain rate, the extent of DRX increases. The volume fraction of HAGBs is 31.39% at the temperature of 1200 °C and the strain rate of $0.1\ \mathrm{s}^{-1}$, and that significantly rises to 89.01% at the strain rate of $0.001\ \mathrm{s}^{-1}$.

(4) Most DRX grains are twin-related to the parent grain. The formation of twins can promote the DRX process. The volume fraction of TBs is only 1.69% at the temperature of 1200 °C and the strain rate of $0.1\ \mathrm{s}^{-1}$, but that increases to 16.93% at the condition of completely DRX.

Acknowledgments: This work is financially supported by the Aeronautical Science Foundation of China (Project Number: 2015ZE53057) and the National key Research and Development Program of China (Project Number: 2016YFB0701303).

Author Contributions: Bin Tang and Jinshan Li conceived and designed the experiments; Lin Xiang performed the experiments; Xiangyi Xue and Lin Xiang analyzed the data; Hongchao Kou contributed reagents/materials/analysis tools; Lin Xiang wrote the paper.

Conflicts of Interest: The authors declare no conflict of interest.

References

1. Harding, R.A.; Wickins, M.; Wang, H.; Djambazov, G.; Pericleous, K.A. Development of a turbulence-free casting technique for titanium aluminides. *Intermetallics* **2011**, *19*, 805–813. [CrossRef]

2. Loria, E.A. Gamma titanium aluminides as prospective structural materials. *Intermetallics* **2000**, *8*, 1339–1345. [CrossRef]

3. Clemens, H.; Kestler, H. Processing and Applications of Intermetallic γ-TiAl-Based Alloys. *Adv. Eng. Mater.* **2000**, *2*, 551–570. [CrossRef]

4. Zhang, W.J.; Lorenz, U.; Appel, F. Recovery, recrystallization and phase transformations during thermomechanical processing and treatment of TiAl-based alloys. *Acta Mater.* **2000**, *48*, 2803–2813. [CrossRef]

5. Clemens, H.; Wallgram, W.; Kremmer, S.; Güther, V.; Otto, A.; Bartels, A. Design of novel β-solidifying TiAl alloys with adjustable β/B2-phase fraction and excellent hot-workabilit. *Adv. Eng. Mater.* **2008**, *10*, 707–713. [CrossRef]

6. Clemens, H.; Chladil, H.F.; Wallgram, W.; Zickler, G.A.; Gerling, R.; Liss, K.D.; Kremmer, S.; Güther, V.; Smarsly, W. In and ex situ investigations of the β-phase in a Nb and Mo containing γ-TiAl-based alloy. *Intermetallics* **2008**, *16*, 827–833. [CrossRef]

7. Appel, F.; Oehring, M.; Wagner, R. Novel design concepts for gamma-base titanium aluminide alloys. *Intermetallics* **2000**, *8*, 1283–1312. [CrossRef]

8. Appel, F.; Oehring, M.; Paul, J. Nano-scale design of TiAl alloys based on beta phase decomposition. *Adv. Eng. Mater.* **2006**, *8*, 371–376. [CrossRef]

9. Tang, B.; Cheng, L.; Kou, H.C.; Li, J.S. Hot forging design and microstructure evolution of a high Nb containing TiAl alloy. *Intermetallics* **2015**, *58*, 7–14. [CrossRef]
10. Zghal, S.; Thomas, M.; Naka, S.; Finel, A.; Couret, A. Phase transformations in TiAl-based alloys. *Acta Mater.* **2005**, *53*, 2653–2664. [CrossRef]
11. Cui, N.; Kong, F.T.; Wang, X.P.; Chen, Y.Y.; Zhou, H.T. Hot deformation behavior and dynamic recrystallization of a β-solidifying TiAl alloy. *Mater. Sci. Eng. A* **2016**, *652*, 231–238. [CrossRef]
12. Huang, L.; Liaw, P.K.; Liu, Y.; Huang, J.S. Effect of hot-deformation on the microstructure of the Ti-Al-Nb-W-B alloy. *Intermetallics* **2012**, *28*, 11–15. [CrossRef]
13. Zhang, S.Z.; Kong, F.T.; Chen, Y.Y.; Liu, Z.Y.; Lin, J.P. Phase transformation and microstructure evolution of differently processed Ti-45Al-9Nb-Y alloy. *Intermetallics* **2012**, *31*, 208–216. [CrossRef]
14. Niu, H.Z.; Chen, Y.Y.; Kong, F.T.; Lin, J.P. Microstructure evolution, hot deformation behavior and mechanical properties of Ti-43Al-6Nb-1B alloy. *Intermetallics* **2012**, *31*, 249–256. [CrossRef]
15. Li, J.B.; Liu, Y.; Liu, B.; Wang, Y.; Cao, P.; Zhou, C.X.; Xiang, C.J.; He, Y.H. High temperature deformation behavior of near γ-phase high Nb containing TiAl alloy. *Intermetallics* **2014**, *52*, 49–56. [CrossRef]
16. Cheng, L.; Chang, H.; Tang, B.; Kou, H.C.; Li, J.S. Characteristics of metadynamic recrystallization of a high Nb containing TiAl alloy. *Mater. Lett.* **2013**, *92*, 430–432. [CrossRef]
17. Zhang, S.Z.; Zhang, C.J.; Du, Z.X.; Hou, Z.P.; Lin, P.; Kong, F.T.; Chen, Y.Y. Deformation behavior of high Nb containing TiAl-based alloy in α + γ two phase field region. *Mater. Des.* **2016**, *90*, 225–229. [CrossRef]
18. Cheng, L.; Chang, H.; Tang, B.; Kou, H.C.; Li, J.S. Deformation and dynamic recrystallization behavior of a high Nb containing TiAl alloy. *J. Alloy. Compd.* **2013**, *552*, 363–369. [CrossRef]
19. Wei, D.X.; Koizumi, Y.C.; Nagasako, M.; Chiba, A. Refinement of lamellar structures in Ti-Al alloy. *Acta Mater.* **2017**, *125*, 81–97. [CrossRef]
20. Li, B.H.; Chen, Y.Y.; Hou, Z.Q.; Kong, F.T. Microstructure and mechanical properties of as-cast Ti-43Al-9V-0.3Y alloy. *J. Alloy. Compd.* **2009**, *473*, 123–126. [CrossRef]
21. Chen, Y.Y.; Yang, F.; Kong, F.T.; Xiao, S.L. Microstructure, mechanical properties, hot deformation and oxidation behavior of Ti-45Al-54V-3.6Nb-0.3Y alloy. *J. Alloy. Compd.* **2010**, *498*, 95–101. [CrossRef]
22. Cheng, L.; Xue, X.Y.; Tang, B.; Kou, H.C.; Li, J.S. Flow characteristics and constitutive modeling for elevated temperature deformation of a high Nb containing TiAl alloy. *Intermetallics* **2014**, *49*, 23–28. [CrossRef]
23. Herzig, C.; Przeorski, T.; Mishin, Y. Self-diffusion in γ-TiAl: An experimental study and atomistic calculations. *Intermetallics* **1999**, *7*, 389–404. [CrossRef]
24. Mishin, Y.; Herzig, C. Diffusion in the Ti-Al system. *Acta Mater.* **2000**, *48*, 589–623. [CrossRef]
25. Sprengel, W.; Oikawa, N.; Nakajima, H. Single-phase interdiffusion in TiAl. *Intermetallics* **1996**, *4*, 185–189. [CrossRef]
26. Witusiewicz, V.T.; Bondar, A.A.; Hecht, U.; Velikanova, T.Y. The Al-B-Nb-Ti system IV. Experimental study and thermodynamic re-evaluation of the binary Al-Nb and ternary Al-Nb-Ti systems. *J. Alloy. Compd.* **2009**, *472*, 133–161. [CrossRef]
27. Niu, H.Z.; Kong, F.T.; Chen, Y.Y.; Yang, F. Microstructure characterization and tensile properties of β phase containing TiAl pancake. *J. Alloy. Compd.* **2011**, *509*, 10179–10184. [CrossRef]
28. Gourdet, S.; Montheillet, F. An experimental study of the recrystallization mechanism during hot deformation of aluminium. *Mater. Sci. Eng. A* **2000**, *283*, 274–288. [CrossRef]
29. Miura, H.; Sakai, T.; Mogawa, R.; Jonas, J.J. Nucleation of dynamic recrystallization and variant selection in copper bicrystals. *Philos. Mag.* **2007**, *87*, 4197–4209. [CrossRef]

Article

Springback Estimation in the Hydroforming Process of UNS A92024-T3 Aluminum Alloy by FEM Simulations

Cristina Churiaque [1], Jose Maria Sánchez-Amaya [1,*], Francisco Caamaño [2], Juan Manuel Vazquez-Martinez [3] and Javier Botana [1]

[1] Department of Materials Science and Metallurgical Engineering and Inorganic Chemistry, LABCYP, Faculty of Engineering, University of Cádiz, Av. Universidad de Cadiz 10, 11519 Puerto Real, Spain; cristina.churiaque@uca.es (C.C.); javier.botana@uca.es (J.B.)

[2] Titania, Ensayos y Proyectos Industriales S.L., 11500 El Puerto de Santa María, Spain; francisco.caamano@titania.aero

[3] Department of Mechanical Engineering & Industrial Design, Faculty of Engineering, University of Cádiz, Av. Universidad de Cadiz 10, 11519 Puerto Real, Spain; juanmanuel.vazquez@uca.es

* Correspondence: josemaria.sanchez@uca.es; Tel.: +34-956-483339

Received: 2 May 2018; Accepted: 27 May 2018; Published: 1 June 2018

Abstract: The production of metal parts manufactured through the hydroforming process is strongly affected by the difficulty in predicting the elastic recovery (springback) of the material. In addition, the formation of wrinkles and crack growth should be avoided. Manual cold work is widely employed in industry to obtain the final shape of the manufactured parts. Therefore, an accurate springback estimation is of high interest to reduce the overall time of manufacturing and also to decrease the manual rectification stage. A working procedure based on finite element simulations (FEM) was developed to estimate the elastic recovery and predict the final morphology of UNS A92024-T3 aluminum alloy pieces after forming. Experimental results of real hydroformed parts were compared with the results obtained in simulations performed with PAM-STAMP software. The influence of different experimental parameters on the forming processes was also analyzed, such as the material properties, the rolling direction of sheet metal, or the hardening criteria employed to characterize the plastic region of the alloy. Results obtained in the present work show an excellent agreement between real and simulated tests, the maximum morphology deviations being less than the thickness of parts (2.5 mm). FEM simulations have become a suitable and mature tool that allows the prediction of the pieces springback, a precise material characterization being required to obtain reliable results.

Keywords: hydroforming; springback; FEM simulation; UNS A92024-T3; hardening criteria

---

1. Introduction

The continuous growth of the aerospace industry, especially due to an important increase in the demand of new aircraft, involves some challenges that companies in the sector may face to be more competitive. Among other investments, efforts should be focused towards the improvements of the engineering design, in order to achieve a better development of aerospace products. At the same time, the development of manufacturing processes should be as simple, straightforward, and robust as possible, also maintaining profit margins and creating a consolidated working base for the future. Another key to improve operational performance are the investment in new services and the transformation of manufacturing programs to make them more effective.

Hydroforming in fluid cell presses is a known widespread process applied in different industrial sectors, since it allows one to obtain pieces with different morphologies from flat metal sheets.

This process has the main advantage of being relatively quick, being also carried out at room temperature. These two factors make it a relatively cheap and convenient process. However, it also presents several inherent technical disadvantages, such as the development of defects (cracks or wrinkles) in areas with high plastic deformation, and the elastic recovery (springback) after the application of pressure. The elastic recovery must be estimated and taken into account, as the final shape of the formed part should be as close as possible to the design part [1]. Therefore, in order to manufacture parts with curved shapes, the dies used must have a more pronounced morphology than the design part, with the aim of compensating the springback. In many cases, the dies geometry is estimated taking into account the intuition and practical experience of the operators, or by means of trial–error tests, which constitute significant losses of time and material, decreasing the overall production rate.

As stated before, the hydroforming process involves the generation of residual stresses, high elastic recovery, and the possibility of defects formation, such as wrinkles and cracks. The uncertainty associated with the low precision in the springback estimation usually generates critical variations between machined and design parts, which leads to subsequent manual adjustment of the geometry of the workpiece/part. Most of the studies working on this topic are focused on flange forming [2–4]. One of the emerging methodologies to reduce the possible defects caused in the manufacture of parts by this process is the use of Finite Element Method (FEM) simulations. Making use of these tools, most researches perform a comparison between experimental and simulation studies of the springback effect on flange bending. For example, Lei Chen et al. [5] focused their research on straight jogged flange forming, studying the influence of different parameters on the process, such as: sheet thickness, radius of curvature, flanging height, press-time cycle of press, as well as the influence on the results when using different rubber materials in the press (Shore A70 rubber and Shore A80 polyurethane). The influence of the lubricant and the flexible rubber in the hydroforming process of aluminum components was studied in by Paunoui et al. [6]. These authors claimed that a flexible rubber assures uniform pressure toward the blank, and additionally, that the lubricant influences the process of deformation.

A numerical simulation parametric study of rubber-pad forming process was developed in [7]. This research was focused on the analysis of forming parameters, such as the hardness and type of rubber, or the blank thickness, on the elastic recovery. An interesting result obtained is that the springback angle increases as the initial aluminum blank thickness increases, using the same hardness of the flexible punch with polyurethane and natural rubber. Another remarkable finding is that the use of polyurethane rubber as soft punch is recommended to minimize the springback of aluminum sheet when compared to the natural rubber. Besides, the employment of polyurethane rubber, with less hardness value, may delay the occurrence of damage.

In [8], an analysis of the reduction of time and costs was performed in the sheet metal forming process, making use of FEM simulations. In particular, simulations were carried out to minimize the deviations of the stamping and to anticipate the detection of problems associated with the effects of the springback. This methodology facilitates the perception and monitoring of the piece at different stages of the forming process.

Most recently, advanced simulations of the metal forming process analyze the influence of the die on the elastic deformation [9,10]. Thus, an engineering methodology for the structural assessment of the stamping tooling and the die-face designs during the sheet metal forming processes were presented in [10]. This study compares the simulation results of the metal forming process considering the die as Rigid Solid (common approach in FEM models) with those obtained considering it as Deformable Solid. The deformable solid model was successfully developed, presenting better springback estimation in the forming process of a high strength steel of moderate thickness [10].

On the other hand, an enhancement of dimple formability in sheet metals by two-step forming was proposed in [11] to reduce stamping flaws in curved parts. It presents several FEM simulation models in 2D and 3D, exposing the differences found between the results of the one-step and two-step metallic forming. According to the FEM model, the following outputs were compared in [11]: sheet metal formability and deviations from the objective geometry, thickness distribution and damage percentage, and comparison of strains. The two-step forming process was confirmed to provide better formability, reducing flaws and cracks.

As mentioned in [12], an adequate material characterization is crucial to obtain a reasonable fitting between simulation and experimental data. This implies that the material sheet involved in the forming process should be mechanically tested in order to minimize the comparative errors between numerical results obtained from the FEM simulations (according to the plasticity criteria and selected hardening laws), and the experimental results obtained from the real manufacturing test. Key for the success of simulations of forming processes is the constitutive model used for the description of the plastic behavior. The global plastic deformation model requires the estimation of two factors: the anisotropic behavior and the work-hardening law. Both issues can be estimated independently in the optimization procedure; in a first stage, the material parameters related to the anisotropic behavior can be measured, while in a second stage, the best suitable isotropic work-hardening law can be identified. Given the high importance of understanding both factors, it is considered necessary to revise in the next sections the most common available plasticity criteria and hardening laws.

1.1. Anisotropic Plasticity Criteria

Hill's (1948) quadratic yield criterion is one of the most widely used anisotropic yield criterion. The Hill's 48 yield function is also easy to formulate, and therefore, has been widely used to investigate the effect of anisotropy on springback, especially in sheet metal forming [13]. The material parameters of the Hill's 48 yield function are mainly obtained from Lankford values at angles of $0°$, $45°$, and $90°$ to the rolling direction. The Lankford coefficients needed depend on the type of anisotropy characterized. A type of orthotropic anisotropy (employed in the present study) considers anisotropy in three directions, and accordingly, three values are needed: r_0, r_{45}, and r_{90}. Lankford's coefficient along the "α" direction can be estimated as with Equation (1):

$$r_\alpha = \frac{\varepsilon_w(\alpha)}{\varepsilon_t(\alpha)} \tag{1}$$

where ε_w(width strain) and ε_t(thickness strain). are measured on the specimens in tensile tests.

If Hill's coefficients are known, but not the Lankford's coefficients, it is possible to define them and to transform them into Lankford's coefficients using the following equations. The anisotropic parameters F, G, H, and N can be formulated in terms of the r-values (r_0, r_{45}, r_{90}) as follows in Equations (2)–(5) [14]:

$$F = \frac{r_0}{r_{90}(1 + r_{90})} \tag{2}$$

$$G = \frac{1}{(1 + r_0)} \tag{3}$$

$$H = \frac{r_0}{(1 + r_0)} \tag{4}$$

$$N = \frac{(r_0 + r_{90})(1 + 2r_{45})}{2r_{90}(1 + r_0)} \tag{5}$$

1.2. Hardening Law

During cold work hardening, mechanical deformation brings the material into plastic domain and the material behaves nonlinearly. This behavior exhibits a nonlinear relationship between stress σ and strain ε. In cold forming processes, such as hydroforming the material is subjected to biaxial deformation conditions, and therefore higher strain values than the ones provided by uniaxial tensile test (in which constriction cannot be reached) are needed [15]. Therefore, it is necessary to extrapolate the plastic stress–strain curve to higher values of deformation by different hardening model equations, such as the Krupkowsky Law, Hollomon's Law, or Power-Law. In any case, a common approach to identify the material's work-hardening parameters, is to assume isotropic behavior. The Krupkowsky law follows Equation (6):

$$\sigma = K \cdot (\varepsilon_0 + \varepsilon_p)^n,$$

(6)

In which K is the work hardening coefficient, and n, the work hardening exponent. σ, ε, and ε_0 are the equivalent stress, equivalent strain, and the offset (elastic) strain, respectively [16,17]. Alternatively, the Power-Law can also be employed to model the plastic region, as indicated in Equation (7), in which σ, ε_p, and n are the equivalent stress, the equivalent plastic strain, and the work hardening exponent, respectively.

$$\sigma = a + b \cdot \varepsilon_p^n,$$

(7)

Another common mathematical description of the work hardening phenomenon for an isotropic material is the Hollomon's stress–strain equation [18], shown in Equation (8).

$$\sigma = K \cdot \varepsilon^n,$$

(8)

where K and n are two material constants commonly referred to as strength coefficient and strain hardening exponent respectively.

It is worth mentioning that a comparison of models is provided in [15], where the Krupkowsky law is seen to be preferable to the classical Hollomon law because it better fits to the experimental results and leads to a finite σ/ε_p slope at initial yielding.

1.3. Objective

The main objective of this study was to predict the springback of UNS A92024-T3 aluminum alloy parts with double curvature, manufactured by the hydroforming process with fluid cell press. For this purpose, an experimental procedure based on finite element simulations was developed. In order to obtain reliable results, the influence of different process parameters was determined, as the mechanical properties of the material being formed, the rolling direction of the blank (to take into account the anisotropy), and the hardening model for deformation of the elastoplastic material. After estimating the correct values of these parameters, they were used as input for the FEM model. Finally, results of the simulation stage were compared with experimental tests of the hydroforming process. This methodology may allow precise estimation of the final geometry of the parts and the possible appearance of wrinkles and cracks.

2. Materials and Methods

In this work, flat sheets of UNS A92024-T3 aluminum alloy, composition detailed in Table 1, were hydroformed to generate double curvature probes. The size of the parts is: 215.601 mm × 226.677 mm, with a thickness of 2.5 mm. Two different parts (pieces) were formed, encoded as Part 1 and Part 2, changing only the curvature of the employed dies. The design of Part 1 is depicted in Figure 1. The formed elements acquire double curvature, presenting a maximum difference in height between the corners of the sheet (lower level) and the central area of the piece (maximum level) of 14.94 mm in Part 1 and 10.71 mm in Part 2, as shown in Figure 2. Note that the hydroforming process was performed without drilling the holes marked in Figure 1.

Table 1. Chemical composition (wt %) of the UNS A92024-T3 aluminum alloy.

Material	Si	Fe	Cu	Mn	Mg	Cr	Zn	Ti	Other Each	Other Total	Al
UNS A92024-T3	0.038	0.091	4.46	0.62	1.41	0.005	0.083	0.031	<0.05	<0.15	Bal.

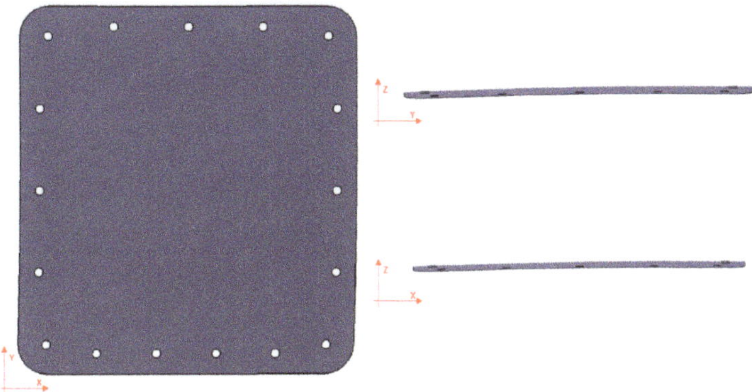

Figure 1. Design of UNS A92024-T3 aluminum alloy hydroformed Part 1.

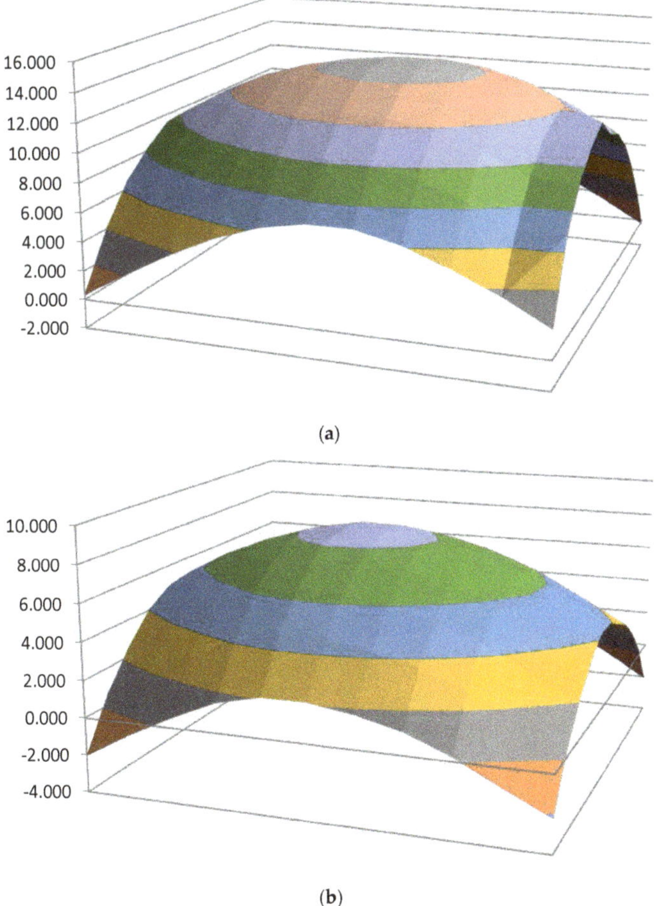

(a)

(b)

Figure 2. (a) Three dimensional surface chart of Part 1 morphology, in mm; (b) Three dimensional surface chart of Part 2 morphology, in mm.

2.1. Material Properties and Experimental Characterization Tests

According to our preliminary experience and to the previous results reported by other authors [12], an adequate material characterization is crucial in order to accurately estimate the actual springback of the hydroformed pieces and therefore, to obtain reliable simulations. In order to evaluate the influence of the mechanical properties of the material being forming, different Material Properties, encoded as MP1 to MP6, have been taken as input of the simulation model. MP1 and MP2 were materials parameters obtained from experimental tensile tests, while MP3 to MP6 were materials parameters taken from minimum requirements of aerospace standards. Table 2 summarizes the source employed to obtain these Material Characterization parameters.

Table 2. Material Properties employed in hydroforming simulations of UNS A92024-T3.

Material Properties	Source	Description of Material Properties (MPs)
MP1	Experimental tests	Tensile tests performed on UNS A92024-T3511 with the same thickness as the hydroformed sheet (2.5 mm)
MP2	Experimental tests	15 tensile tests performed on UNS A92024-T3 samples extracted from the same material batch used in the hydroforming process (2.5 mm)
MP3	Standards data	Minimum MPs values requested by the aerospace standard SAE-AMS-QQ-A-250/5 [19]
MP4	Standards data	Minimum MPs values requested by the aerospace standard SAE-AMS-QQ-A-200/3 [20]
MP5	Standards data	MPs based on SAE-AMS-QQ-A-200/3 [21], with higher elongation value (20%)
MP6	Standards data	MPs based on SAE-AMS-QQ-A-200/3 [21], with higher elongation value (30%)

In order to obtain material parameters MP1 and MP2, standard uniaxial tensile tests, according to ISO 6892-1:2016 [21], were performed on the samples, using a SHIMADZU MTS universal testing machine, with a maximum Load Capacity of 100 kN. The results obtained from the tensile test, as elastic limit, tensile strength, % elongation, Poisson's coefficient and Young's Modulus, were incorporated as input data in the simulation software. MP1 data were obtained from the average values of 3 experimental tensile tests performed on UNS A92024-T3511 samples with the same thickness as the hydroformed sheet (2.5 mm). MP2 data were obtained from a set of 15 standard uniaxial tensile tests performed to 2.5 mm thick samples of UNS A92024-T3 extracted from the same material batch used in the real hydroforming tests. In order to obtain precise materials parameters in MP2, different rolling directions were considered. Thus, 5 specimens with 0° rolling direction, 5 specimens with 45° rolling direction and 5 remaining specimens at 90° were tested. Biaxial strain gauges were installed on specimens to obtain the biaxial yield stress and biaxial anisotropic plasticity coefficients which were derived from the developed curve. The mechanical properties and material constants determined for MP2 are summarized in Table 3.

Table 3. Material properties MP2 of tested material: UNS A92024-T3.

Test Direction	Young's Modulus (MPa)	Yield Strength (MPa)	Ultimate Tensile Stress (MPa)	Elongation (%)	*R*-Value
Rolling direction (0°)	72,809	332	491	19.98	0.49
Diagonal direction (45°)	74,711	294	458	25.25	0.59
Transverse direction (90°)	71,803	337	481	20.86	0.50

The third material properties considered, encoded as MP3, correspond to the minimum material properties values requested by the aerospace specification SAE-AMS-QQ-A-250/5 [19]. This standard is applied to UNS A92024-T3 sheets with thicknesses ranging from 1.25 mm and 6.5 mm, reporting the following minimum values for the Ultimate Tensile Strength (UTS), Yield Strength (YS), and Strain to Rupture (Elongation %): UTS min = 63 ksi (434 MPa); YS min = 42 ksi (290 MPa); Elongation = 15%. Similarly, MP4 to MP6 contain the reference values of material properties indicated in SAE-AMS-QQ-A-200/3 [20], providing the specific material requirements for aluminum alloy 2024 bar, rod, shapes, tube, and wire produced by extrusion. In this standard, the minimum values of UTS, YS and Elongation %, for 2024-T3 with lower thickness than 6.3 mm, are: UTS min = 57 ksi (393 MPa); YS min = 42 ksi (290 MPa); Elongation = 12%. These minimum values were adopted for MP4. MP5 and MP6 have the same material parameters as MP4, with a unique difference imposed of a higher Elongation value. Thus, in MP5, the Elongation was fixed at 20%; while in MP6, the Elongation was 30%. These latter three MPs allowed the analysis of the influence of the ductility.

2.2. Experimental Hydroforming and Geometry Measurement

The fluid cell press forming process is based on the filling of a fluid cell by oil, producing the expansion of an elastic element called a diaphragm. Fluid pressure produces the sheets deformation adopting the geometry of a specific matrix, as shown in Figure 3. In the present study, the experimental hydroforming process was executed using a Flexform Fluid Cell Presses of Quintus Technologies, Type QFC, with rectangular forming trays, and a maximum draw depth of 300 mm. The working pressure was 80 MPa. Polyurethane rubber plates with hardness of 90 Shore A were used in experimental tests.

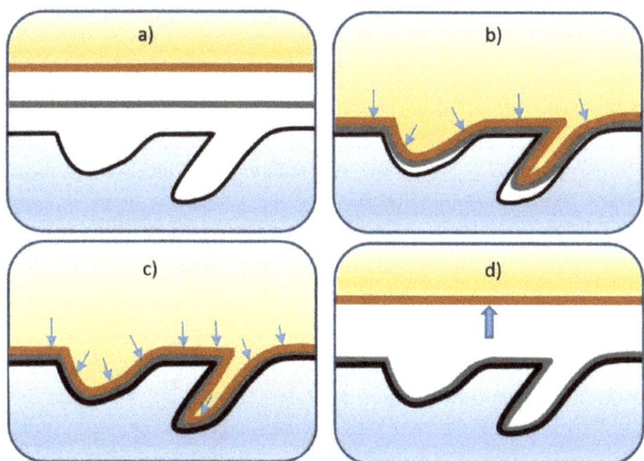

Figure 3. Schematic steps of Flexform process: (**a**) Initial pressure application; (**b**) Sheet partially hydroformed; (**c**) Sheet completely hydroformed; (**d**) Elevation of rubber.

The dimensional evaluation of the geometry generated in the hydroformed elements was carried out using a Helios electronic comparator set, positioning the specimen on a reference flatness surface. The electronic comparator set employed (Helios Messtechnik GmbH & Co., Dörzbach, Germany) has a capacity of 0–12 mm, a resolution of 0.0001 mm, and a measurement uncertainty of 0.0004 mm. To perform the measurement, a grid of 20 mm × 20 mm squares was designed on the shaped pieces, measuring the height of the nodes of the mesh, as illustrated in Figure 4.

The coordinate values obtained in the measurement were used to generate a point cloud using CATIA V5-6R2013 software. Subsequently, a surface containing all the points of the measurement corresponding to the morphology of the real hydroformed piece was generated. This geometry was imported into the simulation software to make a comparison with the pieces obtained in a computational way and assess the reliability of the FEM model.

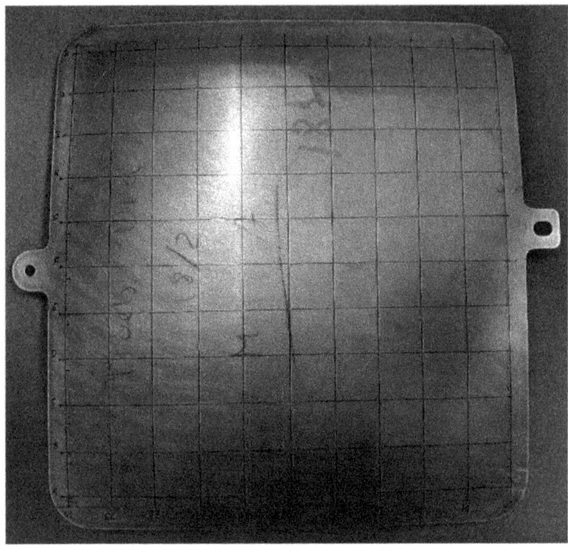

Figure 4. Real hydroformed Piece 1, with a marked surface grid, required to measure the height of different nodes.

2.3. Simulation

An FEM model was developed under the PAM-STAMP 2G simulation package. The different components taking part in the forming process (base, die, metal flat sheet, fixation elements, rubber, etc.) were defined in this software before running the FEM simulations. The load (80 MPa) was applied using a fluid cell forming process. The die geometry was imported from CAD files. The meshes of the involved tools in the forming process are automatically generated, the blank mesh being defined inside the window attributes "Blank editor", where the size and type of elements can be established in PAM-STAMP. The blank mesh is mainly formed by elements of 4 nodes, excepting the areas close to the pin locator holes and the part outline, where elements have 3 nodes. These elements allow the adjustment of the blank mesh to the geometry of the design parts. These pin locators were the fixing elements of the blank during the whole forming process. The total number of elements related to the deformable blank was 13,143, and the average characteristic length (average size of elements) was 1931 mm. The boundary conditions, materials characteristics, and properties, and the 2 processes steps (Stamping and Springback) were also subsequently defined. The boundary conditions in the forming stage are restricted by locator pins. The translation and rotation movement of the blank at locator pins was locked in X, Y, and Z. The boundary conditions set in the springback stage were isostatic locking points. An implicit advanced simulation was carried out at the springback stage. Standard contact was used for advanced implicit springback, available for the tool-blank contact pair. Standard contact is based on a penalty method and it supports friction. The main output of the simulations was the sheet geometry after the hydroforming process. The simulations also allowed us to estimate the elastic recovery (springback) and the prevention of defects formation, such as wrinkles or cracks. In a final step, the geometries obtained in real hydroforming tests were compared with those predicted by the simulations, allowing the estimation of the deviation degree between the FEM estimated and real parts, as presented in one example in Figure 5.

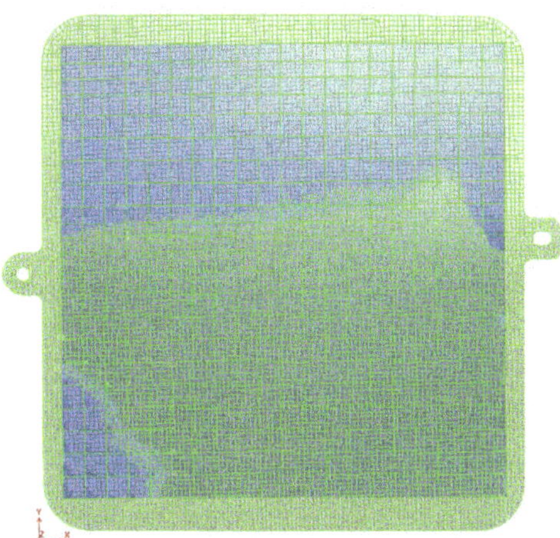

Figure 5. Example of surface superposition of simulated (green mesh) and real hydroformed piece (blue mesh), employed to measure the distance (deviation) between nodes of the grid.

3. Results and Discussion

Numerical simulations of the hydroforming process were carried out with different material parameters for the UNS A92024-T3 alloy. As example of the process, Figure 6 shows the results of hydroformed Part 1 in isometric view (a) and YZ view (b), before (red blank) and after (blue blank) springback. The influence of some variables, such as the mechanical properties, the rolling direction of the blank, or the hardening model, on the process springback was investigated.

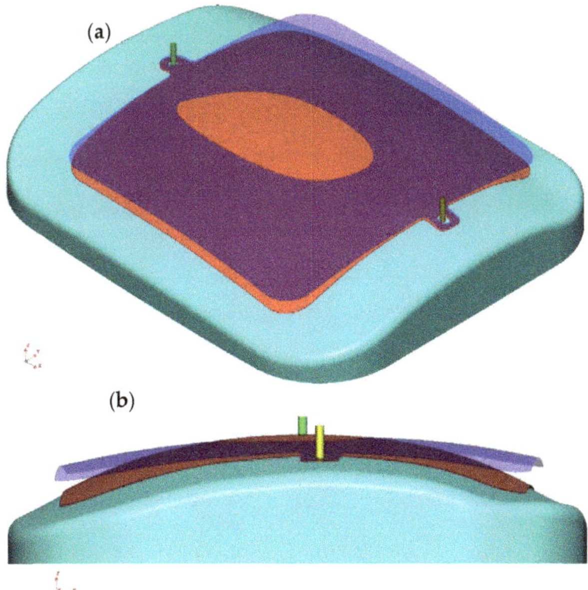

Figure 6. (**a**) Isometric view and (**b**) YZ view, of hydroformed Part 1, before springback (red blank) and after springback (blue blank).

3.1. Influence of Material Properties

The influence of the Material Properties MP1 to MP6 detailed in Section 2.1 was studied with the objective of reproducing the real hydroforming process, and subsequently, being able to estimate the final geometry by means of the FEM simulations. The sources used to obtain the different MPs are reported in Table 2. The first three rows of Table 4 include the values of the main mechanical properties estimated: UTS, YS and elongation. Different FEM simulations of hydroforming process of Part 1 (Piece 1) were solved taking into account these input parameters values. The geometry predicted by these simulations was later compared with the real geometry obtained in the plant for the Part 1. This actual real geometry was obtained from the CATIA point cloud measured with the electronic comparator set. As example of these comparisons, Figure 7 depicts the distance between the real geometry of Part1 and the predicted geometry of the simulations considering material properties MP1 to MP6. The best fitting was found in the case of MP2. As a summary of the comparison between the real and predicted geometries by simulations, two parameters were considered: Maximum distance between the simulated and real Part, (encoded as *DMAX*, measured in millimeters), and Percentage of surface with distance between simulated and real Part lower than 0.5 mm (in %, indicated as *%SURF*). These proposed parameters allow an adequate comparison between the real measured geometry and the geometries estimated by simulations. The fitting is much better (meaning better simulations) as *DMAX* (linear graph) is minimized and *%SURF* (Bar Chart) is maximized. These parameters results are depicted in the last two rows of Table 4. For better visualization, these results have been included in Figure 8.

Figure 8 shows that MP2 is the material parameter data set that better reproduces the morphology of the real hydroformed Part 1. In fact, when MP2 is employed as input parameters, the simulation provides a geometry quite close to the real measured one, giving the highest *%SURF* and the lowest *DMAX* values. Thus, the simulation run with MP2 as input parameters generated a surface with lower deviation than 0.5 mm in a 92.5% of the real piece. In addition, the maximum distance measured between the simulated and real part was 2 mm. When comparing MP1 and MP2 results (tested materials), it is also interesting to note that MP2, with higher Elongation value and lower UTS and YS values, provided a simulation with lower springback, and therefore, with higher curvature than MP1. This behavior is related to the fact that the MP1 was obtained from a cold hardened sheet (related to the bending process of T3511 treatment), in comparison with MP2, a sheet not previously subjected to cold work hardening. As a consequence, MP1 provides higher springback and therefore, lower curvature than MP2. The better approximation obtained with MP2 was expected, provided that these values were obtained from a complete experimental tensile tests program performed with samples of the same material being formed, allowing also the estimation of the anisotropy values (only available for MP2).

Although it is clear that MP1 (batch subjected to cold-rolling + bending) has higher anisotropy than MP2 (only subjected to previous cold-rolling), the hardening process (related to the last bending step of MP1) is thought to have a much higher influence on the springback results than the anisotropy. In fact, mechanical values, included in Table 4, show that MP1 presents higher UTS and YS, and lower elongation ductility than MP2. Therefore, although MP2 presents lower anisotropy than MP1, it seems that the hardening effect has a higher influence on results than the anisotropy. This assumption is dealt with in continuation (results of MP4–MP6) and in Section 3.2 (influence of rolling direction). In any case, it should be clarified that the influence of the anisotropy on the springback results was not explicitly investigated in this work, being a very interesting issue to address in future research.

Table 4. Summary of Material Properties (MPs) and results of forming simulations with different MPs.

Summary of Mechanical Properties and FEM Results	MP1	MP2	MP3	MP4	MP5	MP6
Ultimate Tensile Strength, UTS (MPa)	535	477	421	393	393	393
Yield Strength, YS (MPa)	396	321	276	290	290	290
Elongation (%)	11	22	15	12	20	30
Maximum distance between simulated and real Part (mm)	3.5	2.0	2.8	2.6	3.3	3.6
Surface with distance between simulated and real Part < 0.5 mm (%)	24.4	92.5	50.6	45.4	25.8	20.7

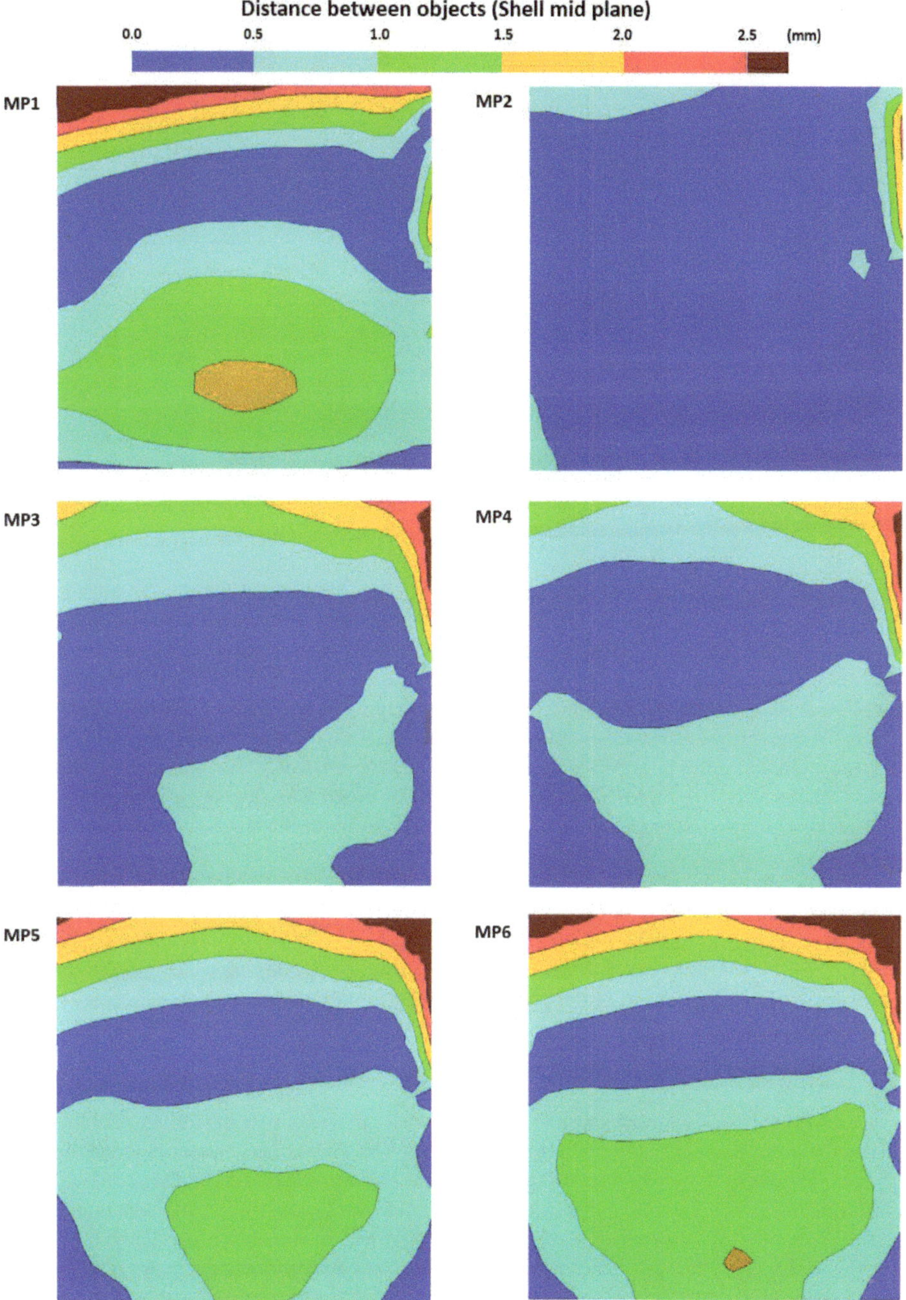

Figure 7. Distance between the real hydroformed piece (Part 1) and simulations employing the six different material properties (MP1 to MP6) summarized in Table 4.

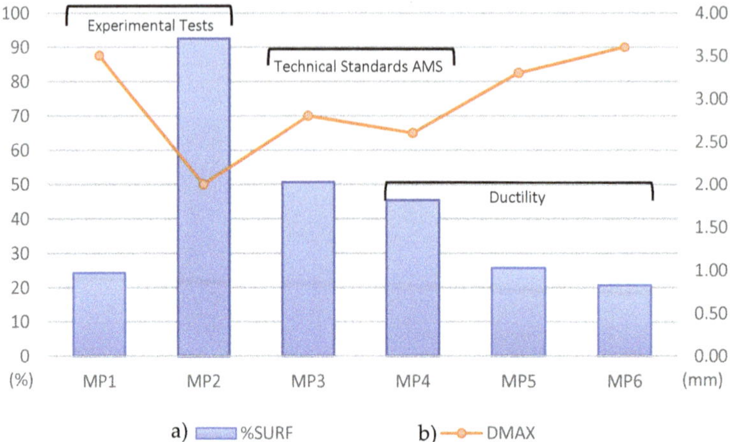

Figure 8. Results of simulations performed with different mechanical properties of the material: (**a**) Bar Chart with Surface % with distance between Real and Simulation Parts < 0.5 mm (*%SURF*); (**b**) Linear Graph with maximum distance between Real and Simulation Piece (*DMAX*).

MP2 was also compared with material properties of requirements SAE-AMS-QQ-A-250/5 (MP3) and SAE-AMS-QQ-A-200/3 (MP4). The fittings provided by MP3 and MP4 were acceptable, although they were worse than that obtained with MP2 even though, both MP3 and MP4 gave better results than MP1. These results are related to the relatively low values of elongation, UTS, and YS of both MP3 and MP4. These properties were observed to lead to simulations with moderated springback values. In order to evaluate the influence of the sheet ductility on the hydroforming process, MP5 and MP6 were considered as input parameters, having the same properties of MP4, excepting the elongation values (12% for MP4, 20% for MP5, and 30% for MP6). The progressive fluence increase (higher plastic deformation) imposed from MP4 to MP6, leads to lower springback values, and therefore to higher curvature, in agreement with [12,22].

The overall results obtained in this section demonstrate the big influence of the material parameters on the reliability of the hydroforming simulations. An appropriate material characterization test program is seen to be essential to estimate the correct parameters values of the material being hydroformed, which is required to obtain reliable springback estimations. In general, hardened materials (with low ductility and high UTS and YS values) lead to high springback recovery values, and therefore, to low curvature hydroformed pieces [23].

3.2. Influence of the Rolling Direction

A new set of hydroforming simulations was carried out to evaluate the influence of the rolling direction. For this study, only the material characterization MP2 was considered, comparing the results achieved when the rolling direction was parallel to the X-axis (as indicated in Figure 9a) with those results obtained when the rolling direction was parallel to the Y-axis (Figure 9b). Note that in all the simulations reported in the previous section, the rolling direction was always parallel to the X-axis.

The geometry of the real hydroformed Part 1 was compared with the geometry obtained from simulations with rolling direction parallel to the X-axis (RD1X) and with rolling direction parallel to the Y-axis (RD1Y). The distances between Part 1 and geometries obtained with those simulations are summarized in Figure 10, in which the results of both previously employed parameters (*%SURF* and *DMAX*) are depicted.

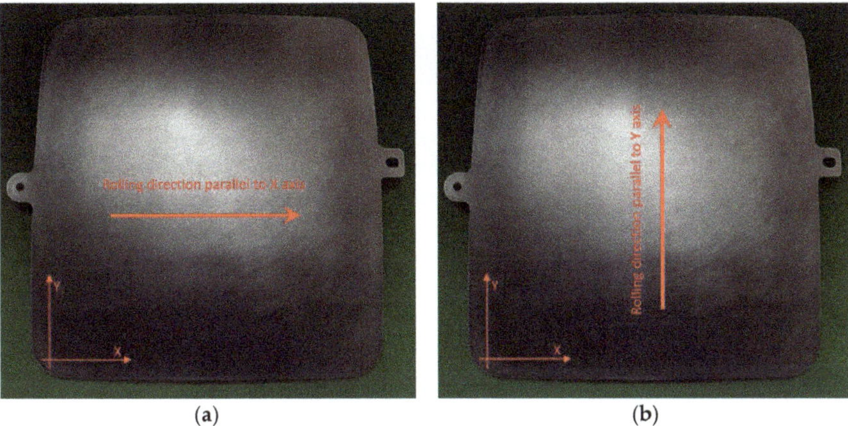

(**a**)   (**b**)

Figure 9. (**a**) Rolling direction of the simulations and real hydroformed sheet encoded as Part 1 (parallel to the X-axis); (**b**) rolling direction considered for the simulation of the hydroformed sheet (parallel to the Y-axis).

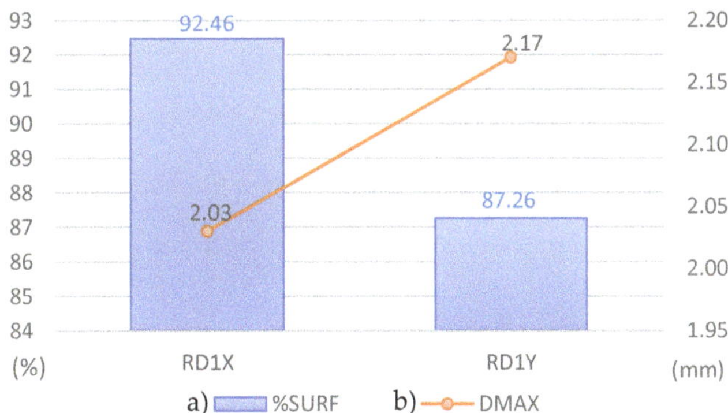

Figure 10. Comparison between the real hydroformed Part 1 with simulations considering rolling direction parallel to X (RD1X) and rolling direction parallel to Y (RD1Y): (**a**) Bar Chart with Surface % with distance between Real and Simulation Parts < 0.5 mm; (**b**) Linear Graph with maximum distance between Real and Simulation Piece.

It is clearly observed in Figure 8 that the geometry in RD1Y has a greater deviation than in RD1X. It is experimentally observed that the formed sheet presents greater recovery when positioned with RD1Y (Figure 9b), adopting therefore, less overall curvature. This difference is associated with the asymmetric behavior of the sheet being formed. Nevertheless, the differences found when varying the rolling direction were relatively low when compared with the influence of the material parameters. Therefore, it can be concluded in this section that the rolling direction is a variable with relatively low influence on the simulations results of the studied piece. This finding is of high importance at a manufacturing industrial level, as in the real hydroforming process performed in the plant, it is not usual to consider the rolling direction before cutting and mechanizing the hydroforming blank. In summary, although the rolling direction does change the results of the springback, and consequently, the results of the final geometry, it is not a critical factor.

3.3. Influence of the Hardening Model

Many published studies have investigated the influence of constitutive equations on simulated springback prediction [5,14,22]. For example, Ki-Young Seo et al. [14] performed a springback evaluation of TRIP1180 steel sheet. The constitutive equations Hill's 48 and Yld2000-2d were used to describe its yield behavior. Moreover, isotropic and kinematic hardening models based on the Yoshida-Uemori model were adopted to express hardening behavior. In this case, the material constants of TRIP1180 for the constitutive equations were obtained from uniaxial tension, tension—compression, loading—unloading, and hydraulic bulging tests.

The influence of the hardening model on the precision of the hydroforming simulation results is analyzed in this section. The morphology obtained from the real hydroforming experiment of Piece 1 was compared with the geometries predicted by the simulations when applying three different hardening criteria: Krupkowsky, Power Law and Hollomon (according to Equations (6)–(8), respectively). An Excel VBA (Visual Basic for Applications) tool was created to obtain the coefficients of each hardening law. The procedure consisted of fitting the three hardening curves to the plastic region of the true stress–strain curves, considering MP2 as the materials properties. These fittings, plotted in Figure 11, allowed the estimation of the coefficients of the three hardening curves (Krupkowsky, Power Law, and Hollomon).

The real hydroformed geometry of Piece 1 was compared with those obtained by simulations employing Krupkowsky (HC1K), Power Law (HC1P) and Hollomon (HC1H) as hardening criteria. Figure 12 summarizes the results of the already-mentioned parameters *%SURF* and *DMAX*. The figure reveals the relatively high correlation values obtained in the three cases, as all *%SURF* values were always higher than 90% and *DMAX* lower than the sheet thickness (2.5 mm). This means that the three criteria allow a more than reasonable fitting of the simulated curves to the true stress–strain curves. HC1K results (simulations with Krupkowsky coefficients) present the highest *%SURF*, being HC1H (Hollomon Law) the one providing the lowest *DMAX*. Considering *%SURF* as the prevalent parameter, it can be concluded that Krupkowsky criterion provides more adjusted results in the simulations of the sheet metal hydroforming process of this material. Therefore, according to these results, the most precise hydroforming simulations for the UNS A92024-T3 alloy were obtained when employing the Krupkowsky coefficients as hardening criteria.

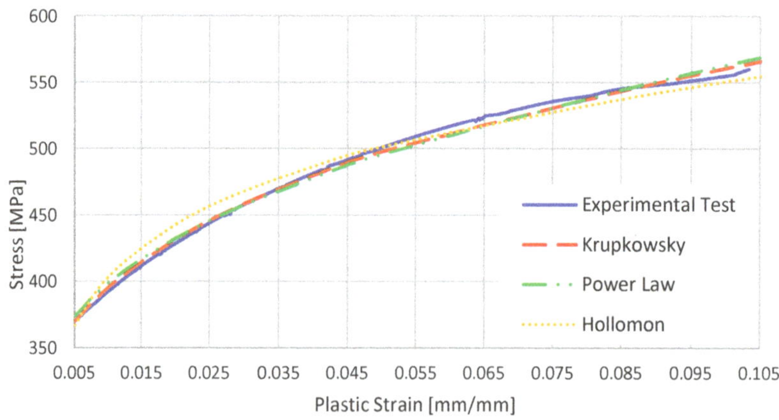

Figure 11. Hardening curves of the material UNS A92024-T3 aluminum alloy according to Krupkowsky, Power Law, and Hollomon criteria.

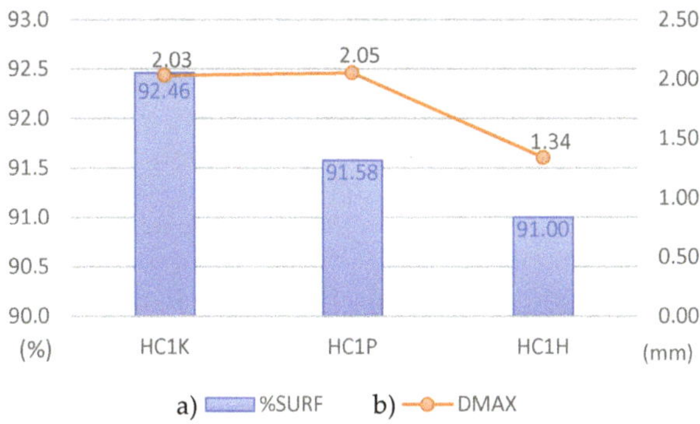

Figure 12. Results of simulations performed with different hardening criteria: Krupkowsky (HC1K), Power Law (HC1P) and Hollomon (HC1H). (**a**) Bar Chart with % Surface with distance between Real and Simulation Parts < 0.5 mm; (**b**) Linear Graph with maximum distance between Real Piece and Simulation Piece.

3.4. Validation of the Model

From the above-mentioned results comparing the simulations and the real hydroforming process of Piece 1, it was determined that the better fittings were obtained with the material properties MP2, with the rolling direction RD1X, and with Krupkowsky Law (HCK) as hardening criteria. In order to validate all the above mentioned experimental and simulation parameters, a second hydroforming experiment was performed to provide the Piece 2, similar to Piece 1, but with lower curvature. This Piece 2 was formed with a new Die2 tool. A new hydroforming simulation with this Die2 and the previously optimized parameters (MP2, RD1X, and HCK) was launched. The predicted morphology was compared with the measured geometry of Piece 2. The results of this comparison are plotted in Figure 13. An excellent geometric adjustment was found, the values of *%SURF* and *DMAX* being 93.79% and 1.63 mm, respectively. The fitting of Piece 2 is therefore even better than the one of Piece 1 (Figure 10), as *%SURF* is higher and *DMAX* is lower than in Piece 1.

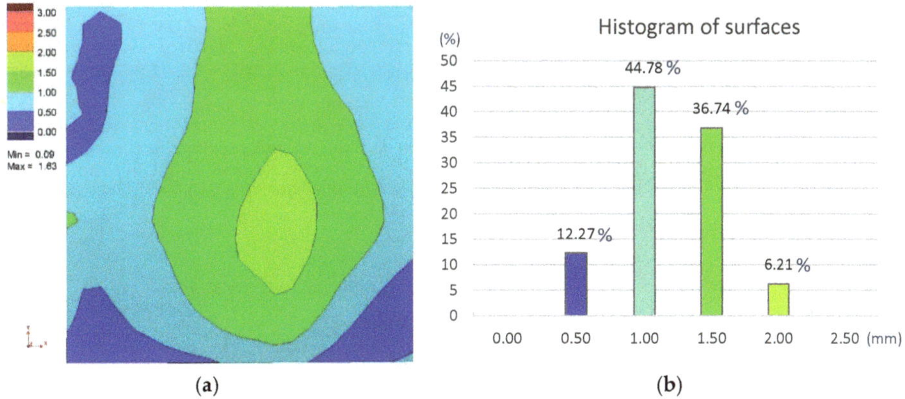

Figure 13. Simulation of Piece 2: (**a**) Distance between Real Part and Simulated Part with Die2, material characterization MP2, rolling direction parallel to the X-axis and hardening law according to Krupkowsky. (**b**) Histogram bar chart depicting the surface percentage with lower deviations than 0.5, 1.0, 1.5, and 2.0 mm, measured between the real and simulated blank.

The results of this second hydroforming experiments allowed us to confirm that the developed simulation process provides a reliable springback estimation, and therefore, an accurate geometry approximation.

The overall results obtained in this investigation demonstrate the high influence of some experimental variables of the hydroforming process on the springback results. Among these variables, the material properties of the blank are observed to play an essential role, as different batches of sheets (MP1 and MP2) lead to different springback values. Therefore, an appropriate methodology to estimate the hydroforming springback is highly advisable, such as the FEM procedure developed in the present work. These tools can clearly benefit the manufacturing industry of sheet metal production, as they can make the hydroforming a much more robust, accurate, predictable, efficient, and productive process.

4. Conclusions

In the present paper, a working procedure based on finite element simulations was developed to predict the elastic recovery (springback) of double curvature UNS A92024-T3 pieces after the hydroforming process, estimating therefore the final geometry of the parts. Experimental results of real formed parts were compared with the results obtained in simulations. The influence of different parameters was analyzed, such as the material properties, the rolling direction and the hardening criteria.

The material parameters showed a high effect on the results of the hydroforming simulations. Thus, an appropriate material characterization is observed to be required to obtain reliable springback estimations. It is highly recommended to use material properties obtained from the same batches than the material being hydroformed. In general, hardened materials lead to high springback recovery values, and therefore, to low curvature hydroformed pieces. The rolling direction is observed to induce little change on the springback, being considered a low influencing parameter. Krupkowsky, Power Law, and Hollomon hardening models of the material being hydroformed were observed to lead to similar reliable results, all of them providing a good fitting to the experimental stress strain curves. Krupkowsky criterion seems to provide slightly more adjusted results.

It is highly recommended to follow a systematic simulation procedure to optimize the results of the hydroforming process, as it allows a better fitting between simulated and real results. The developed FEM strategy was validated taking into account the real geometry of two different hydroformed pieces.

Author Contributions: The present paper is part of the MSc Thesis defended by C.C. C.C. optimized and performed the FEM simulations, analyzed the overall data, and wrote the paper; J.M.S.-A. supervised the whole work, analyzed and discussed the data, and revised the paper. F.C. performed at Titania facilities the tensile tests required to obtain MP1 and MP2. J.M.V.-M. performed the geometry measurements of the hydroformed parts; and J.B. discussed the data and revised the paper.

Acknowledgments: The authors would like to thank Antonio Aragón and María Cruz Jimena, of Airbus D&S (CBC), for providing the 2024-T3 material, sharing their hydroforming process expertise, and performing the experimental hydroforming tests at the CBC oil hydraulic press equipment. Authors would like also thank the student Alberto Baro for his help in fitting the hardening curves with the Excel VBA tool.

Conflicts of Interest: The authors declare no conflict of interest.

References

1. Sala, G. A numerical and experimental approach to optimise sheet stamping technologies: Part II—Aluminium alloys rubber-forming. *Mater. Des.* **2000**, *22*, 299–315. [CrossRef]
2. Chen, L.; Chen, H.; Guo, W.; Chen, G.; Wang, Q. Experimental and simulation studies of springback in rubber forming using aluminium sheet straight flanging process. *Mater. Des.* **2014**, *54*, 354–360. [CrossRef]
3. Sinke, J. Spring back of Curved Flanges of Rubber Formed Aluminium parts. *Key Eng. Mater.* **2013**, *554–557*, 1851–1855. [CrossRef]
4. Asnafi, N. On stretch and shrink flanging of sheet aluminium by fluid forming. *Mater. Process. Technol.* **1999**, *96*, 198–214. [CrossRef]

5. Chen, L.; Chen, H.; Wang, Q.; Li, Z. Studies on wrinkling and control method in rubber forming using aluminium sheet shrink flanging process. *Mater. Des.* **2015**, *65*, 505–510. [CrossRef]

6. Paunoiu, V.; Teodor, V.; Susac, F. Researches regarding the hydroforming process of aluminum components. *Mod. Technol. Ind. Eng.* **2015**, *95*, 1–6. [CrossRef]

7. Belhassen, L.; Koubaa, S.; Wali, M.; Dammak, F. Numerical prediction of springback and ductile damage in rubber-pad forming process of aluminum sheet metal. *Int. J. Mech. Sci.* **2016**, *117*, 218–226. [CrossRef]

8. Gomes, T.; Silva, F.J.G.; Campilho, R.D.G.S. Reducing the simulation cost on dual-phase steel stamping process. *Procedia Manuf.* **2017**, *11*, 474–481. [CrossRef]

9. Neto, D.M.; Coër, J.; Oliveira, M.C.; Alves, J.L.; Manach, P.Y.; Menezes, L.F. Numerical analysis on the elastic deformation of the tools in sheet metal forming processes. *Int. J. Solids Struct.* **2016**, *100–101*, 270–285. [CrossRef]

10. Firat, M. Computer aided analysis and design of sheet metal forming processes: Part III: Stamping die-face design. *Mater. Des.* **2007**, *28*, 1311–1320. [CrossRef]

11. Kim, M.; Bang, S.; Lee, H.; Kim, N.; Kim, D. Enhancement of dimple formability in sheet metals by 2-step forming. *Mater. Des.* **2014**, *54*, 121–129. [CrossRef]

12. Banabic, D. Phenomenological Constitutive Models Parameters Optimization. In *Advanced Methods in Material Forming*; Springer: Berlin/Heidelberg, Germany, 2007; p. 39.

13. Hill, R. A theory of the yielding and plastic flow of anisotropic metals. *Proc. R. Soc. Lond. Ser. A Math. Phys. Sci.* **1948**, *A193*, 281–297. [CrossRef]

14. Seo, K.-Y.; Kim, J.-H.; Lee, H.-S.; Kimm, J.H.; Kim, B.-M. Effect of Constitutive Equations on Springback Prediction Accuracy in the TRIP1180 Cold Stamping. *Metals* **2017**, *8*, 18. [CrossRef]

15. Atluri, S.N.; Yagawa, G.; Cruse, T.A. Computational Mechanics '95: Vol. 1 and Vol. 2. Theory and Applications. In Proceedings of the International Conference on Computational Engineering Science, Atlanta, GA, USA, 10–14 April 1988; Springer: Berlin/Heidelberg, Germany, 1995; p. 1326.

16. Abe, T.; Tsuta, T. Pergamon, Amsterdam-Oxford-New York-Tokyo. In Proceedings of the Asi-PAcific Symposium on Advances in Engineering Plasticity and Its Applications (AEPA'96), Hiroshima, Japan, 21–24 August 1996; p. 686.

17. Zhao, B.; Zhang, S.; Lu, X.; Dong, Q. Cyclic tangential loading of a power-law hardening elastic–plastic spherical contact in pre-sliding stage. *Int. J. Mech. Sci.* **2017**, *128–129*, 652–658. [CrossRef]

18. Elgindi, M.B.; Wei, D.; Liu, Y.; Kamran, K.; Xu, H. Buckling and deformation of Hollomon's power-law tubes. *Thin-Walled Struct.* **2014**, *74*, 213–221. [CrossRef]

19. AMS. *Aluminium Alloy 2024, Plate and Sheet: SAE-AMS-QQ-A-250/4B (2015)*; Aerospace Material Specification, SAE International; AMS: Warrendale, PA, USA, 2015.

20. AMS. *Aluminum Alloy 2024, Bar, Rod, Shapes, Tube, and Wire, Extruded: SAE-AMS-QQ-A-200-3 (2007)*; Aerospace Material Specification, SAE International; AMS: Warrendale, PA, USA, 2015.

21. ISO. *Metallic Materials. Tensile Testing. Part 1: Method of Test at Room Temperature*; International Standard ISO-6892-1:2016; International Organization for Standard: Geneva, Switzerland, 2016.

22. Asnafi, N. On springback of double-curved autobody panels. *Int. J. Mech. Sci.* **2001**, *43*, 5–37. [CrossRef]

23. Bruni, C.; Celeghini, M.; Geiger, M.; Gabrielli, F. A study of techniques in the evaluation of springback and residual stress in hydroforming. *Int. J. Adv. Manuf. Technol.* **2007**, *33*, 929–939. [CrossRef]

 *metals*

Article

Effects of Platform Pre-Heating and Thermal-Treatment Strategies on Properties of AlSi10Mg Alloy Processed by Selective Laser Melting

Riccardo Casati [1], Milad Hamidi Nasab [1], Mauro Coduri [2], Valeria Tirelli [3] and Maurizio Vedani [1,*]

[1] Department of Mechanical Engineering, Politecnico di Milano, 20156 Milan, Italy; riccardo.casati@polimi.it (R.C.); milad.hamidi@polimi.it (M.H.N.)
[2] ESRF, European Syncrotron Radiation Facility, 38043 Grenoble, France; mauro.coduri@esrf.fr
[3] Aidro S.r.l, 21020 Taino, Italy; valeria.tirelli@aidro.it
* Correspondence: maurizio.vedani@polimi.it; Tel.: +39-02-2399-8230

Received: 12 October 2018; Accepted: 13 November 2018; Published: 15 November 2018

Abstract: The AlSi10Mg alloy was processed by selective laser melting using both hot- and cold-build platforms. The investigation was aimed at defining suitable platform pre-heating and post-process thermal treatment strategies, taking into consideration the peculiar microstructures generated. Microstructural analyses, differential scanning calorimetry, and high-resolution diffraction from synchrotron radiation, showed that in the cold platform as-built condition, the amount of supersaturated Si was higher than in hot platform samples. The best hardness and tensile performance were achieved upon direct aging from cold platform-printed alloys. The hot platform strategy led to a loss in the aging response, since the long processing times spent at high temperature induced a substantial overaging effect, already in the as-built samples. Finally, the standard T6 temper consisting of post-process solution annealing followed by artificial aging, resulted in higher ductility but lower mechanical strength.

Keywords: selective laser melting; AlSi10Mg alloy; processing temperature; aging treatment

---

1. Introduction

Due to large freedom in shaping and high material usage efficiency, selective laser melting (SLM) is attracting a great interest for a large number of industrial applications requiring near-net shape manufacturing of light components. SLM allows generating parts by selectively scanning powdered metals with a laser beam, so as to produce layer by layer solid volumes after melting and solidification [1,2]. AlSi10Mg and AlSi7Mg alloys are amongst the most popular materials for light structures built by SLM. They are selected among the traditional foundry alloys featuring good castability and well-known solidification behavior [3–5].

The trend toward increased lightness and higher load-bearing capacity is now pushing for higher material strength, which can be achieved either by alternative Al alloy formulations or by optimized thermal treatments of current materials. High-strength precipitation hardenable Al-Zn-Mg-Cu and Al-Cu-Mg-Si alloys offer large opportunities to achieve high mechanical performances, but their processability is often limited due to hot-cracking phenomena occurring during the last stages of the rapid solidification induced by SLM [6,7]. An alternative approach, aimed at raising the strength of the above easily processable Al-Si-Mg alloys, is to optimize their post-process thermal treatments, trying to exploit, as much as possible, the supersaturation of the solid solution and the fine cellular structure brought about by the rapid solidification and cooling induced by SLM.

From the recent literature, there is evidence that several kinds of post-process heat treatments can be considered. Stress relieving is often carried out on as-built AlSi10Mg parts at temperatures of 300 °C for 2 h [8–11], but investigations showing the effectiveness of annealing treatments at temperatures as low as 200 °C are also available [10,12]. Such treatments lead to a substantial increase in alloy ductility and a concurrent drop in strength.

Solution-annealing treatments and solution annealing followed by aging (T6-like thermal treatments) have been also investigated on AlSi10Mg and AlSi7Mg alloys processed by SLM, in order to modify the properties of the as-built materials [3,5,9,11,13–19]. It was shown that the standard solution annealing, followed by water quenching and artificial aging, replaces the fine α-Al cells, and the Si network segregated at their boundaries into coarser grains and globular Si particles. Evidence of laser tracks, and of the related heat-affected zones, also disappeared during solution annealing [3,9,13,15,19]. Also, in this case, fracture elongation significantly improves, but hardness, yield strength, and ultimate tensile strength drop down to values lower than those of the as-built material.

An interesting database about the mechanical properties available in literature for the AlSi10Mg alloy, processed by SLM, was given in a recent paper by Tang and Pistorious [20], who showed a large variability of the achieved performance depending on sample orientation, processing parameters, and heat treatment condition. Maamoun and co-workers [21] published a research study on thermal post-processing of AlSi10Mg alloy after SLM, and supplied conclusive suggestions in the form of a microstructure/microhardness map on possible options for post-processing treatments. However, it must be considered that the adoption of the aging treatment was evaluated only after the standard solution-annealing treatment, according to conventional T6-like treatment used in cast and wrought Al alloys.

The aging response of an A357 (AlSi7Mg) alloy produced by SLM was specifically investigated in [18]. Differential scanning calorimetry (DSC) and microhardness analyses showed that SLM-processed samples feature the same precipitation sequence of the corresponding cast alloy, by the sequential formation of Mg-Mg, Si-Si, and Mg-Si co-clusters, β'', β', and β (Mg_2Si) phases, along with the precipitation of Si [22–24]. Moreover, it was demonstrated that the as-built and the solution-annealed and quenched samples experienced, substantially, the same precipitation sequence when subjected to direct aging. The as-built condition can, therefore, be considered as fully appropriate for an effective dispersion strengthening process by direct artificial aging, without the need of any post-SLM solution-annealing treatment.

Further information on aging behavior of supersaturated alloys generated by rapid solidification can be obtained from a recent work published by Marola and co-authors [25]. Their comparative study on an AlSi10Mg alloy, processed by SLM, copper mold casting, and melt spinning, showed that extensive Si supersaturation takes place in rapidly solidified alloys. At the early stages of aging, the supersaturated Si was able to precipitate, leading to broad DSC signals that overlap to those originally assigned to β'' formation. These results are consistent with XRD data supplied by Li and co-workers in their study on effects of heat treatments of AlSi10Mg alloy produced by SLM [13].

The above observations suggest that the peculiar microstructure of Al alloys after the rapid solidification and cooling conditions inherited by SLM could be considered as a profitable starting point for the design of optimized aging treatments in additively manufactured Al alloys. The present paper is, therefore, aimed at evaluating the benefits of different possible heat treatment strategies on an AlSi10Mg alloy. Opportunities offered by SLM processing on room-temperature building platforms, rather than on hot building platforms, are also investigated. In the first scenario, the supersaturated solid solution of as-built material is exploited as the starting condition for subsequent aging whereas, in the latter, the strengthening action of a possible in situ aging treatment is investigated.

2. Materials and Methods

An EOS M290 SLM system was adopted to process gas-atomized AlSi10Mg powders according to the parameters supplied in Table 1. The table also shows that a set of samples, labeled as HP, was produced by depositing powder layers on a preheated build platform at 160 °C, whereas a second set, labeled as CP, was printed without any preheating. The specific temperature level of 160 °C, for the hot platform, was selected in order to reproduce the usual aging temperature adopted for cast AlSi10Mg alloys. It should be considered that information about the build platform temperature are not always supplied in the experimental descriptions of published investigations. Examples of 35 °C [8,10,26–28], 100 °C [13], and 200 °C [21] platform processing are available in literature.

Cubes of 20 mm side, and cylindrical samples with a diameter of 10 mm and length of 100 mm, were built using a scanning strategy involving a rotation of 67° between consecutive layers. Sample density was checked by Archimedes' method, assuming a theoretical density of the alloy of 2.67 g/cm^3. Five distinct measurements were performed.

Table 1. Process parameters adopted for the production of the AlSi10Mg samples.

Sample Code	Power (W)	Hatch Distance (mm)	Scan Rate (mm/s)	Layer Thickness (mm)	Platform Temperature (°C)
CP	340	0.2	1300	0.03	RT
HP					160

Different thermal treatment conditions were considered for the investigation, according to data collected in Table 2. In particular, samples produced by the cold platform were investigated either in the as-built (AB), as-built and directly aged (T5 temper), and solution-annealed, water-quenched, and aged (T6 temper) conditions. Samples produced by the hot platform procedure were only considered in the as-built condition, in an attempt to investigate the possible in situ aging effects induced by the high-temperature holding times during SLM processing.

Scanning electron microscope (SEM) (EVO 50, Zeiss, Germany) and optical microscope (Leitz Aristomet, Germany) were used to investigate alloy microstructure after chemical etching with Keller's and Weck's solutions. DSC analyses were performed in an Ar atmosphere using a heating rate of 30 °C/min to investigate aging behavior of the alloy treated according to the different conditions (i.e., as-built from cold platform, as-built from hot platform, after solution annealing and water quenching).

Table 2. Investigated tempers of the AlSi10Mg samples (w.q.: water quenching).

Sample Code	Platform Temperature	Solution Treatment	Aging
CP AB	RT	None	None
CP T5	RT	None	160 °C 4 h
CP sol	RT	540 °C 1 h, w.q.	None
CP T6	RT	540 °C 1 h, w.q.	160 °C 4 h
HP AB	160 °C	None	None

XRD investigations were carried out at the ID22 beamline of the European Synchrotron Radiation Facility (ESRF) in Grenoble (France), using a high-resolution setup, i.e., with nine scintillators preceded by Si 111 crystal analyzers, which guarantee accurate determination of lattice parameters, as well as suppression of parasitic scattering. The specimens were cut with a cylindrical geometry and diameter of about 1 mm, and rotated during acquisition to increase statistical averaging of crystallites. Incident wavelength was set to 0.35434 Å (~35 keV). XRD patterns were collected, summing different scans with 45 min total acquisition time for each sample, from 0° to 35° in 2 theta. For data analysis, the amounts of Si phase and the estimated standard deviations were computed by Rietveld refinements based on the relative intensities of Si and Al Bragg reflections. Moreover, the crystallite size of Si (when in nanocrystalline form) was evaluated on the basis of the broadening of Bragg peaks, using the Williamson-Hall approach [29].

Tensile tests were performed according to EN ISO 6892-1:2016 standard at room temperature, with a crosshead speed of 0.5 mm/min, using an MTS Alliance RT/100 universal testing frame equipped with an extensometer. Specimens having a gauge length of 30 mm and a diameter of 6 mm, machined from bars built along horizontal (powder bed layers parallel to longitudinal axis of the specimens) and vertical positions, were adopted. At least three specimens for each condition were tested.

3. Results

3.1. Microstructure

Representative images of the as-built microstructure of samples, produced by cold and hot platform, are depicted in Figures 1 and 2, respectively. As expected, in both conditions, the distinct tracks of the melt pools left by the laser scanning are visible, especially from lateral views (Figures 1 and 2). No significant differences are detectable in terms of size and shape of the melt pools, when comparing hot and cold platform processing.

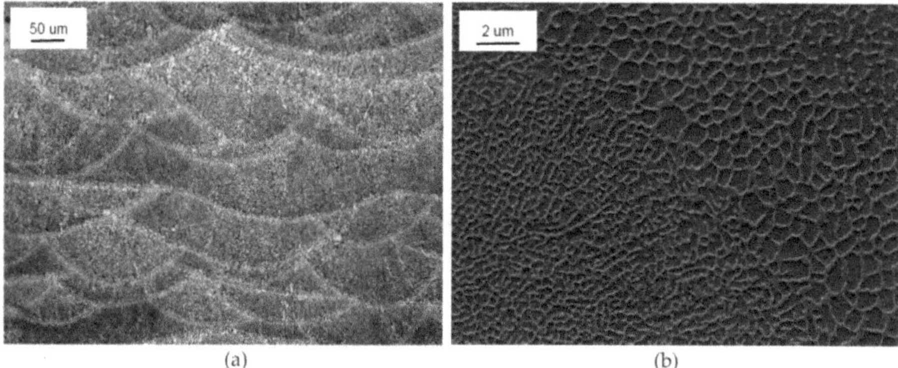

(a)   (b)

Figure 1. Microstructure of the AlSi10Mg alloy produced by selective laser melting (SLM) on a cold platform. (**a**) Low magnification optical microscope view (section plan parallel to building direction) of the solidified melt pools left by laser tracks; (**b**) SEM image of the cellular structure.

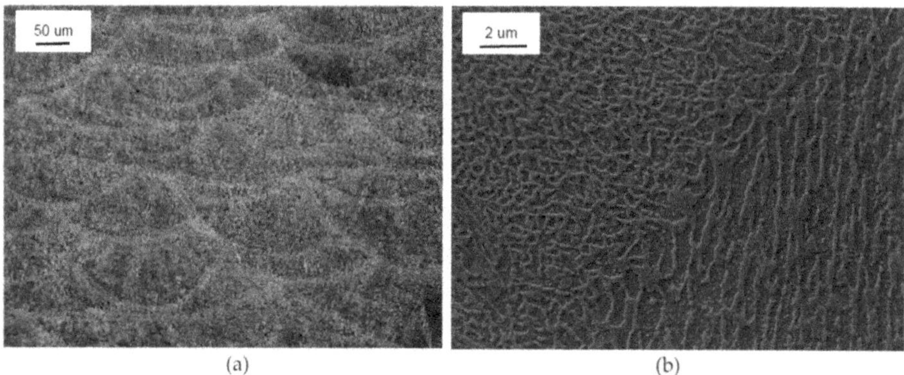

(a)   (b)

Figure 2. Microstructure of the AlSi10Mg alloy produced by SLM on a hot platform. (**a**) Low magnification optical microscope view (section plan parallel to building direction) of the solidified melt pools left by laser tracks; (**b**) SEM image of the cellular structure.

In both cases, a very limited number of defects (oxides, gas pores, unmelted particles, lack of fusion at melt pool boundaries) was detected. The material density, as measured by Archimedes' method, supplied an average value of 99.467% and a standard deviation of 0.004.

High magnification observations revealed, in both conditions, the presence of a very fine cellular solidification structure consisting in submicrometric primary α-Al cells decorated by a network of Si. Figures 1 and 2 also show that, when crossing the boundary of the melt pools, the size of the cells is subjected to a clear change, due to reheating effects produced by the overlaying passes [4], and to local changes in solidification conditions [13]. The comparison between cold and hot platform microstructure, proposed in Figure 3, highlights, by higher magnification views, that the Si network is substantially continuous in the CP samples (Figure 3a), while it becomes broken, combined with an additional intracellular precipitation of tiny particles, in HP samples (Figure 3b). It is presumed that the longer holding period at high temperature results in higher diffusion, leading to spheroidization phenomena in the Si network and precipitation of supersaturated Si atoms at cell interiors. This hypothesis is in good agreement with the results published by Li and co-authors, who investigated the precipitation of Si atoms into nano- or micrometric particles from supersaturated solid solutions in a SLM-processed AlSi10Mg alloy [13].

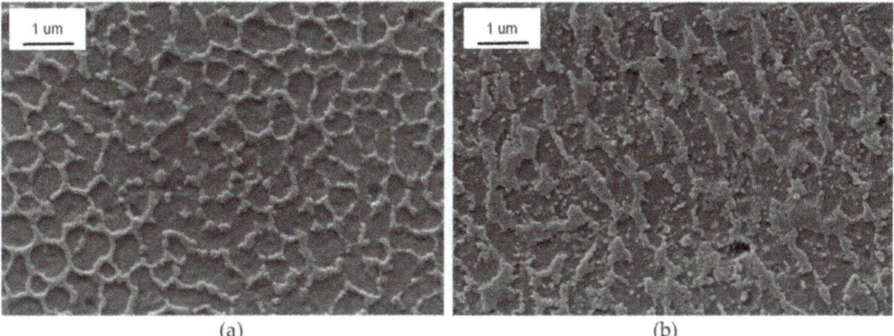

(a) (b)

Figure 3. High magnification SEM views of the as-built microstructure detected in cold platform (**a**) and hot platform (**b**) samples. The Si network surrounding the α-Al cells is the lighter phase, discrete intracellular Si precipitates are observed in panel b as spheroidal-shaped particles located inside the α-Al cells.

The effects of thermally activated diffusion become more evident after the solution-annealing treatment in CP sol and CP T6 samples. Figure 4 depicts the coarse grain structure formed after holding the alloy at 540 °C for 1 h. It is revealed that the cellular structure fully transforms into a set of equiaxed grains having a size of the order of 10 μm. The Si particles also coarsen, reaching an average size of 2.8 μm and a more rounded shape.

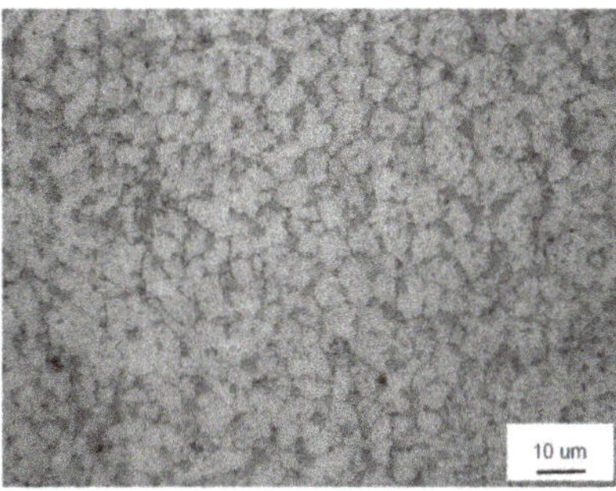

Figure 4. Representative microstructure of the solution-annealed (CP T6) sample.

3.2. X-ray Diffraction

High-resolution diffraction from synchrotron radiation was mainly used to investigate the evolution of the Si phase, based on phase fraction and lattice parameters. Only Al and Si phases were observed within the detection limit of the instrument. Figure 5 depicts the low-angle region of the patterns of the investigated samples treated according to the different tempers, while Table 3 summarizes the lattice parameters of the α-Al and Si phases.

Both the solution-annealing treatment (sample CP sol and CP T6) and the hot platform solidification mode (sample HP AB) promote higher fraction of precipitated Si. In CP AB sample, a significantly lower amount of Si is detected, but it readily increases after aging (sample CP T5). Consistently, the lattice parameters of the α-Al show larger distortions with increasing degree of Si supersaturation. It is worth remarking that the reference lattice parameters of the α-Al phase in the AlSi10Mg alloy under equilibrium is 0.40515 nm and, by increasing the fraction of solute Si, it is expected to decrease according to the smaller atomic radius of Si, compared to that of Al [25]. As the high resolution of the instrument allows a very accurate determination of lattice parameter of microcrystalline Al, its evolution confirms the trend observed for the Si phase fractions reported in Table 3.

The XRD data also show that the size of precipitates increases when moving from CP AB to CP T5 and HP AB samples, according to a presumable dependence on heating time. Considering the two material conditions involving high-temperature solution treatment (CP sol and CP T6), the breadth of the Si peaks becomes similar to that of α-Al, confirming that crystals coarsened to a size within the micrometer scale, too big to be accurately quantified through powder diffraction. The size of 2–8 μm, already evaluated by microscopy, should be considered for these conditions.

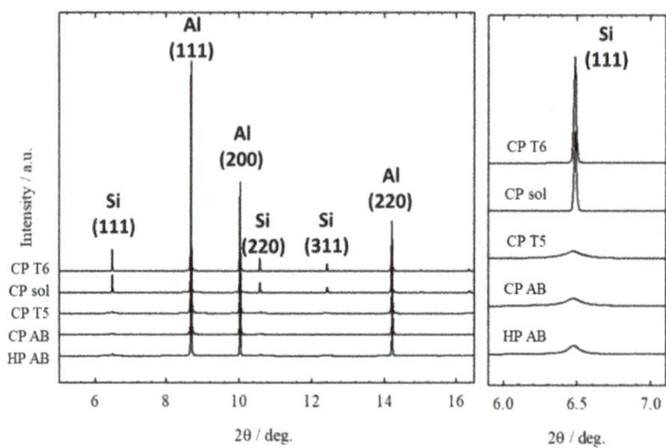

Figure 5. XRD patterns of the AlSi10Mg samples treated according to the different tempers. The panel on the right better highlights the changes in peak shape of Si (111).

Table 3. Al and Si lattice parameters and features of Si phase detected by XRD (the standard deviation of measures is given in brackets).

Sample Code	Lattice Parameters (nm)		Si Phase (wt %)	Size Range of Si Crystallites
	α-Al	Si		
CP AB	0.404833	0.54361	8.77 (0.05)	8 nm
CP T5	0.405011	0.54386	10.03 (0.05)	8 nm
CP sol	0.405182	0.54276	10.16 (0.03)	micrometric
CP T6	0.405155	0.54286	10.23 (0.03)	micrometric
HP AB	0.405079	0.54333	10.21 (0.05)	15 nm

3.3. Thermal Analysis and Aging Curves

DSC curves of as-built and of solution-annealed samples have been collected in order to evaluate the aging potential of the AlSi10Mg alloy, starting from different conditions. Figure 6 displays representative profiles of the heating ramps within the temperature range, where precipitation of strengthening phases is expected. The solution-annealed sample (CP sol) shows the sequence of precipitation peaks according to accepted literature [18,22–25]. The DSC curve of the as-built sample produced on a cold platform (CP AB) shows a similar shape, suggesting that most of the strengthening potential of the alloy can be exploited by thermal aging, starting from the as-built state, without the need of any prior solution-annealing treatment. On the contrary, the curve of the as-built sample produced on hot platform (HP AB) only shows a very limited response to the thermal ramp, confirming that most of the precipitation already occurred during the hot-stage manufacturing process.

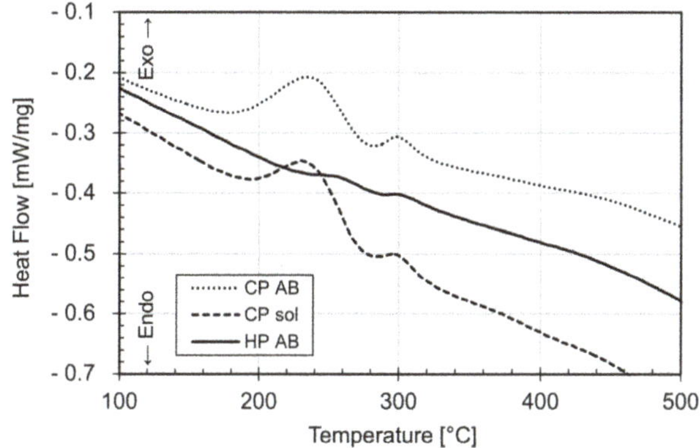

Figure 6. DSC scan profiles collected on heating of the SLM-processed AlSi10Mg alloy within the temperature range of interest for the precipitation of the strengthening phases.

The evolution of Vickers' microhardness of as-built samples, as a function of holding time at the aging temperature of 160 °C, is depicted in Figure 7. In full agreement with DSC data, it is reported that aging starting from CP AB condition produces a clear hardening, with an increase in microhardness from 116 to 134 HVn. Conversely, aging of the hot platform-processed samples led to a slight loss in hardness, from 117 to 113 HVn, after 4 h of aging at 160 °C. Thus, it can be reasonably inferred that further aging of HP AB alloy can only induce stress relaxation and/or overaging effects, without any improvement in mechanical properties.

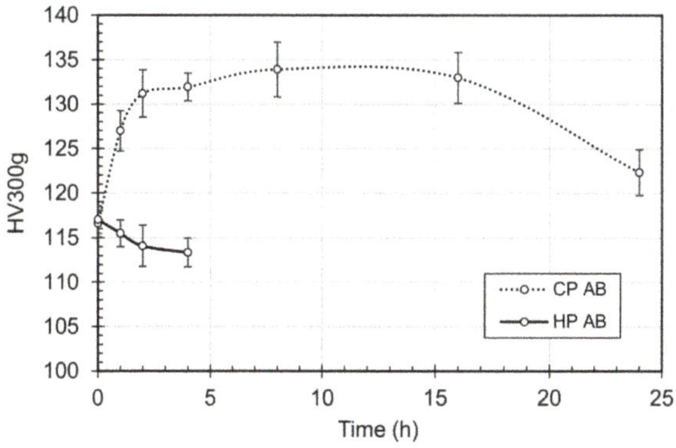

Figure 7. Aging curves at 160 °C of as-built SLM-processed AlSi10Mg alloy.

3.4. Mechanical Properties

Tensile tests were performed on the investigated alloy to evaluate the influence of the different tempers. In addition, the effect of sample orientation was considered by testing samples machined from both horizontal and vertical bars, namely, specimens with longitudinal axis normal and parallel to the building direction, respectively. Table 4 summarizes the main tensile properties achieved, while Figure 8 shows the recorded tensile curves.

The collected data clearly show that the best strength is achieved by optimally aging the cold platform as-built samples (CP T5), thus preserving the ultrafine cellular structure and inducing the expected precipitation of strengthening phases. In spite of its large aging potential, highlighted by the DSC analyses, the solution-treated samples (CP T6) attained the lowest strength, but took benefit from a remarkable improvement in fracture elongation. Finally, the performance of samples built on hot platform (HP AB) showed an intermediate condition in terms of strength and ductility. It is suggested that, in this temper, the material could benefit from the fine cellular structure, but could not fully exploit its strengthening potential, due to the lack of a tailored aging stage.

Finally, as far as the effect of specimen orientation is concerned, there is evidence from tensile data that CP T5 and HP AB samples tested in the horizontal direction always feature higher yield strength and fracture elongation, but lower ultimate tensile strength. On the contrary, solution annealing and aging (CP T6) led to the best yield and ultimate strength values in the horizontal samples, but comparable ductility for the two directions.

Table 4. Average tensile data (UTS: ultimate tensile strength; YS: yield strength) of the SLM-processed AlSi10Mg alloy heat-treated to the different tempers. Standard deviation is given in brackets.

Sample Code	Sample Orientation	UTS (N/mm^2)	0.2 YS (N/mm^2)	Strain at Fracture (%)
CP T5	horizontal	471 (0.8)	321 (1.8)	8.6 (0.5)
	vertical	493 (0.6)	292 (0.6)	6.0 (0.6)
CP T6	horizontal	323 (0.0)	243 (0.0)	15.3 (2.4)
	vertical	302 (1.4)	223 (2.8)	16.0 (2.5)
HP AB	horizontal	386 (2.6)	248 (1.7)	8.6 (1.4)
	vertical	412 (5.5)	228 (4.1)	7.0 (0.1)

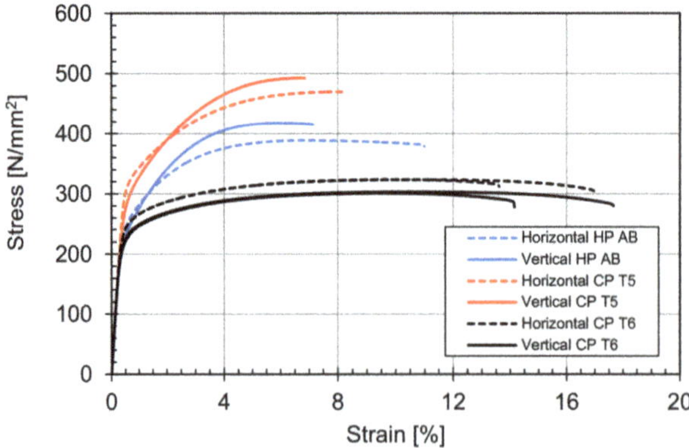

Figure 8. Tensile curves of the SLM-processed AlSi10Mg alloy heat-treated to the different tempers.

4. Discussion

The present results clearly demonstrate that optimization of thermal treatment of SLM-processed Al alloy can led to a remarkable improvement in strength over the as-built condition, and that the standard thermal treatment procedures (i.e., T6 temper) might be inadequate when trying to improve the tensile strength. Skipping the solution-annealing stage and promoting precipitation aging, starting right from the supersaturated, rapidly solidified and cooled, as-built material, already revealed itself to be a promising strategy for other alloys [30–32]. Jägle and co-workers [30] deeply investigated the effects of the thermal cycles experienced in various age-hardenable alloys during SLM and laser metal deposition. Their study highlighted desired and undesired precipitation reactions that could occur on the processing of several alloys, during intrinsic heat treatments generated by subsequent deposition passes (i.e., in already consolidated material) and during the following aging treatments. It was concluded that laser additive manufacturing can be used to produce supersaturated alloys that can lead to precipitation of second phases during a following aging stage. The present findings confirm that, also in the AlSi10Mg alloy, the precipitation reaction can be effectively exploited starting from the as-built condition, and it can lead to an appreciable increase in mechanical strength, reaching values of 471 MPa and 321 MPa for the ultimate tensile strength and the yield strength of the CP T5 samples, respectively. When compared to the tensile data survey given in the literature [20], the achieved performance ranks in the highest positions, also considering the favorable trade-off with fracture elongation. Conversely, the inappropriateness of a conventional T6 heat treatment on SLM-processed alloys is highlighted, owing to extensive grain growth induced by the high-temperature solution annealing and coarsening of the Si particles. A careful observation of the tensile curves reveals that the samples tested in their original SLM solidification structure (i.e., CP T5 and HP AB samples) show distinctive shapes of the stress vs strain curves. Specimens machined from horizontal bars have higher yield strength, and lower work-hardening rate and ultimate tensile strength, combined with improved ductility (see Table 4). Such behavior is supposed to be due to peculiar crystal orientation of the elongated solidification grains (containing the above described cellular substructure) that produce an anisotropic effect in material behavior [4,19], and to possible preferential location of small defects at laser track boundaries, leading to easier crack growth normal to loading direction, in vertical specimens.

A further issue to be considered when comparing cold and hot platform-printing strategies is the effectiveness of the in situ aging process in the latter situation. From DSC and microhardness results, it is inferred that HP AB samples do not show any significant response to aging under DSC heating ramps, and are even subjected to a slight loss in hardness on isothermal aging (see Figure 7). This suggests that overaging effects could take place in the samples, and it is reasonable to speculate that the actual aging effect promoted by hot platform processing would depend on printing time and on the position of the reference volume along the part height. Indeed, material volumes close to the build platform are subjected to longer processing times than those located close to the upper surfaces of the part. The volume of printed parts should also play a role, since thin parts featuring a lower re-melted area and higher fraction of relatively cold powder surrounding it would experience shorter holding times at high temperature than bulky shapes. A simple experimental test was designed in this perspective, measuring the microhardness profile along a sectioned vertical cylindrical bar (10 mm in diameter, 100 mm in length) produced by the hot platform strategy. The profile, reported in Figure 9, shows that no significant trend can be envisaged, at least for the bar size and the alloy investigated here. A similar experiment was also performed by Maamoun and co-authors [21], who showed, for the same AlSI10Mg alloy, an increasing hardness trend with increasing distance from platform. However, their samples were only 10 mm high along the z direction, and the SLM process was performed by a build platform temperature of 200 °C, instead of the 160 °C used in the present investigation.

It must also be considered that hot platform processing is of paramount importance for the control of residual stresses and distortions for difficult-to-process alloys and complex geometries [10,33–35]. Therefore, the hot platform processing method could represent a profitable alternative in the case of large parts with critical shapes, also considering cost savings related to the absence of any thermal post-processing stage.

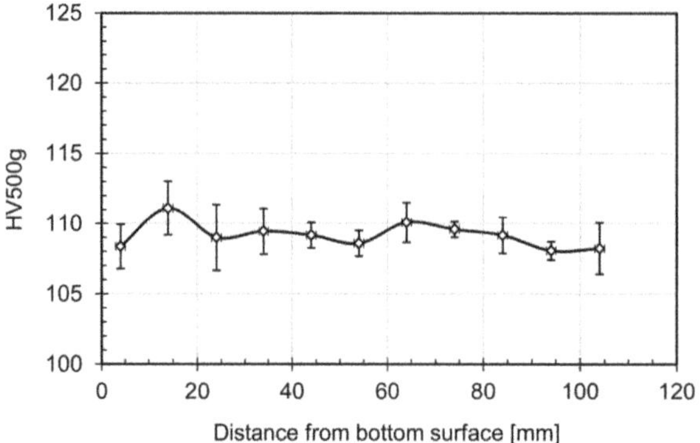

Figure 9. Microhardness profile measured in a HP AB sample as a function of position along build direction. Data refers to average and standard deviation values out of five microhardness readings per position.

5. Conclusions

Hot and cold platform processing were considered, among other strategies, to assess the aging response of the AlSi10Mg alloy processed by SLM, in order to achieve improved mechanical properties. In particular, hot platform was devised to test the ability of the alloy to age harden "in situ", namely, during the printing process itself. On the contrary, cold platform was used in order to investigate the possibility of directly aging the alloy right after the SLM process, without the need of a pre-solution-annealing treatment. The achieved results allowed for drawing the following conclusions.

The microstructure of the alloy produced by SLM on a cold platform consists of ultrafine α-Al cells surrounded by a network of Si phase. High-resolution XRD analysis demonstrated that, in the cold platform as-built condition, the amount of supersaturated Si is higher than in hot platform and in solution-annealed samples.

Upon aging, the rapidly solidified cold platform-processed alloy could be directly strengthened starting from as-built condition, supplying the highest level of hardness and tensile strength, together with fairly good levels of fracture elongation.

Hot platform processing led to a significant loss of the aging response with a concurrent drop of mechanical properties. It is supposed that the rather long holding times spent at high temperature during processing could induce a substantial overaged temper in the HP samples.

The standard solution annealing, followed by water quenching and artificial aging, resulted in the lowest achievable strength, thus demonstrating the inappropriateness of a conventional T6 heat treatment on SLM-processed AlSi10Mg alloys when high-strength alloys are required.

Author Contributions: Conceptualization by R.C., M.H.N., M.V.; V.T. supplied the SLM samples, M.C. conducted XRD experiments at ESFR, and performed data analysis; M.H.N. performed microstructural analyses, M.V. wrote the manuscript. All authors discussed the results and approved the final manuscript.

Funding: The present research work has been partially funded by Regione Lombardia within the frame of the project "NUVOLE" (id 187033) for Politecnico di Milano and within the project INNODRIVER-S3—2017 for Aidro S.r.l.

Acknowledgments: Research activities have been supported by the use of the interdepartmental Laboratory AMALA at Politecnico di Milano. The European Synchrotron Radiation Facility in Grenoble (France) is acknowledged for provision of beamtime.

Conflicts of Interest: The authors declare no conflict of interest.

References

1. Sames, W.J.; List, F.A.; Pannala, S.; Dehoff, R.R.; Babu, S.S. The metallurgy and processing science of metal additive manufacturing. *Int. Mater. Rev.* **2018**, *61*, 1–46. [CrossRef]
2. DebRoy, T.; Wei, H.L.; Zuiback, J.S.; Mukherjee, T.; Elmer, J.W.; Milewski, J.O.; Beese, A.M.; Wilson-Heid, A.; De, A.; Zhang, W. Additive manufacturing of metallic components—Process, structure and properties. *Prog. Mater. Sci.* **2018**, *92*, 112–224. [CrossRef]
3. Bradl, E.; Heckenberger, U.; Holzinger, V.; Buchbinder, D. Additive manufactured AlSi10Mg samples using selective laser melting: Microstructure, high cycle fatigue, and fracture behavior. *Mater. Des.* **2012**, *34*, 159–169. [CrossRef]
4. Thijs, L.; Kempen, K.; Kruth, J.-P.; Van Humbeeck, J. Fine-structured aluminium products with controllable texture by selective laser melting of pre-alloyed AlSi10Mg powder. *Acta Mater.* **2013**, *61*, 1800–1819. [CrossRef]
5. Kimura, T.; Nakamoto, T. Microstructures and mechanical properties of A357 (AlSi7Mg0,3) aluminium alloy fabricated by selective laser melting. *Mater. Des.* **2016**, *89*, 1294–1301. [CrossRef]
6. Zhang, H.; Zhu, H.; Qi, T.; Hu, Z.; Zeng, X. Selective laser melting of high strength Al-Cu-Mg alloys: Processing microstructure and mechanical properties. *Mater. Sci. Eng. A* **2016**, *656*, 47–54. [CrossRef]
7. Montero Sistiaga, M.L.; Mertens, R.; Vrancken, B.; Wang, X.; Van Hooreweder, B.; Kruth, J.P.; Van Humbeeck, J. Changing the alloy composition of Al7075 for better processability by selective laser melting. *J. Mater. Proc. Technol.* **2016**, *238*, 437–445. [CrossRef]
8. Rosenthal, I.; Stern, A.; Frage, N. Strain rate sensitivity and fracture mechanism of AlSi10Mg parts produced by selective laser melting. *Mater. Sci. Eng. A* **2017**, *682*, 509–517. [CrossRef]
9. Fousová, M.; Dvorsky, D.; Michalková, A.; Vojtěch, D. Changes in microstructure and mechanical properties of additively manufactured AlSi10Mg alloy after exposure to elevated temperatures. *Mater. Charact.* **2018**, *137*, 119–126. [CrossRef]
10. Rosenthal, I.; Schneck, R.; Stern, A. Heat treatment effect on the mechanical properties and fracture mechanism in AlSi10Mg fabricated by additive manufacturing selective laser melting. *Mater. Sci. Eng. A* **2018**, *729*, 310–322. [CrossRef]
11. Zang, C.; Zhu, H.; Liao, H.; Cheng, Y.; Hu, Z.; Zeng, X. Effect of heat treatment on fatigue property of selective laser melting AlSi10Mg. *Int. J. Fatigue* **2018**, *116*, 513–522. [CrossRef]
12. Fiocchi, J.; Tuissi, A.; Bassani, P.; Biffi, C.A. Low temperature annealing dedicated to AlSI10Mg selective laser melting products. *J. Alloys Compd.* **2017**, *695*, 3402–3409. [CrossRef]
13. Li, W.; Li, S.; Liu, J.; Zhang, A.; Zhou, Y.; Wei, Q.; Yan, C.; Shi, Y. Effect of heat treatment on AlSi10Mg alloy fabricated by selective laser melting: Microstructure evolution, mechanical properties and fracture behaviour. *Mater. Sci. Eng. A* **2016**, *663*, 116–125. [CrossRef]
14. Aboulkhair, N.T.; Tuck, C.; Ashcroft, I.; Everitt, N.M. On the precipitation hardening of selective laser melted AlSi10Mg. *Metall. Mater. Trans.* **2015**, *46A*, 3337–3341. [CrossRef]
15. Aboulkhair, N.T.; Maskery, I.; Tuck, C.; Ashcroft, I.; Everitt, N.M. The microstructure and mechanical properties of selectively laser melted AlSi10Mg: The effect of a conventional T6-like heat treatment. *Mater. Sci. Eng. A* **2016**, *667*, 139–146. [CrossRef]
16. Tradowski, U.; White, J.; Ward, R.M.; Reimers, W.; Attallah, M.M. Selective laser melting of AlSi10Mg: Influence of post-processing on the microstructural and tensile properties development. *Mater. Des.* **2016**, *105*, 217–222. [CrossRef]
17. Wang, L.F.; Sun, J.; Yu, X.L.; Shi, Y.; Zhu, X.G.; Cheng, L.Y.; Liang, H.H.; Yan, B.; Guo, L.J. Enhancement in mechanical properties of selectively laser melted AlSi10Mg aluminium alloys by T6-like heat treatment. *Mater. Sci. Eng. A* **2018**, *734*, 299–310. [CrossRef]
18. Casati, R.; Vedani, M. Aging response of an A357 Al alloy processed by selective laser melting. *Adv. Eng. Mater.* **2018**, 1–7. [CrossRef]
19. Yang, K.V.; Rometsch, P.; Davies, C.H.J.; Huang, A.; Wu, X. Effect of heat treatment on the microstructure and anisotropy in mechanical properties of A357 alloy produced by selective laser melting. *Mater. Des.* **2018**, *154*, 275–290. [CrossRef]
20. Tang, M.; Pistorius, P.C. Oxides, porosity and fatigue performance of AlSi10Mg parts produced by selective laser melting. *Int. J. Fatigue* **2017**, *94*, 192–201. [CrossRef]

21. Maamoun, A.H.; Elbestawl, M.; Dosbaeva, G.K.; Veldhuis, S.C. Thermal post-processing of AlSi10Mg parts produced by selective laser melting using recycled powder. *Addit. Manuf.* **2018**, *21*, 234–247. [CrossRef]

22. Chen, W.C.; Lee, S.L.; Tan, A.H. Effect of Pre-Ageing on the precipitation behaviors and mechanical properties of Al-7Si-Mg Alloys. *J. Mater. Sci. Chem. Eng.* **2018**, *6*, 55–67. [CrossRef]

23. Edward, G.A.; Stiller, K.; Dunlop, G.L.; Couper, M.J. The precipitation sequence in Al-Mg-Si alloys. *Acta Mater.* **1998**, *46*, 3893–3904. [CrossRef]

24. Tadeka, M.; Ohkubo, F.; Shirai, T.; Fukui, K. Stability of metastable phases and microstructures in the ageing process of Al-Mg-Si ternary alloys. *J. Mater. Sci.* **1998**, *33*, 2385–2390.

25. Marola, S.; Manfredi, D.; Fiore, G.; Poletti, M.G.; Lombardi, M.; Fino, P.; Battezzati, L. A comparison of Selective Laser Melting with bulk rapid solidification of AlSi10Mg alloy. *J. Alloys Compd.* **2018**, *742*, 271–279. [CrossRef]

26. Brandão, A.D.; Gumpinger, J.; Gschweitl, M.; Seyfert, C.; Hofbauer, P.; Ghidini, T. Fatigue properties of additively manufactured AlSi10Mg—Surface treatment effects. *Proc. Struct. Int.* **2017**, *7*, 58–66. [CrossRef]

27. Shi, Y.; Yang, K.V.; Kairy, S.K.; Palm, F.; Wu, X.; Rometsch, P.A. Effect of platform temperature on the porosity, microstructure and mechanical properties of an Al-Mg-Sc-Zr alloy fabricated by selective laser Melting. *Mater. Sci. Eng. A* **2018**, *732*, 41–52. [CrossRef]

28. Yang, K.V.; Shi, Y.; Palm, F.; Wu, X.; Rometsch, P. Columnar to equiaxed transition in Al-Mg(-Sc)-Zr alloys produced by selective laser melting. *Scripta Mater.* **2018**, *145*, 113–117. [CrossRef]

29. Williamson, G.K.; Hall, W.H. X-ray line broadening from filed aluminium and wolfram. *Acta Metall.* **1953**, *1*, 22–31. [CrossRef]

30. Jägle, E.A.; Sheng, Z.; Wu, L.; Lu, L.; Kisse, J.; Weisheit, A.; Raabe, D. Precipitation reactions in age-hardenable alloys during laser additive manufacturing. *JOM* **2016**, *68*, 943–949. [CrossRef]

31. Casati, R.; Lemke, J.N.; Zanatta Alarcon, A.; Vedani, M. Aging behavior of high-strength Al Alloy 2618 produced by selective laser melting. *Metall. Mater. Trans. A* **2017**, *48A*, 575–579. [CrossRef]

32. Casati, R.; Lemke, J.N.; Tuissi, A.; Vedani, M. Aging behaviour and mechanical performance of 18-Ni 300 steel processed by selective laser melting. *Metals* **2016**, *6*, 218. [CrossRef]

33. Li, C.; Liu, Z.Y.; Fang, X.Y.; Guo, Y.B. Residual stresses in metal additive manufacturing. *Procedia CIRP.* **2018**, *71*, 348–353. [CrossRef]

34. Mukherjee, T.; Zhang, W.; DebRoy, T. An improved prediction of residual stresses and distortion in additive manufacturing. *Comput. Mater. Sci.* **2017**, *126*, 360–372. [CrossRef]

35. Bartlett, J.L.; Croom, B.P.; Burdick, J.; Henkel, D.; Li, X. Revealing mechanisms of residual stress development in additive manufacturing via digital image correlation. *Addit. Manuf.* **2018**, *22*, 1–12. [CrossRef]

Article

Microstructure and Fatigue Properties of AlZn6Mg0.8Zr Alloy Subjected to Low-Temperature Thermomechanical Processing

Aleksander Kowalski [1], Wojciech Ozgowicz [1], Adam Grajcar [1,*], Marzena Lech-Grega [2] and Andrzej Kurek [3]

[1]   Silesian University of Technology, Institute of Engineering Materials and Biomaterials, 44-100 Gliwice, Poland; kowalski.polsl@gmail.com (A.K.); wojciech.ozgowicz@polsl.pl (W.O.)
[2]   Institute of Non-Ferrous Metals in Gliwice, Light Metals Division, 32-050 Skawina, Poland; mlechgrega@imn.skawina.pl
[3]   Opole University of Technology, Faculty of Mechanical Engineering, 45-271 Opole, Poland; a.kurek@po.opole.pl
*   Correspondence: adam.grajcar@polsl.pl; Tel.: +48-32-237-2940

Received: 22 August 2017; Accepted: 12 October 2017; Published: 21 October 2017

Abstract: The paper presents results of the investigations on the effect of the low-temperature thermomechanical treatment on the microstructure of AlZn6Mg0.8Zr alloy (7003 alloy) and the relationships between microstructure and fatigue properties and fractography of fractured samples. Fatigue life has been determined in a mechanical test at a simple state of loading under conditions of bending as well as torsion. The development of fatigue cracking has been described based on fractography investigations of the fractured samples making use of a scanning electron microscope (SEM). It was found that the factors determining the fatigue strength of the tested alloy are the microstructure as well as the type and size of the cyclic stresses. These factors determine the fractography of fatigue samples.

Keywords: aluminum alloy; 7003 alloy; fatigue properties; thermomechanical treatment; fractography

---

1. Introduction

Heat and thermomechanical processing of light metal alloys, particularly aluminum alloys for plastic working, is an effective way to increase their operating properties. Accelerated ageing process, connected with the impact of increased temperature and dependent on the rate of plastic deformation, mainly determines the microstructure of these alloys as well as the morphology and distribution of precipitations. Deformation of the material before ageing causes the precipitation becomes a competitive process between bulk precipitation and precipitation on dislocations. Moreover, microstructure evolution and mechanical properties depend on the sequence of the ageing and dynamics of phases formation after predeformation. It is important that the precipitates grow more slowly in the bulk than on dislocations. Therefore, it is necessary to consider each case individually and then it is possible to assess their synergistic effect during ageing. [1–3]. On the other hand, the grain size of the matrix of these alloys and the presence of strengthening intermetallic phases assure significant enhancement of their mechanical properties, and often also their ductile, fatigue and corrosion properties [4–6]. The attempts to prepare 7055 and 7075 aluminum alloys with such complex set of properties were noted in the work of Zuo et al. [7] and Das et al. [8].

The Al–Zn–Mg alloys (7000 series) are a relatively recent group of aluminum alloys applied in shipbuilding for construction of ships and displacement crafts, as well as high-speed surface effect vehicles, capable of transmitting high dynamic loads. The use of new constructional solutions and analytical methods for calculations of strength, based on technical stereomechanics for individual assemblies and ship structures, make modern high-speed fleet hulls approach more aircraft constructions rather than traditional naval solutions [9–11]. The 7000 series alloys, similarly to the 2000 series (Al–Li alloys) [12], are characterized by the highest strength among all aluminum alloy grades. However, this is dependent on properly performed heat and thermomechanical treatment as well as the obtained dispersion of precipitations both inside and on grain boundaries.

The knowledge of the fatigue characteristics of AlZn6Mg0.8Zr alloy (7003) is important due to varying loads of external forces, hydrodynamic loads in particular, and those caused by longitudinal bending moment, which are transferred by the bottom plating of high-speed surface effect vehicles. Many of these characteristics have not been reported in the literature, and most often there are data on the results of fatigue tests under cyclic tensile-compressive loads. Park, Jo et al. [13,14] determined the interaction of manganese-rich particles on the propagation of fatigue cracking of Al–Zn–Mg alloy during low-cycle fatigue. The influence of ageing conditions on fatigue properties of Al alloy with small concentration of Fe and Si admixtures (alloy 7475) has been investigated in [15]. Deng et al. [16] defined the fatigue strength of joints produced by the friction stir welding (FSW) of the 7050 aluminum alloy, while Effertz et al. [17] researched fatigue strength of joints of this alloy produced with friction spot welding (FSpW). The impact of secondary phase particles in the 7075 aluminum alloy, used for airplane wings components, as origins for fatigue cracking initiation as a function of the number of cycles was reported by Payne, Weiland et al. [18,19]. The attempts to simulate cracking initiation on the basis of these data were developed by Li et al. [20]. Zhao et al. [21] have tested the effects of stress ratio (R), overloading, underloading, and high-low sequence loading on fatigue crack growth rate of the 7075 commercial alloy. Moreover, Zhao et al. [22] performed extensive studies on multiaxial fatigue load for the same alloy. Available literature is focused mostly on the 7000 series aluminum alloys with Cu addition, which leads to the increased corrosion of these alloys. Moreover, there are not many publications concerning bending and torsional fatigue loads. Therefore the purpose of undertaken studies was to determine the influence of diversified loading conditions during low- and high-cycle oscillatory bending and double-sided torsion on fatigue life and fatigue cracking resistance of the 7003 aluminum alloy without the addition of Cu, after low temperature thermomechanical processing.

2. Materials and Methods

2.1. Materials and Heat Treatment

The study material was a metal sheet, cold-rolled from Al–Zn–Mg (7000 series) commercial aluminum alloy with a chemical composition is shown in Table 1.

Table 1. Chemical compositions of the investigated alloy.

Denotation of the Alloy	Chemical Composition (Mass %)												
	Zn	Mg	Mn	Fe	Cr	Si	Zr	Cu	Ga	Ti	Ni	Pb	Al
EN AW-7003 EN AW-Al Zn6Mg0.8Zr	6.134	0.742	0.291	0.197	0.167	0.121	0.080	0.036	0.005	0.005	0.004	0.003	rest

The investigated alloy was subjected to a low-temperature thermomechanical treatment as shown in Figure 1:

- Preheated up to 500 °C and soaked for one hour,
- Cooled down in water,
- Cold-rolled with a degree of deformation of: 10%, 20% and 30%,
- Aged for 12 h at a temperature of 150 °C, then cooled down in the open air.

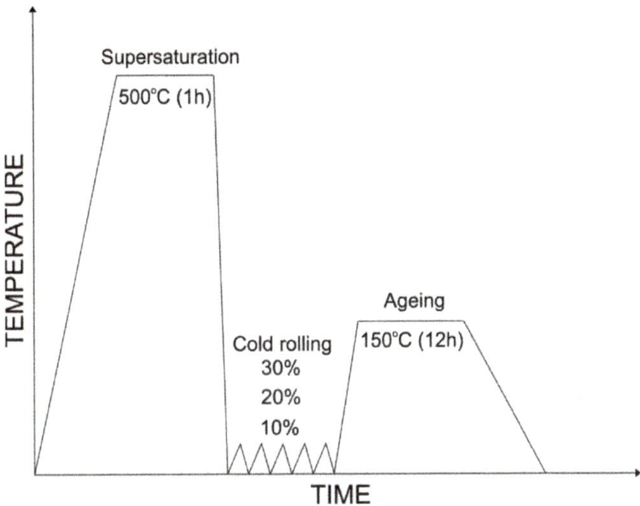

Figure 1. Scheme of low-temperature thermomechanical treatment of AlZn6Mg0.8Zr alloy.

Mechanical properties after the low-temperature thermomechanical treatment are presented in Table 2.

Table 2. Mechanical properties of the investigated alloy after the low-temperature thermomechanical treatment.

Degree of Deformation [%]	Mechanical Properties			
	$\overline{R_{p0.2}}$ [MPa]	$\overline{R_m}$ [MPa]	$\overline{A}$ [%]	$\overline{Z}$ [%]
10%	256	321	10.2	49.5
20%	294	341	8.2	47.3
30%	300	347	9.1	40.4
Supersaturation 500 °C/1 h/water	174	313	19.8	25.0

2.2. Fatigue Tests

The fatigue strength was tested on a test-stand MZGS-100 (Opole University of Technology, Faculty of Mechanical Engineering, Opole, Poland) [23] (Figure 2a) using un-notched standard samples (Figure 2b). The specimens were tested in two different stress conditions: oscillatory (cycling) bending and double-sided torsion under sinusoidally changing loading, both with a stress ratio R = −1 and a constant amplitude of bending or torsion moment (σ_{max} = | σ_{min} |). The stress (σ_a, τ_a) amplitudes were calculated in the specimen area with a smallest diameter (10 mm) as a bending/torsion moment ratio to bending/torsional strength index, respectively. The stress concentration factor for the specimen geometry, especially for torsion, was calculated at a level of 1.03 and considered irrelevant for further studies.

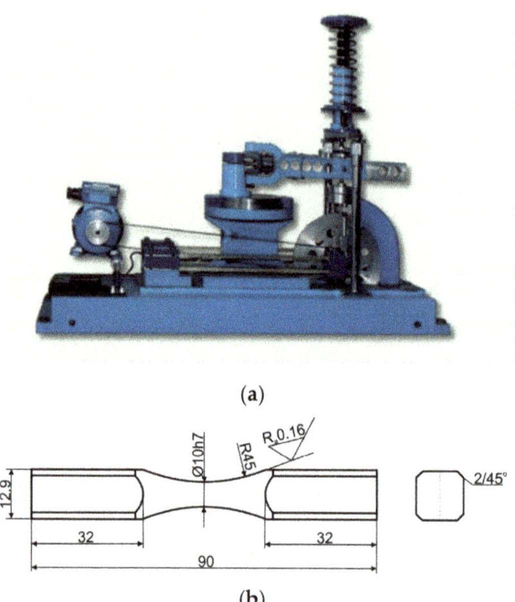

(a)

(b)

Figure 2. (**a**) Test stand MZGS-100 and (**b**) specimen for fatigue strength tests.

The tests were carried out at diversified intensity of stresses, ensuring low- and high-cycle fatigue of specimens. The results were plotted using the binary logarithmic system diagram $\log(\sigma_a)$ vs. $\log(N_f)$ or $\log(\tau_a)$ vs. $\log(N_f)$. The regression model was described with the following dependence:

$$Y = A + mX \tag{1}$$

where: Y—$\log(\sigma_a)$—logarithm of normal or shearing stress—$\log(\tau_a)$, X—$\log(N_f)$—logarithm of fatigue life, m—directional coefficient.

The experimental results were approximated by the regression Equation (1) for bending and torsion in the form:

$$\log(N_f) = A_\sigma + m_\sigma \log(\sigma_a) \tag{2}$$

and for double-sided torsion in the form:

$$\log(N_f) = A_\tau + m_\tau \log(\tau_a) \tag{3}$$

where: m_σ, m_τ—slope of regression line coefficients, A_σ, A_τ—constant terms.

Fatigue tests of AlZn6Mg0.8Zr alloy were carried out on thermoplastically processed specimens, taking into consideration the number of fatigue cycles from approx. 2×10^4 to approx. 10^7 and stress amplitude from 128 to 193 MPa in case of oscillatory bending, and from 74 to 122 in case of double-sided torsion. The influence of low-temperature thermomechanical processing (LTMP) parameters on fatigue strength of the examined alloy was analyzed in relation to the supersaturated state. The basic parameter of the LTMP used in the comparative study of the temporary fatigue strength of examined alloy was the degree of cold deformation after supersaturation during LTMP.

2.3. Metallographic Examinations

Metallographic tests were carried out on microsections of longitudinal samples (Figure 3). The preparation of the microsections comprised standard operations of submerging the samples in chemohardened resin, grinding and mechanical polishing on a Struers LaboPol-21 machine (Struers Inc., Cleveland, OH, USA). The specimens for grain size observation were electrolytically etched in Barker's reagent. Observations of metallographic specimens using the polarized light were performed in the OLYMPUS GX-71 (Microscope Systems Limited, Glasgow, UK) light microscope, at magnifications of 500×.

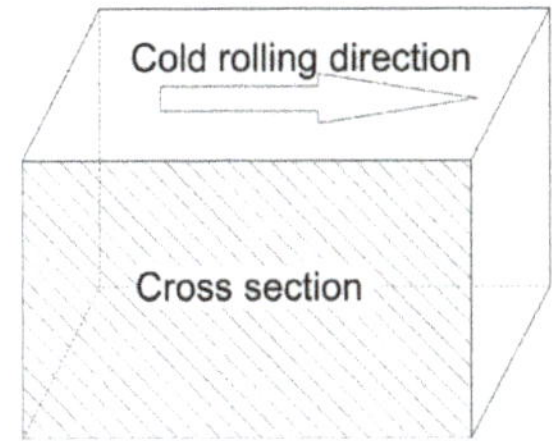

Figure 3. Cross section of longitudinal samples used in metallographic examinations.

2.4. Fractographic Examinations

Samples for fractographic investigations after bending and torsion tests were performed (Figure 4). For this purpose an electron-scanning microscope (SEM) of the ZEISS Supra 25-type (Carl Zeiss AG, Jena, Germany) was applied with an electron device GEMINI (Carl Zeiss AG, Jena, Germany) at a voltage of 20 kV. Comparisons of AlZn6Mg0.8Zr alloy fractures were done mainly due to the type of cyclically changing loads.

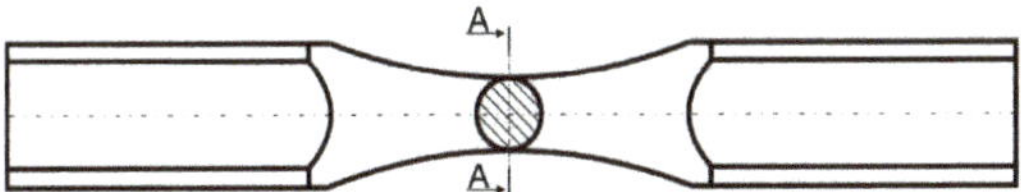

Figure 4. Cross section of fatigue samples used in fractographic observations.

3. Results and Discussion

3.1. Fatigue Properties

The values of coefficients in Equations (2) and (3) $m_{B,T}$ and $A_{B,T}$, are summarized in Table 3.

Table 3. Equation coefficients of regression lines for AlZn6Mg0.8Zr alloy after various stages of low-temperature thermomechanical processing (LTMP) and cyclic load tests.

Stage of LTMP	Coefficient Value/Type of Load			
	Bending		Torsion	
Material State	m_B	A_B	m_T	A_T
Supersaturation 500 °C	−23.810	58.310	−20.833	46.208
Plastic deformation [%]				
10	−27.778	67.694	−15.625	36.500
20	−19.231	48.212	−17.241	39.655
30	−38.462	91.192	−20.833	46.854

The AlZn6Mg0.8Zr alloy after supersaturation in water from the temperature of 500 °C in the LTMP, reveals the highest temporary fatigue strength for the analyzed range of load cycles in the oscillatory bending test (Figure 5). Loading of studied alloy with maximum bending stresses (σ_a) of 174 MPa and torsional stresses (τ_a) of 97 MPa results in the failure of the sample at approx. 1.79×10^5 and 1.61×10^5, respectively. The high parallelism of regression lines was found for oscillatory bending and double-sided torsion in case of specimens supersaturated in water from the temperature of

500 °C ($m_B = -23.810$, $m_T = -20.833$), suggesting a similar change of fatigue life along with increasing bending and torsion fatigue stresses (Table 3). The experimental data, obtained in case of fatigue loading of studied alloy with bending stresses after supersaturation from the temperature of 500 °C, reveal small scatter of results (SD = 0.023), which demonstrates the high stability behavior of the tested alloy during stress changes σ_a as a function of the analyzed fatigue load cycles.

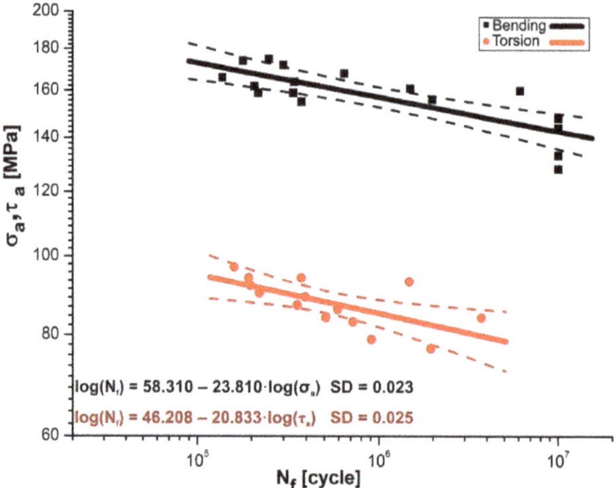

Figure 5. Basquin's characteristics of regression lines with confidence interval (α = 0.05) for AlZn6Mg0.8Zr alloy after supersaturation in water from temperature of 500 °C.

The course of temporary fatigue strength changes for tested alloy, subjected to LTMP with the deformation degree in the range (10–30)% as a function of number of cycles and the load type (Figure 6), is similar to the one observed in case of supersaturation of the alloy in water from the temperature of 500 °C (Figure 5). The alloy subjected to the deformation of 10% during LTMP exhibits fatigue life equal approx. 6.40×10^4 cycles at maximum bending stresses of 193 MPa, whereas in case of 30% deformation, loading of tested alloy with maximum bending stresses of 174 MPa results in obtaining fatigue life of approx. 1.44×10^5 cycles. In case of torsional loads, the fatigue life for the alloy is equal about 2.70×10^4 cycles at maximum bending stresses of 122 MPa for deformation of 10% and approx. 8.90×10^4 cycles at maximum stresses of 101 MPa for 30% deformation during LTMP. Das et al. [8], while investigating the alloy of the same series (7075), characterized by higher concentration of Mg and the presence of Cu, have noted an increase in fatigue life up to 70% for tensile-compressive loads along with increasing plastic strain at cryogenic temperature. The initial material, without deformation, revealed fatigue life for this type of load equal approx. 1.68×10^5 at maximum stresses of 240 MPa.

Minimal dispersion of experimental results was noted in case of cyclic bending loads. The standard deviation of regression lines (SD) for this type of load is the smaller the higher is the degree of deformation in the LTMP process. The value of SD for the degree of deformation of 10% to 30% changes from approximately 0.018 to about 0.005, respectively (Figure 6). It was found that the value coefficients of regression lines m_B and m_T in case of the studied alloy subjected to LTMP with 20% degree of deformation (Figure 6b) are similar (in contrast to LTMP with 10% and 30% deformation degree) and they are −19.231 and −17.241, respectively (Table 3). This allows to state that AlZn6Mg0.8Zr alloy, cold-rolled with 20% deformation degree during LTMP, is characterized by a similar course of temporary fatigue strength under analyzed cyclically changing loads conditions. Moreover, regression lines in case of double-sided torsion of the examined alloy reveal high parallelism, irrespective of the degree of plastic deformation in the LTMP (Table 3).

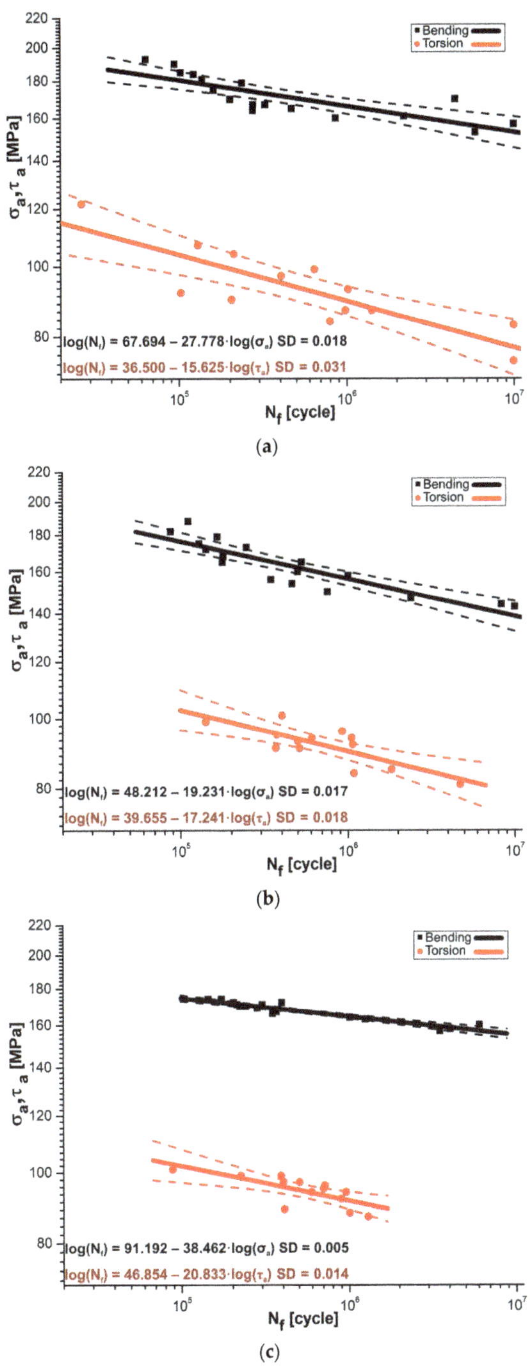

Figure 6. Basquin's characteristics of regression lines with confidence interval (α = 0.05) for AlZn6Mg0.8Zr alloy subjected to cold rolling in LTMP with degree of deformation of: (a) 10%; (b) 20% and (c) 30% after supersaturation from temperature of 500 °C.

3.2. Microstructure after Low-Temperature Thermomechanical Processing

Microstructure of AlZn6Mg0.8Zr alloy, subjected to supersaturation in water from 500 °C with subsequent rolling using 10% plastic deformation and ageing at the temperature of 150 °C for 12 h, is characterized by a banding arrangement of grains of the α solution, inside which parallel slip bands and precipitations of primary phases have been observed (Figure 7a). Similar microstructure occurs for the 20% deformed alloy (Figure 7b). Plastic deformation of examined alloy with higher degree of cold work (30%) during LTMP leads also to a banding system of grains in the α solution matrix, inside of which an increased amount of cold work symptoms are observed, most often as parallel slip bands (Figure 7c). It was found that the morphology and distribution of observed precipitations do not depend on the conditions of performed LTMP, which suggests that these are precipitations of primary phases. The microanalysis of chemical composition (using SEM) of dark points revealed mainly that these are Fe-rich phases, additionally suggesting that these are primary phases produced during casting process. The same observations were noted recently by Lech-Grega et al. [24], in the case of the 7020 aluminium alloy. The phases they observed contained Fe, Mn and Cr.

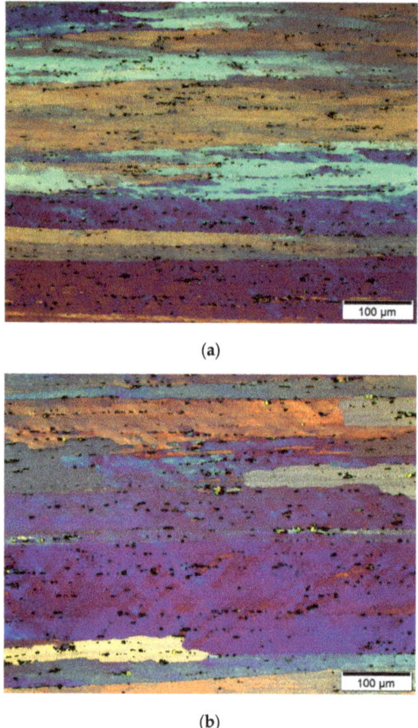

(a)

(b)

Figure 7. *Cont.*

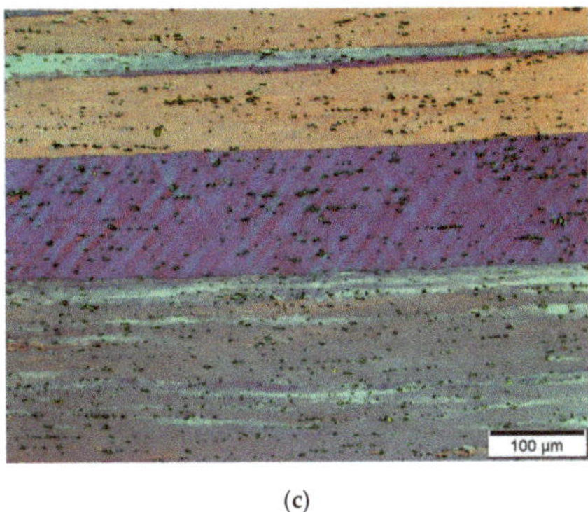

(c)

Figure 7. AlZn6Mg0.8Zr alloy subjected to LTMP with degree of deformation of: (**a**) 10%; (**b**) 20%; (**c**) 30%.

3.3. Fractography after Decohesion during Fatigue Tests

During oscillatory bending tests of AlZn6Mg0.8Zr alloy samples, supersaturated in water from the temperature of 500 °C, one can observe that the fractures analyzed in the range of low-cycle fatigue strength are primarily characterized by the presence of smooth surfaces of jogs of cleavage planes with locally occurring slight traces of plastic deformation (Figure 8a). Secondary fatigue cracks are also visible (Figure 8b) as well as areas with mixed fracture characteristics (Figure 8c). The fractures of examined alloy in the range of high-cycle fatigue strength are quasi-cleavage with smooth surfaces of jogs of cleavage planes and cracks proceeding along these planes (Figure 8d).

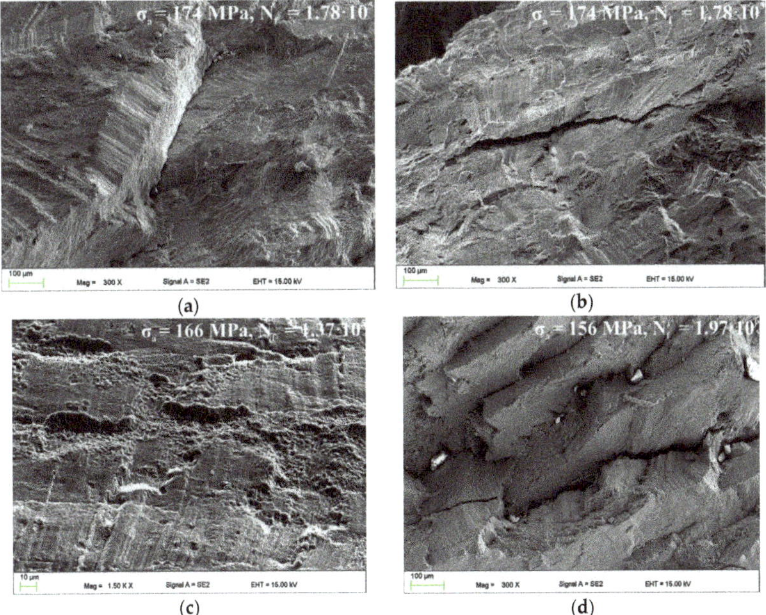

Figure 8. Fractures of fatigue test samples (oscillatory bending) of AlZn6Mg0.8Zr alloy, subjected to supersaturation in water from 500 °C: (**a**) smooth surfaces of jogs of cleavage planes; (**b**) secondary crack in the fracture surface; (**c**) mixed nature of the fracture; (**d**) cracks proceeding along the jogs of cleavage planes.

The fractures of AlZn6Mg0.8Zr alloy, after supersaturation from the temperature of 500 °C, subjected to double-sided torsion within the range of limited fatigue strength, are of the transcrystalline quasi-cleavage type with apparent slip bands systems on the cleavage planes, where secondary fatigue cracks proceeding along these bands can be also observed (Figure 9a). Moreover, differently oriented fatigue crack planes with irregular jogs and secondary fractures systems were revealed (Figure 9b,c). The areas with parallel slip bands on the cleavage planes are also typical (Figure 9d).

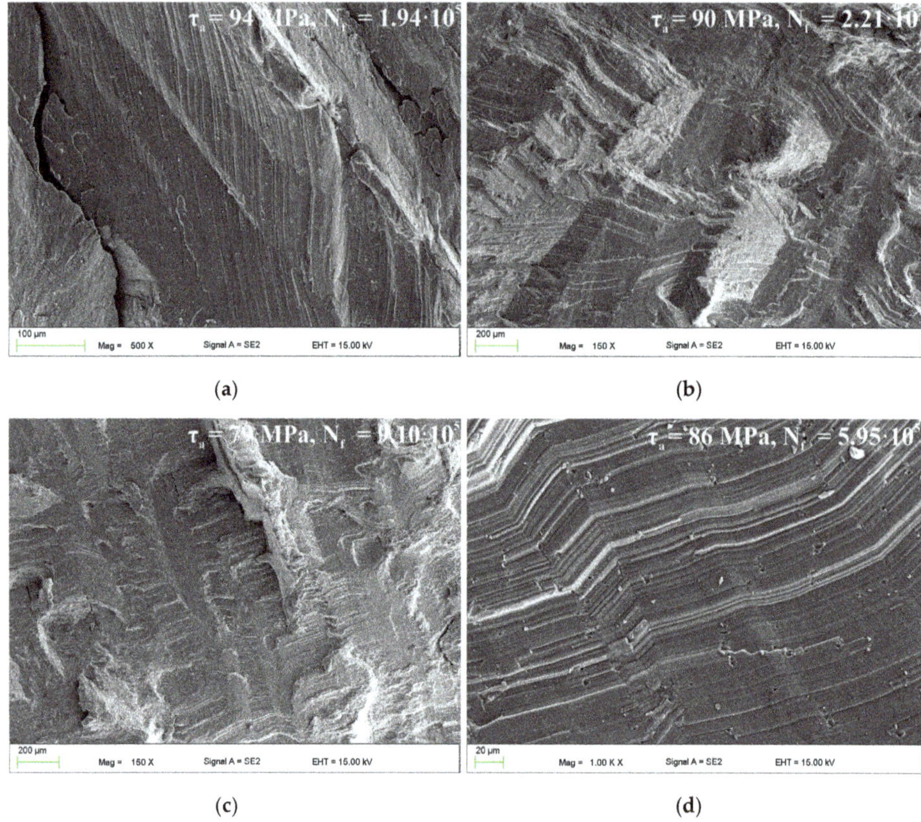

Figure 9. Fractures of fatigue test samples (double-sided torsion) of AlZn6Mg0.8Zr alloy, subjected to supersaturation in water from 500 °C: (**a**) secondary fatigue cracks along slip bands; (**b**) jogs of differently oriented fatigue crack planes; (**c**) jogs systems of cleavage planes; (**d**) parallel slip bands on the cleavage planes.

Fractures of studied alloy specimens after LTMP with the degree of deformation equal to 10%, subjected to oscillatory bending with stress amplitude (σ_a) in the range of low-cycle fatigue strength are of the transcrystalline quasi-cleavage type, with smooth jogs of cleavage planes with traces of plastic strain of the surface (Figure 10a). For specimens loaded with an amplitude of bending stress in the range of high-cycle fatigue strength, the presence of the origins of fatigue fractures with jogs of the transcrystalline-cleavage planes nature and numerous tongue-steps were observed in the surface area (Figure 10b). However, detailed microfractographic analysis of studied samples allows to state that observed transcrystalline quasi-cleavage fatigue fractures reveal local areas of plastic deformation on the jogs surfaces and traces of this deformation on potentially stochastic cleavage planes

(Figure 10c). There are also often visible local material tears and cracks, probably of intercrystalline nature (Figure 10d).

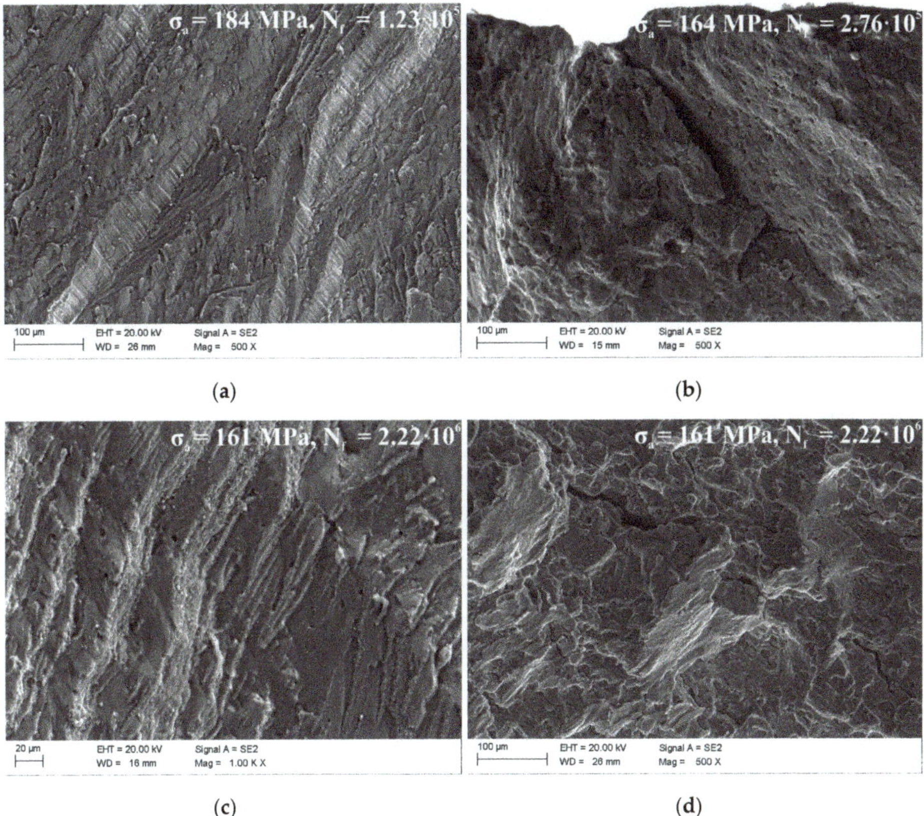

Figure 10. Fractures of fatigue test samples (oscillatory bending) of AlZn6Mg0.8Zr alloy, subjected to LTMP with 10% degree of deformation: (**a**) traces of plastic deformation of the jogs surfaces of fatigue crack planes; (**b**) origins of fatigue fracture in the surface area; (**c**) traces of plastic deformation on the jogs surfaces of cleavage planes; (**d**) local material tears and cracks on the fatigue fracture surface.

Numerous secondary fatigue cracks of various length and depth are present on the fractures of investigated alloy, subjected to LTMP with 10% deformation after double-sided torsion, in the range of low-cycle fatigue strength (Figure 11a). Cracking initiation in these samples occurs in the surface zone where primary fatigue cracks and their origins were observed (Figure 11b). Irregularly arranged jogs of the cleavage planes were observed on the fractures of specimens subjected to double-sided torsion in the range of high-cycle fatigue strength (Figure 11c). Fatigue strips have been localized on surfaces of these jogs (Figure 11d).

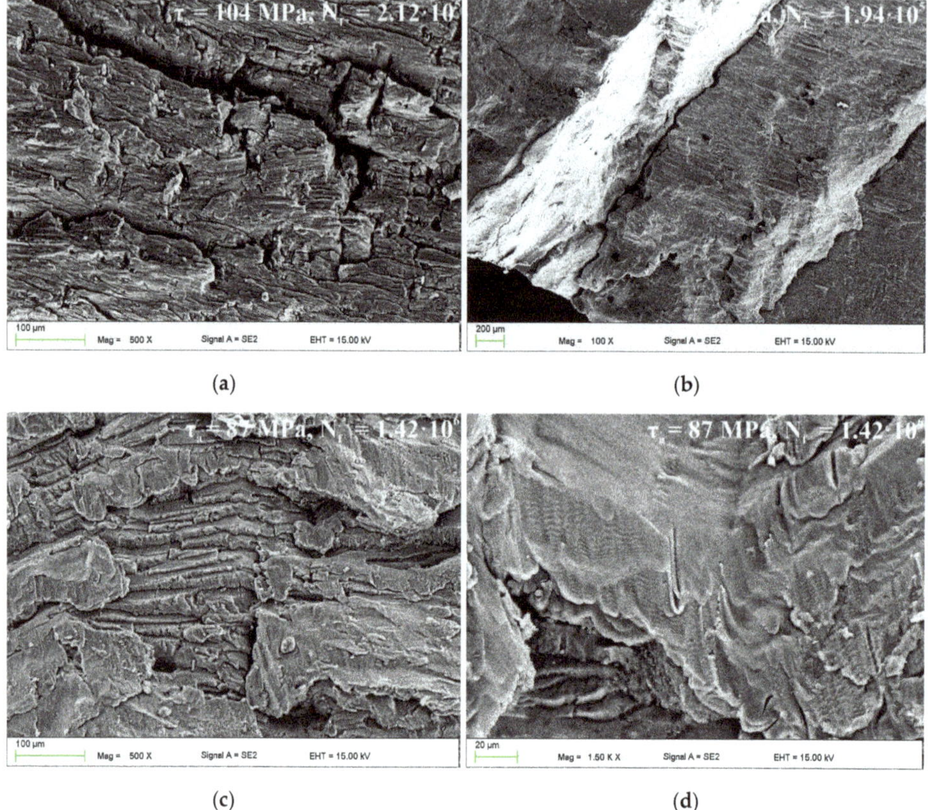

Figure 11. Fractures of fatigue test samples (double-sided torsion) of AlZn6Mg0.8Zr alloy, subjected to LTMP with 10% degree of deformation: (**a**) systems of secondary cracks on the cleavage planes; (**b**) focal point of fracture and primary fatigue crack on the specimen surface; (**c**) varied shape of jogs of the cleavage planes; (**d**) plastic fatigue strips on the jogs surface of crack planes.

The fractures of AlZn6Mg0.8Zr alloy specimens, subjected to LTMP with 30% cold rolling deformation after oscillatory bending with stress amplitude in the range of low-cycle fatigue strength, are of a transcrystalline quasi-cleavage nature (Figure 12a). Areas with the significant plastic deformation are observed in the form of plastic fatigue strips (Figure 12b). The occurrence of such effects of plastic deformation of fatigue fractures of the 7000 series alloy was also reported in [8]. However, Das et al. have observed these effects only in the range of high-cycle fatigue and not as in this case—in the whole examined range. It's possible that this is due to another type of load being applied (in [8] the samples were subjected to tensile-compressive stresses). In the vicinity of these strips, on the cleavage planes, secondary fatigue cracks were detected (Figure 12c). In addition, in the range of high-cycle fatigue strength, investigated alloy specimens exhibit the transcrystalline quasi-cleavage fracture with numerous jogs in the planes of cleavage, forming the systems of characteristic rivers and river basins (Figure 12d).

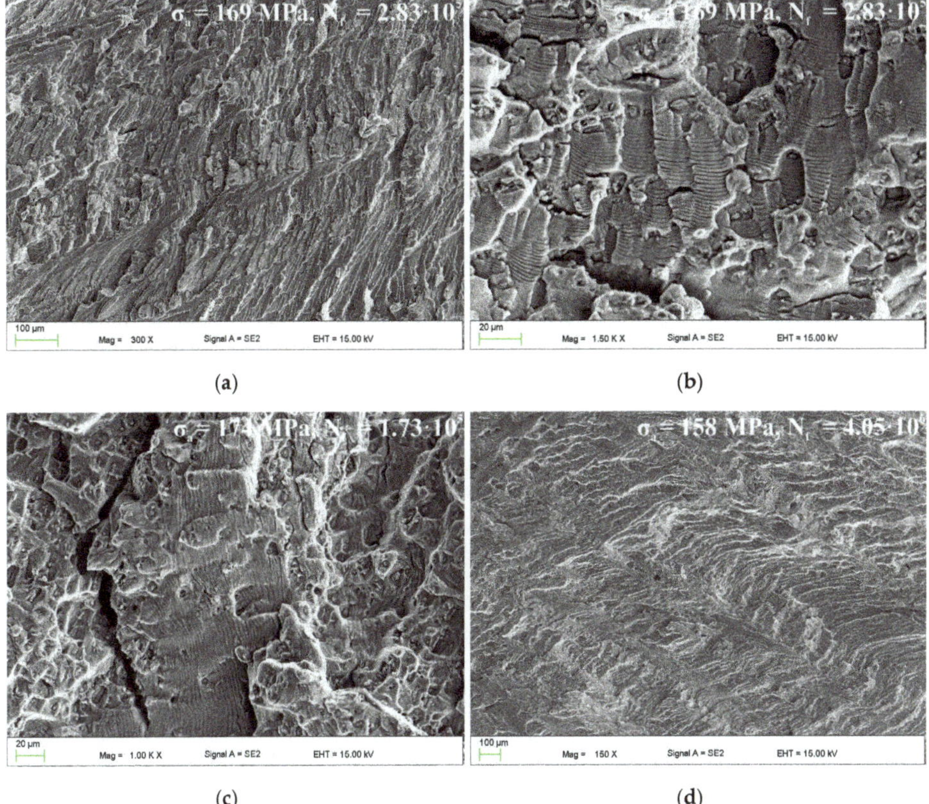

Figure 12. Fractures of fatigue test samples (oscillatory bending) of AlZn6Mg0.8Zr alloy, subjected to LTMP with 30% degree of deformation: (**a**) quasi-cleavage character of the investigated alloy fracture; (**b**) fatigue strip systems on the surface of cleavage planes; (**c**) secondary crack and adjacent fatigue strips (**d**) jogs systems on the fatigue fracture surface of investigated alloy forming rivers and river basins.

The fractures of specimens from the alloy subjected to LTMP with a 30% deformation degree reveal, after double-sided torsion with stress amplitude (τ_a approx. 94–96) in the range of high number of cycles, transcrystalline quasi-cleavage nature with numerous fatigue cracks (Figure 13a) and craters of various size (Figure 13b). For the amplitude of torsional stress (τ_a) in the range of (89–92) MPa, transcrystalline-cleavage fractures with extensive secondary cracks were detected (Figure 13c). Frequently, these cracks are spread in different directions, which may indicate the presence of a probable obstacle, e.g., in the form of precipitations, or the case where paths of cracks propagation run along grain boundaries (Figure 13d). Similar effects were observed by Park, Jo et al. in their works [13,14]. In Al–Zn–Mg alloy with 0.5% concentration of Mn, the paths are different from those observed in alloys without this addition. The Mn-rich dispersive phases act as hindrances and lead to refraction of cracking paths, so that they become more "tortuous". Moreover, they found that fine Mn-rich particles, present in Al–Zn–Mg–(Mn) alloy, cause both increased low-cycle fatigue strength and crack toughness. Payne et al. [18] found that phase particles (e.g., Al_7Cu_2Fe), which are harder than the matrix, become the precursors of cracking and when the number of fatigue cycles increase, they lead to propagation of cracking in the alloy matrix. On the other hand, the particles with hardness smaller than the matrix (e.g., Mg_2Si) are not often privileged areas for initiation of cracks.

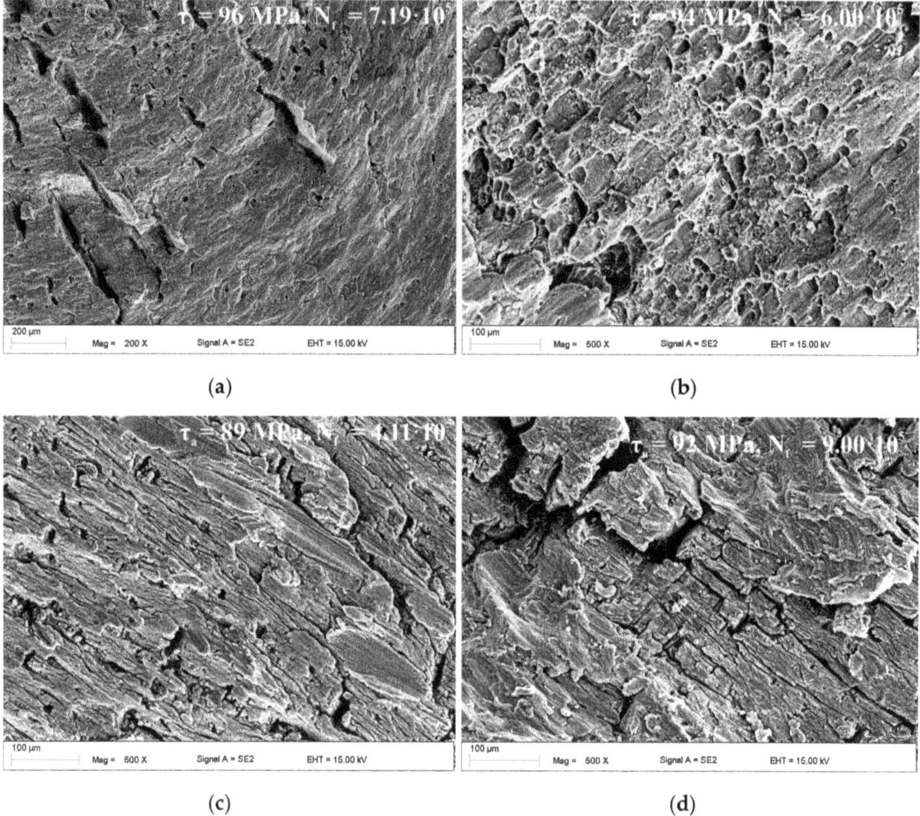

(a) (b)

(c) (d)

Figure 13. Fractures of fatigue test samples (double-sided torsion) of AlZn6Mg0.8Zr alloy, subjected to LTMP with 30% degree of deformation: (**a**) transcrystalline quasi-cleavage fracture with secondary cracks; (**b**) craters with diversified size on the fracture surface; (**c**) secondary cracks systems on the fatigue fracture surface of investigated alloy; (**d**) secondary cracks of deep depth on the fracture surface.

Increasing the degree of plastic deformation from 10% to 30% during LTMP results in increasing number of fracture areas with traces of plastic deformation, especially on the surface of cleavage planes. This is particularly evident in case of oscillatory bending. However, it does not result in an absolute increase of fatigue strength, but it affects the lower slope of the regression line. This observation allows us to assert that plastic deformation has a beneficial effect on the temporary fatigue strength of these alloys. Reducing the slope of the regression lines evidences greater homogeneity of the alloy 7003 during increasing fatigue load, both in the case of bending and torsion stresses.

4. Conclusions

The performed fatigue tests of the 7003 series Al–Zn–Mg alloy and metallographic analysis allow to formulate the following conclusions:

- AlZn6Mg0.8Zr alloy exhibits higher temporary fatigue strength in case of cyclic bending loads compared to torsion loads, regardless of state of the material.

- Investigated alloy, subjected to fatigue stresses from bending and torsion demonstrates more uniform course of temporary fatigue strength in case of oscillatory bending, determined statistically in the analyzed range of loads.
- It has been found that the important factors determining the portion of transcrystalline quasi-cleavage and ductile fractures are mainly: material state, type of applied fatigue load and amplitude of stress, and conditioning the number of cycles leading to failure.
- Increase in the degree of plastic deformation in the range from 10 to 30% results in an increase of the area of ductile fracture, which causes greater homogeneity of the 7003 alloy during increasing bending and torsion fatigue loads.

Acknowledgments: The publication is partially supported by the rector grant in the area of scientific research and development works. Silesian University of Technology, No. 10/010/RGJ17/0143.

Author Contributions: Aleksander Kowalski and Wojciech Ozgowicz conceived and designed the experiments; Aleksander Kowalski performed fractographic observations, analyzed the data and wrote the paper; Wojciech Ozgowicz supervised the work; Adam Grajcar analyzed the data and reviewed the paper; Marzena Lech-Grega performed metallographic experiments and analyzed the results; Andrzej Kurek performed fatigue tests and analyzed the results; all authors discussed the paper.

Conflicts of Interest: The authors declare no conflicts of interest.

References

1. Ferragut, R.; Somoza, A.; Tolley, A.; Torriani, I. Precipitation kinetics in Al–Zn–Mg commercial alloys. *J. Mater. Process. Technol.* **2003**, *141*, 35–40. [CrossRef]
2. Thomson, D.S.; Subramanya, B.S.; Levy, S.A. Quench rate effects in Al–Zn–Mg–Cu alloys. *Metall. Trans.* **1971**, *2*, 1149–1160. [CrossRef]
3. Deschamps, A.; Livet, F.; Bréchet, Y. Influence of predeformation on ageing in an Al–Zn–Mg alloy—I. Microstructure evolution and mechanical properties. *Acta Mater.* **1998**, *47*, 281–292.
4. Deschamps, A.; Bréchet, Y. Influence of quench and heating rates on the ageing response of an Al–Zn–Mg–(Zr) alloy. *Mater. Sci. Eng. A* **1998**, *251*, 200–207. [CrossRef]
5. Chemingui, M.; Khitouni, M.; Jozwiak, K.; Mesmacque, G.; Kolsi, A. Characterization of the mechanical properties changes in an Al–Zn–Mg alloy after a two-step ageing treatment at 70 °C and 135 °C. *Mater. Design* **2010**, *31*, 3134–3139. [CrossRef]
6. Hoyt, J. On the coarsening of precipitates located on grain boundaries and dislocations. *Acta Metall. Mater.* **1991**, *39*, 2091–2098. [CrossRef]
7. Zuo, J.; Hou, L.; Shi, J.; Cui, H.; Zhuang, L.; Zhang, J. Effect of deformation induced precipitation on dynamic aging process and improvement of mechanical/corrosion properties AA7055 aluminum alloy. *J. Alloys Comp.*. [CrossRef]
8. Das, P.; Jayaganthan, R.; Chowdhury, T.; Singh, I.V. Fatigue behaviour and crack growth rate of cryorolled Al 7075 alloy. *Mater. Sci. Eng. A* **2011**, *528*, 7124–7132. [CrossRef]
9. Heinz, A.; Haszler, A.; Keidel, C.; Moldenhauer, S.; Benedictus, R.; Miller, W.S. Recent development in aluminium alloys for aerospace applications. *Mater. Sci. Eng. A* **2000**, *280*, 102–107. [CrossRef]
10. Williams, J.C.; Starke, E.A., Jr. Progress in structural materials for aerospace systems. *Acta Mater.* **2003**, *51*, 5775–5799. [CrossRef]
11. Shizhen, W.; Cuiyun, L.; Ruolin, W.; Chaoli, M. Effect of cold expansion on high cycle fatigue of 7A85 aluminum alloy straight lugs. *Rare Metal Mater. Eng.* **2015**, *44*, 2358–2362. [CrossRef]
12. Dursun, T.; Soutis, C. Recent developments in advanced aircraft aluminium alloys. *Mater. Design* **2014**, *56*, 862–871. [CrossRef]
13. Park, D.S.; Kong, B.O.; Nam, S.W. Effect of Mn-dispersoid on the low-cycle fatigue life of Al–Zn–Mg alloys. *Metall. Mater. Trans. A* **1994**, *25*, 1547–1550. [CrossRef]
14. Jo, B.L.; Park, D.S.; Nam, S.W. Effect of Mn dispersoid on the fatigue crack propagation of Al–Zn–Mg alloys. *Metall. Mater. Trans. A* **1996**, *27*, 490–493. [CrossRef]
15. Gürbüz, R.; Sarioğlu, F. Fatigue crack growth behavior in aluminium alloy 7475 under different aging conditions. *Mater. Sci. Technol. Ser.* **2001**, *17*, 1539–1543. [CrossRef]

16. Deng, C.; Wang, H.; Gong, B.; Li, X.; Lei, Z. Effects of microstructural heterogeneity on very high cycle fatigue properties of 7050-T7451 aluminum alloy friction stir butt welds. *Int. J. Fatigue* **2016**, *83*, 100–108. [CrossRef]

17. Effertz, P.S.; Infante, V.; Quintino, L.; Suhuddin, U.; Hanke, S.; dos Santos, J.F. Fatigue life assessment of friction spot welded 7050-T76 aluminium alloy using Weibull distribution. *Int. J. Fatigue* **2016**, *87*, 381–390. [CrossRef]

18. Payne, J.; Welsh, G.; Christ, R.J., Jr.; Nardiello, J.; Papazian, J.M. Observations of fatigue crack initiation in 7075-T651. *Int. J. Fatigue* **2010**, *32*, 247–255. [CrossRef]

19. Weiland, H.; Nardiello, J.; Zaefferer, S.; Cheong, S.; Papazian, J.; Raabe, D. Microstructural aspects of crack nucleation during cyclic loading of AA7075-T651. *Eng. Fract. Mech.* **2009**, *76*, 709–714. [CrossRef]

20. Li, L.; Shen, L.; Proust, G. Fatigue crack initiation life prediction for aluminium alloy 7075 using crystal plasticity finite element simulations. *Mech. Mater.* **2015**, *81*, 84–93. [CrossRef]

21. Zhao, T.; Zhang, J.; Jiang, Y. A study of fatigue crack growth of 7075-T651 aluminum alloy. *Int. J. Fatigue* **2008**, *30*, 1169–1180. [CrossRef]

22. Zhao, T.; Jiang, Y. Fatigue of 7075-T651 aluminum alloy. *Int. J. Fatigue* **2008**, *30*, 834–849. [CrossRef]

23. Kurek, A.; Wachowski, M.; Niesłony, A.; Płociński, T.; Kurzydłowski, K.J. Fatigue tests and metallographic of explosively cladded steel-titanium bimetal. *Arch. Metall. Mater.* **2014**, *59*, 1565–1570. [CrossRef]

24. Lech-Grega, M.; Hawryłkiewicz, S.; Richert, M.; Szymański, W. Structural parameters of 7020 alloy after heat treatment simulating the welding process. *Mater. Charact.* **2001**, *46*, 251–257. [CrossRef]

Article

Structural and Mechanical Evaluation of A Nanocrystalline Al–5 wt %Si Alloy Produced by Mechanical Alloying

Davood Dayani [1], Ali Shokuhfar [1], Mohammad Reza Vaezi [2], Seyed Reza Jafarpour Rezaei [3] and Saman Hosseinpour [4,*]

[1] Department of Engineering, Karaj Branch, Islamic Azad University, 3148635731 Karaj, Iran; dayanidavood@gmail.com (D.D.); shokuhfar@kntu.ac.ir (A.S.)

[2] Division of Nanotechnology and Advanced Materials, Materials and Energy Research Center, 3177983634 Karaj, Iran; m_r_vaezi@merc.ac.ir

[3] Department of Materials Science and Engineering, School of Engineering, Shiraz University, Zand Avenue, 7134814666 Shiraz, Iran; rezajafarpour2010@gmail.com

[4] Institute of Particle Technology (LFG), University of Erlangen-Nuremberg, Cauerstrasse 4, 91058 Erlangen, Germany

* Correspondence: saman.hosseinpour@fau.de; Tel.: +49-9131-8529580

Received: 7 July 2017; Accepted: 20 August 2017; Published: 29 August 2017

Abstract: High energy mechanical milling followed by hot-pressing consolidation has been used to produce nanostructured Al–5 wt %Si alloy. X-ray diffraction (XRD), scanning electron microscopy equipped with energy dispersive X-ray detector (SEM-EDX), Vickers hardness, and compression measurements were used to examine the effect of milling duration on microstructure and mechanical properties of the nanostructured consolidated alloys. Crystallite sizes and lattice strains were determined by X-ray peak broadening analysis using the Williamson-Hall (W-H) method. Increasing the milling time reduced the crystallite size, and the minimum crystallite size of about 33 nm was achieved for both consolidated and powdered samples after 50 h of milling. Based on the SEM-EDX observations, the best distribution of silicon into Al matrix was obtained after 20 h of milling and remained unchanged afterwards. Hardness of both consolidated and powder samples increased with milling time, which can be attributed to the reduction of crystallite size and the better distribution of silicon in the aluminum matrix. Similarly, increased milling time increased the yield and compressive strengths of consolidated samples.

Keywords: mechanical alloying; Al–Si alloy; mechanical properties; consolidation

---

1. Introduction

Since mechanical alloying (MA) was first used in 1970 by Benjamin to produce iron and nickel base superalloys [1], Ma has attracted a significant amount of attention for synthesizing novel materials with unique properties [2]. Ma is able to produce both equilibrium and non-equilibrium phases and materials with improved properties compared to their conventionally produced counterparts [3,4]. Ma leads to a remarkable improvement in the mechanical properties of alloys through the homogeneous dispersion of particles along with the reduction of grain size to a nanometer scale [5,6]. Furthermore, Ma improves microstructural stability such as creep resistance at elevated temperatures, which is associated with a suitable dispersion of nanoparticles in the microstructure that prevent high temperature coarsening [7]. The advantages of Ma compared to other conventional alloying methods, such as processing in low temperatures, an improved microstructural refinement, uniform chemical composition, and a lack of solidification defects, have turned Ma into a convenient technique for producing aluminum alloys, composites, and coatings involving aluminum and silicon [8–11]. Mechanically produced Al–Si alloys have been widely used in various fields, such as automotive and aerospace industries, due to their high strength-to-weight ratio, excellent wear resistance, and low thermal expansion coefficient [12]. For practical applications, consolidation, which involves sequences of sintering processes, is a necessary step in the production of desired Al–Si alloys. Hot consolidation that requires the simultaneous application of temperature and pressure is the method of choice for various applications since it increases the chance of the fast and full densification of powders. Furthermore, the formation of unwanted phases or microstructural changes can be prevented by hot consolidation [13].

In the present study, we investigate the nanocrystalline structure of Al–5 wt %Si powders produced by Ma as well as the consolidated alloy after the powders are hot-pressed. The mechanical properties of the produced alloys are assessed using compression and hardness tests, and the microstructural changes and their effects on the mechanical properties of alloys are investigated.

2. Materials and Methods

A mixture of commercial aluminum powders (with 99.9% purity and a particle size of <75 µm, Sigma-Aldrich, St. Louis, MO, USA) and silicon powders (with 99.9% purity and a particle size of <20 µm, Alfa Aesar, Haverhill, MA, USA) was milled in a planetary ball mill under an argon atmosphere to achieve an alloy with a composition of Al–5 wt %Si. The following parameters were used in the milling process: a ball-to-powder mass ratio of 12:1, a ball diameter of 10 mm, and a speed of 250 rpm. We added 1.5 wt % of stearic acid ($CH_3(CH_2)_{16}COOH$) to the mixture as a process control agent (PCA) to moderate the cold welding process and to prevent the adhesion of the powder to the balls and interior surface of the milling tank. The mixture of powders and PCA was milled for up to 50 h, and samples were collected after 5, 20, and 50 h of milling. Mechanically alloyed powders were uniaxially cold-pressed in a cylindrical steel die (a diameter of 10 mm) for 30 s at 500 MPa pressure and then hot-pressed for 60 min at 450 °C under 500 MPa to achieve the consolidated samples. Crystallite size, lattice strain, and hardness of the consolidated samples were compared to their powder counterparts.

X-ray diffraction measurements were performed with a wide angle diffractometer in the θ–2θ step scan mode using CuKα radiation. Scans were collected over a 2θ range of 20–90° with a step of 0.01°. The crystallite size (d) and the equivalent lattice strain (ε) were determined using the Williamson-Hall (W-H) line broadening analysis method [14] according to the following:

$$\beta \cos\theta = \frac{k\lambda}{d} + 2\varepsilon \sin\theta \tag{1}$$

where β is the full width at half maximum (FWHM) of the diffraction peak, θ is the diffraction angle, k is a constant whose values is approximately 0.9, and λ is the incident X-ray wavelength (1.54060 Å).

It is clear from Equation (1) that β cosθ vs. sinθ curve exhibits a straight line with the slope of (2ε) and intercept of $k\lambda/d$. Thus, the crystallite size (d) and the lattice strain (ε) can be achieved.

Scanning electron microscope (SEM) equipped with an energy dispersive X-ray spectrometer (EDX, VEGA/TESCAN, Kohoutovice, Czech Republic) was used in characterizing the microstructure of the samples as well as analyzing the elemental distribution. Additionally, to evaluate the hardness of MA-prepared powders, a micro-hardness test was utilized using a Vickers indenter under the load of 10 gf and dwell time of 15 s (MVK-H21, Akashi Co., Hyogo, Japan). For harness measurements of consolidated samples, a load of 25 gf and a dwell time of 15 s was used. Measuring the hardness of powders is a challenging issue and usually results in scattered values dependent on the method of the measurement, on the properties of the mounting resin, and on the "depth" of the particle under indenter [15]. In this study, we used hot epoxy resin to mount the cold-pressed powders prior the micro-hardness measurements. The excess of the resin was removed from the top layer, exposing the surface of powders to air. The micro-hardness of the powders was estimated by the evaluation of the depth of the indentation. The choice of low load for the hardness measurements on powdered samples was to reduce the indenter impression and its depth, allowing reasonable hardness measurements on small powders. a similar method was employed earlier by Abdoli et al. to evaluate the micro-hardness of nanostructured composite powders produced by MA [16]. For better statistics of the hardness values, the hardness tests were performed on three individual points on each consolidated sample and on 10 points for each powdered sample. Moreover, compression testing on consolidated sample was carried out at a cross-head speed of 0.2 mm/min and a strain rate of 0.005 min^{-1}. Prior to the compression tests, both ends of the specimens were polished to make them parallel to each other.

3. Results

Figure 1 shows the changes in the morphology of mixed powders with increasing milling times. As Figure 1 shows, increasing milling time gradually changes the powder morphology from flat flakes to semi-globular structures. As is expected, in preliminary stages of milling (Figure 1a) aluminum particles are still soft, and they undergo plastic deformation while brittle silicon particles are fragmented. The Si phase is believed to accelerate the deformation of the powder through a second hard phase formation in the mill [17]. Increasing the milling time results in the sequential welding of the aluminum particles and the distribution of silicon particles within the aluminum matrix. With the continuation of milling (Figure 1b), various factors such as deformation, welding, and solid dispersion lead to work hardening of aluminum particles, which in turn enhances their tendency to fracture. Thus, aluminum particles break down, and the average particle size slightly decreases. Finally, welding and fracture processes reach an equilibrium (Figure 1c) where randomly orientated boundaries are formed. In this stage, particles becomes semi-globular in shape, and their size distribution becomes more uniform.

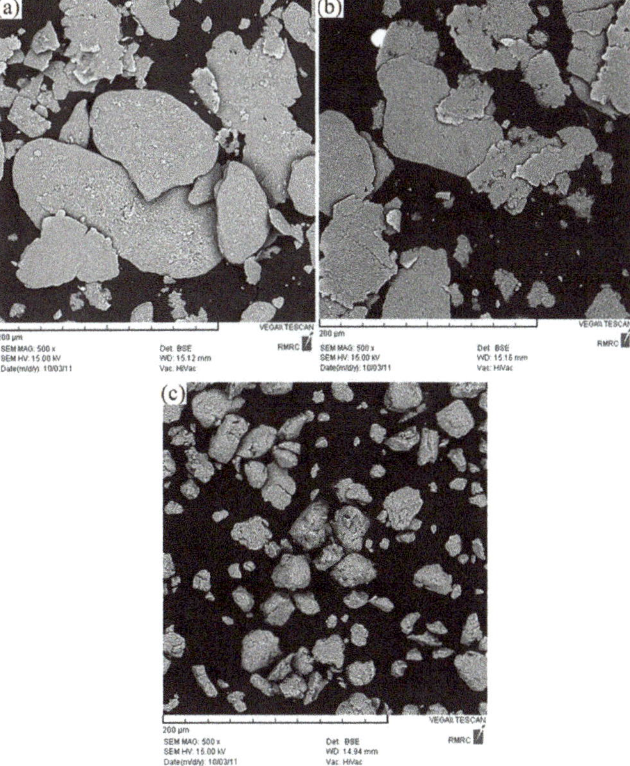

Figure 1. Scanning electron microscope (SEM) images of Al–5 wt %Si powders for different milling durations: (**a**) 5 h, (**b**) 20 h, and (**c**) 50 h.

Figure 2 shows an individual semi-globular particle formed after 50 h of milling. As can be seen from this figure, the surface of the semi-globular particle exhibits some moderate roughness as a consequence of simultaneous competition of welding and fracturing mechanisms. Although Figure 2 exhibits a typical particle shape after 50 h of milling, as can be observed in Figure 1c, there are instances of a transverse crack through the whole particle is observed, most likely due to the increase in the internal stresses and lattice strains within the semi-globular particles, which will be discussed in the following.

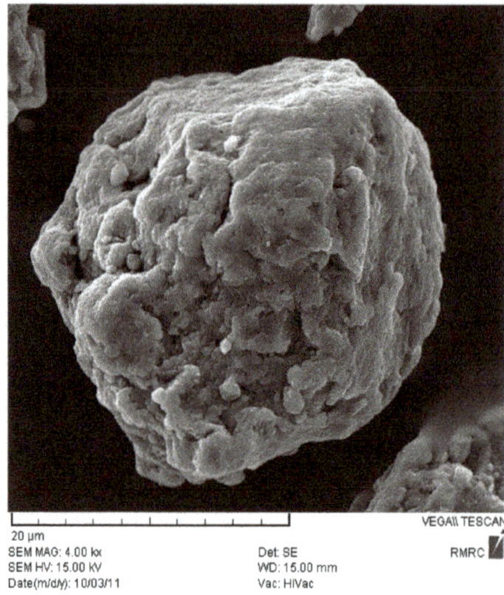

Figure 2. SEM image of semi-globular particle of Al-5 wt %Si formed after 50 h of milling.

From XRD analysis, the crystallite size and lattice strain were measured, and the obtained values are summarized in Tables 1 and 2 for the powder and consolidated samples, respectively. In all cases a decreasing trend in the crystallite size is observed upon increasing the milling time. As proposed by Miraghaei et al. [18], the formation of new defects, especially dislocations, is responsible for the reduction of the crystallite size. Multiple mechanisms were proposed for the accumulation of dislocations. For instance, dense regions of these dislocations can be formed in sub grains, dislocations might pile up at the grain boundaries, or clusters can be accommodated within the crystallites. Overall, a severe plastic deformation during mechanical alloying and consequently the reduction of the crystallite size can contribute to the generation of extra dislocations. Moreover, the results in Tables 1 and 2 indicate that the lattice strain is generally enhanced as the milling time increases. As mentioned before, the severe plastic deformation during the milling process introduces dislocations, vacancies, impurities, and other lattice defects, which, in turn, increase the stress field in the alloys. Similar trends were observed for mechanically produced Al–Mg/Al$_2$O$_3$ nanocomposites, Al/Fe alloys, and alumina-reinforced nanocrystals [19–21].

Table 1. Crystallite size (nm) and lattice strain of powder as functions of milling time (h), determined by the Williamson-Hall (W-H) method.

Time (h)	d (W-H)	ε (W-H)
5	47.8	2.2×10^{-3}
20	33.8	2.8×10^{-3}
30	32.6	2.8×10^{-3}
50	32.6	4.3×10^{-3}

Table 2. Crystallite size (nm) and lattice strain of bulk samples as functions of milling time (h), determined by the W-H method.

Time (h)	d (W-H)	ε (W-H)
5	50.2	1.5×10^{-3}
20	35.43	2.0×10^{-3}
30	33.4	2.1×10^{-3}
50	33.0	2.7×10^{-3}

In the current study, the initial growth of the average particle diameter due to a flattening of the particles, similar to the observations by Milligan et al. [17], is neglected as it is out of the scope of this work. However, a partial crystallite growth is observed for the consolidated samples during the hot pressing stage. Nevertheless, the extensive crystal growth is still retarded due to the pinning effects at the grain boundaries [22]. Explaining such trends for consolidated samples requires a better understanding of the elemental composition distribution, which will be discussed later. As is depicted in Figure 3, the variation of the crystallite size with the increase in milling time exhibits very similar trends for powder and consolidated samples, showing that, after almost 20 h of milling, the size of crystals do not vary any more for either sample.

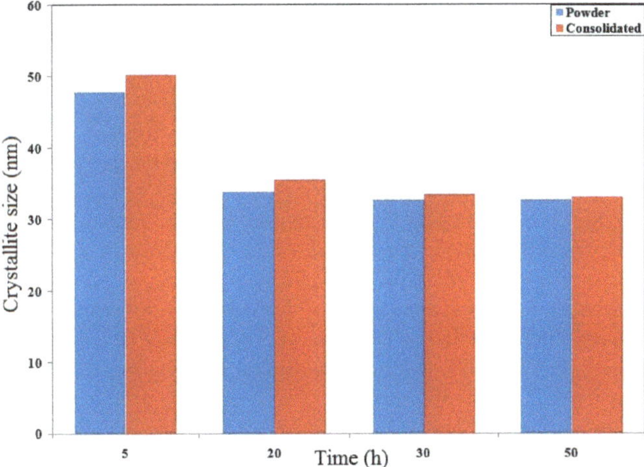

Figure 3. Crystallite size (nm) as a function of milling time (h) for powder and consolidated samples determined by the W-H method.

To understand how the distribution of Si in the aluminum matrix affects the crystallite size and the growth rate by the pinning mechanism, EDX measurements are performed on selected consolidated samples. As can be seen in Figure 4, Si distribution in the Al matrix is more uniform after a prolonged milling time (i.e., 20 h), causing a better pinning of the grain boundaries, which in turn prevents the crystalline growth.

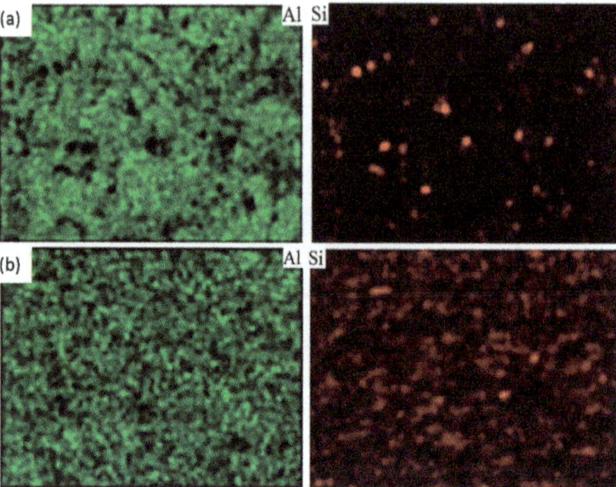

Figure 4. Energy dispersive X-ray spectroscopy (EDX) analysis of consolidated samples with milling time (**a**) 5 h and (**b**) 20 h of milling.

The distribution of Si in the Al matrix as well as the residual internal stress within the crystals affect the hardness of the powders and consolidated samples. Variation in the hardness of the Al–5 wt %Si powders and consolidated samples as a function of milling time is shown in Figure 5. Both crystallite size reduction and elemental dispersion strengthening affect the hardness of samples. Based on these results, the measured hardness for both powdered and consolidated samples is almost monotonically enhanced when the milling time is increased, reaching its maximum value at around 50 h of milling time. However, the consolidated samples show greater absolute hardness values compared to their powder counterparts most likely due to the distribution of the internal stress within their structure (see Tables 1 and 2). It is worth noting that microhardness measurements performed on powder particles are inherently prone to systematic errors. Therefore, microhardness value comparisons between the powder specimens and the consolidated specimens should be considered approximations at best. Despite almost a constant grain size after 20 h of milling (results from Figure 3), overall hardness values increase when the milling time is increased. These phenomena point to the fact that, in addition to grain reinforcement, certain mechanisms, such as work hardening of the fragile fragments in the initial stages of milling, the homogeneous distribution of Si atoms in the Al matrix, and its possible oversaturation, lattice microstrains, as well as the formation and distribution of defects within newly formed crystals, affect the final mechanical properties of the alloys. However, quantification of the fractal contributions of these mechanisms is out of the scope of the current manuscript.

Figure 6 presents yield and compressive strengths of consolidated samples for various milling time. Similar to hardness test results (results from Figure 5), strength values increase with milling time reaching a maximum yield strength after 50 h of milling. It is known that the grain boundaries result in a higher yield stress than the matrix itself, since grain boundaries are able to prevent dislocation movement [21]. In other words, dislocations pile up behind grain boundaries resulting in a stress concentration. Since increasing the milling time reduces the grain size, it will also strengthen the material, resulting in an increase in its yield stress. Furthermore, a better distribution of silicon in the aluminum matrix as well as work hardening can contribute to the increased yield strength of consolidated samples as the milling time increases [23,24].

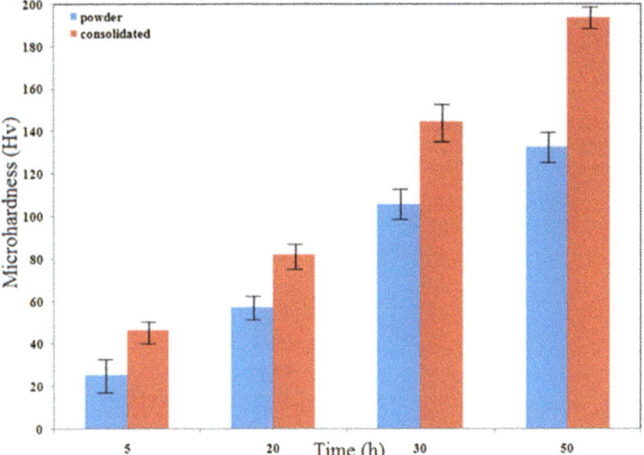

Figure 5. Microhardness values (Hv) of Al–5 wt %Si powder and consolidated samples as a function of the milling time (h).

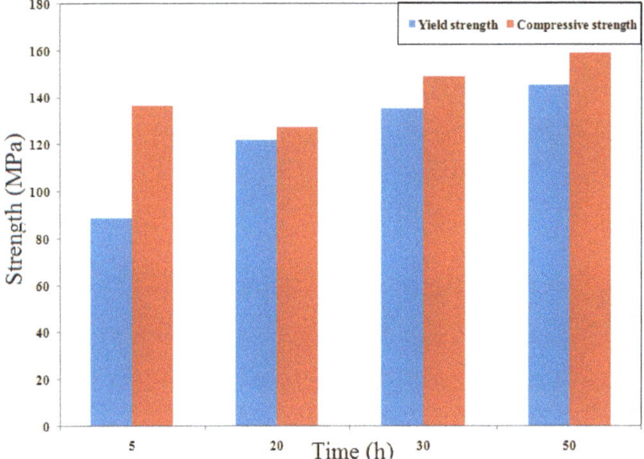

Figure 6. Yield and compressive strengths (MPa) of consolidated samples as a function of the milling time (h).

4. Conclusions

In the present study, nanocrystalline Al–5 wt %Si powder was successfully synthesized by mechanical alloying under an argon atmosphere followed by consolidation using a hot pressing process. Morphology changes of powders revealed that milling, welding of the soft aluminum particles, and the consequent fracture of brittle structures are major competing mechanisms in the first 50 h of milling, leading to an equilibrium semi-globular structure for particulate samples. Evaluation of the crystallite size through analysis of the XRD patterns for both consolidated and powder samples indicated a nanometric crystallite size, which decreases when the milling duration is increased. Consolidation in the hot pressing process led to only a minute crystallite size growth. SEM-EDX studies of the consolidated samples showed that the distribution of Si in the Al matrix is uniform throughout the sample after 20 h of milling, which brings about minimal crystallite size growth. It was also observed that mechanical properties of the consolidated samples such as hardness values and compression strength enhance as the milling time increases as a consequence of uniform distribution of elements and work hardening.

Acknowledgments: The corresponding author would like to acknowledge the FAU library for their support for covering the 25% of the costs to publish in open access.

Author Contributions: Davood Dayani and Seyed Reza Jafarpour Rezaei conceived, designed, and performed the experiments; Davood Dayani wrote a part of the paper; Mohammad Reza Vaezi, Ali Shokuhfar, and Saman Hosseinpour supervised the work; Saman Hosseinpour wrote the paper.

Conflicts of Interest: The authors declare no conflict of interest.

References

1. Benjamin, J.S. Dispersion strengthened superalloys by mechanical alloying. *Metall. Trans.* **1970**, *1*, 2943–2951. [CrossRef]
2. Lü, L.; Lai, M.O. *Mechanical Alloying*; Springer: New York, NY, USA, 1998.
3. Wang, W.; Zhai, H.; Chen, L.; Zhou, Y.; Huang, Z.; Bei, G.; Greil, P. Sintering and properties of mechanical alloyed Ti3AlC2-Cu composites. *Mater. Sci. Eng. A* **2017**, *685*, 154–158. [CrossRef]
4. Shokrollahi, H. The magnetic and structural properties of the most important alloys of iron produced by mechanical alloying. *Mater. Des.* **2009**, *30*, 3374–3387. [CrossRef]
5. Polkin, I.S.; Borzov, A.B. New materials produced by mechanical alloying. *Adv. Perform. Mater.* **1995**, *2*, 99–109. [CrossRef]
6. Chaubey, A.K.; Scudino, S.; Mukhopadhyay, N.K.; Khoshkhoo, M.S.; Mishra, B.K.; Eckert, J. Effect of particle dispersion on the mechanical behavior of Al-based metal matrix composites reinforced with nanocrystalline Al–Ca intermetallics. *J. Alloys Compd.* **2012**, *536*, S134–S137. [CrossRef]
7. Caballero, E.S.; Cintas, J.; Cuevas, F.G.; Montes, J.M.; Herrera-García, M.; Oral, A.Y.; Bahşi, Z.B.; Özer, M.; Sezer, M.; Aköz, M.E. High temperature behavior of nanostructured Al powders obtained by mechanical alloying under NH3 flow. *AIP Conf. Proc.* **2015**, *1653*, 20025. [CrossRef]
8. Fogagnolo, J.; Velasco, F.; Robert, M.; Torralba, J. Effect of mechanical alloying on the morphology, microstructure and properties of aluminium matrix composite powders. *Mater. Sci. Eng. A* **2003**, *342*, 131–143. [CrossRef]
9. Kang, W.-K.; Yılmaz, F.; Kim, H.-S.; Koo, J.-M.; Hong, S.-J. Fabrication of Al–20 wt %Si powder using scrap Si by ultra high-energy milling process. *J. Alloys Compd.* **2012**, *536*, S45–S49. [CrossRef]
10. Khan, A.S.; Farrokh, B.; Takacs, L. Effect of grain refinement on mechanical properties of ball-milled bulk aluminum. *Mater. Sci. Eng. A* **2008**, *489*, 77–84. [CrossRef]
11. Chen, C.; Lu, C.; Feng, X.; Shen, Y. Effects of annealing on Al–Si coating synthesised by mechanical alloying. *Surf. Eng.* **2017**, *33*, 548–558. [CrossRef]
12. Kainer, K.U. *Metal Matrix Composites: Custom-Made Materials for Automotive and Aerospace*; John Wiley & Sons: Hoboken, NJ, USA, 2006.
13. Bose, A.; Eisen, W.B. *Hot Consolidation of Powders and Particulates*; Metal Powder Industries Federation: Princeton, NJ, USA, 2003.
14. Williamson, G.; Hall, W. X-ray line broadening from filed aluminium and wolfram. *Acta Metall.* **1953**, *1*, 22–31. [CrossRef]
15. Hryha, E.; Zubko, P.; Dudrová, E.; Pešek, L.; Bengtsson, S. An application of universal hardness test to metal powder particles. *J. Mater. Proc. Technol.* **2009**, *209*, 2377–2385. [CrossRef]
16. Abdoli, H.; Salahi, E.; Farnoush, H.; Pourazrang, K. Evolutions during synthesis of Al–AlN-nanostructured composite powder by mechanical alloying. *J. Alloys Compd.* **2008**, *461*, 166–172. [CrossRef]
17. Milligan, J.; Vintila, R.; Brochu, M. Nanocrystalline eutectic Al–Si alloy produced by cryomilling. *Mater. Sci. Eng. A* **2009**, *508*, 43–49. [CrossRef]
18. Miraghaei, S.; Abachi, P.; Madaah-Hosseini, H.R.; Bahrami, A. Characterization of mechanically alloyed $Fe_{100-x}Si_x$ and $Fe_{83.5}Si_{13.5}Nb_3$ nanocrystalline powders. *J. Mater. Process. Technol.* **2008**, *203*, 554–560. [CrossRef]

19. Safari, J.; Akbari, G.H.; Shahbazkhan, A.; Delshad Chermahini, M. Microstructural and mechanical properties of Al-Mg/Al$_2$O$_3$ nanocomposite prepared by mechanical alloying. *J. Alloys Compd.* **2011**, *509*, 9419–9424. [CrossRef]

20. Mhadhbi, M.; Khitouni, M.; Azabou, M.; Kolsi, A. Characterization of Al and Fe nanosized powders synthesized by high energy mechanical milling. *Mater. Charact.* **2008**, *59*, 944–950. [CrossRef]

21. Sivasankaran, S.; Sivaprasad, K.; Narayanasamy, R.; Satyanarayana, P.V. X-ray peak broadening analysis of Aa 6061$_{100-x}$–x wt % Al$_2$O$_3$ nanocomposite prepared by mechanical alloying. *Mater. Charact.* **2011**, *62*, 661–672. [CrossRef]

22. Alizadeh, A.; Taheri-Nassaj, E. Mechanical properties and wear behavior of Al–2 wt %Cu alloy composites reinforced by B4C nanoparticles and fabricated by mechanical milling and hot extrusion. *Mater. Charact.* **2012**, *67*, 119–128. [CrossRef]

23. Hall, E.O. The deformation and ageing of mild steel: III discussion of results. *Proc. Phys. Soc. Sect. B* **1951**, *64*, 747. [CrossRef]

24. Poirier, D.; Drew, R.A.; Trudeau, M.L.; Gauvin, R. Fabrication and properties of mechanically milled alumina/aluminum nanocomposites. *Mater. Sci. Eng. A* **2010**, *527*, 7605–7614. [CrossRef]

Article

A Unified Physical Model for Creep and Hot Working of Al-Mg Solid Solution Alloys

Stefano Spigarelli * and Chiara Paoletti

DIISM, Università Politecnica delle Marche, via Brecce Bianche, 60131 Ancona, Italy; chiarella251290@gmail.com
* Correspondence: s.spigarelli@univpm.it; Tel.: +39-071-2204-746

Received: 27 November 2017; Accepted: 22 December 2017; Published: 27 December 2017

Abstract: The description of the dependence of steady-state creep rate on applied stress and temperature is almost invariably based on the Norton equation or on derived power-law relationships. In hot working, the Norton equation does not work, and is therefore usually replaced with the Garofalo (sinh) equation. Both of these equations are phenomenological in nature and can be seldom unambiguously related to microstructural parameters, such as dislocation density, although early efforts in this sense led to the introduction of the "natural power law" with exponent 3. In an attempt to overcome this deficiency, a recent model with sound physical basis has been successfully used to describe the creep response of fcc metals, such as copper. The main advantage of this model is that it does not require any data fitting to predict the strain rate dependence on applied stress and temperature, which is a particularly attractive peculiarity when studying the hot workability of metals. Thus, the model, properly modified to take into account solid solution strengthening effects, has been here applied to the study of the creep and hot-working of simple Al-Mg single phase alloys. The model demonstrated an excellent accuracy in describing both creep and hot working regimes, still maintaining its most important feature, that is, it does not require any fitting of the experimental data.

Keywords: creep; hot working; constitutive equations; solid solution hardening

---

1. Introduction

Although single-phase Al-Mg alloys have limited industrial relevance as creep-resistant alloys, their creep response has been analysed in many papers. A non-exhaustive picture of the extensive coverage of creep of Al-Mg alloys can be obtained, for example, from the relevant chapters in [1–3]. The cause for this interest can be traced back to the simple nature of these materials, since their microstructure constitutes a good case study to identify the main creep mechanisms and separate their relative contributions to creep response in solid solutions.

In most cases, the secondary creep rate ($\dot{\varepsilon}$) dependence on applied stress (σ) and the temperature (T) of single-phase alloys has been described by the conventional power-law and Arrhenius equations, in the form

$$\dot{\varepsilon} = A \frac{D_0 G b}{kT} \left(\frac{\sigma}{G}\right)^n \exp\left(-\frac{Q}{RT}\right) \tag{1}$$

where A is a material parameter, k is the Boltzmann constant, G is the shear modulus, b is the length of the Burgers vector, and R is the gas constant (all the symbols used, and, where relevant, their values, are listed in Table 1). D_0 in Equation (1) is the pre-exponential factor in the Arrhenius form and Q is the activation energy for the relevant diffusional mechanism (self-diffusion or diffusion of Mg atoms in Al). In pure Aluminium the stress exponent n is close to 4.4–5; however, above a certain stress level, power-law breakdown occurs, and the slope of the curve describing the strain rate dependence on applied stress in double-log coordinates increases progressively with the stress level applied.

This behaviour, which is associated to a stress exponent in the power-law regime close to 5 and an activation energy equivalent to that for vacancy diffusion, identifies "class M" (Metal) materials. The addition of Mg results in a more complex dependence of the secondary creep rate on applied stress. The stress exponent in the low stress regime is close to 4–5, it then becomes 3 in an intermediate stress range, and again, 4–5 before power-law breakdown [4–6]. This behaviour ("class A") is generally interpreted by invoking the fact that, since glide and climb of dislocations occur in sequence during high-temperature deformation, the slower is rate controlling. In pure metals, glide is always faster than climb; therefore, the latter is, without exceptions, the rate controlling mechanism. In class A materials, glide is substantially slowed down by the formation of clouds of solute atoms around dislocations, and is, consequently, the rate controlling factor in the intermediate stress regime, leading to $n = 3$ and to an activation energy that is equivalent to the activation energy for diffusion of Mg atoms in Al. It is only when climb is very slow (in the low stress regime) or when the solute atoms no longer play any role, since dislocations have broken away from their atmospheres (in high stress regime), that the stress exponent is 4–5 and a class-M behaviour is apparent.

Table 1. List of fundamental symbols.

b	Burgers vector	2.86×10^{-10} m
c	concentration of Mg in solid solution	[at %]
C_L	work hardening constant	86
D_{0sd}	pre-exponential factor in equation for self-diffusion	8.34×10^{-6} m$^2 \cdot$s^{-1}
D_{0Mg}	pre-exponential factor in equation for diffusion of Mg in Al	1.9×10^{-5} m$^2 \cdot$s^{-1}
G	shear modulus at the testing temperature	$(3.022 \times 10^{10}$–1.6×10^7 $T)$ Pa
k	Boltzmann constant	1.38×10^{-23} J$\cdot$K^{-1}
L	mean dislocation free path	[m]
m	Taylor factor	3.06
M_c	climb mobility of dislocations	[m$^2 \cdot$N$^{-1} \cdot$s^{-1}]
M_{cg}	climb and glide mobility of dislocations	[m$^2 \cdot$N$^{-1} \cdot$s^{-1}]
n	stress exponent in power-law equation	
Q_{sd}	activation energy for vacancy diffusion (self-diffusion)	122×10^3 J$\cdot$mol^{-1}
Q_{dMg}	activation energy for diffusion of Mg in Al	119×10^3 J$\cdot$mol^{-1}
R	universal gas constant	[J$\cdot$mol$^{-1} \cdot$K^{-1}]
R_{max}	maximum back stress	[Pa]
R_{UTS}	ultimate tensile strength	[Pa]
T	absolute temperature	[K]
v_d	velocity of dislocations	[m$\cdot$s^{-1}]
α	material constant in Taylor equations	0.3
δ_{Mg}	volume atomic misfit	
ε	strain	
$\dot{\varepsilon}$	strain rate	[s^{-1}]
ν	Poisson's ratio	0.3
ρ	free dislocation density	[m^{-2}]
ρ_a	free dislocation density in annealed state	[m^{-2}]
σ	stress (creep or constant strain rate experiments)	[Pa]
σ_{ba}	break-away stress	[Pa]
σ_i	internal stress	[Pa]
σ_{ss}	solid solution strengthening stress	[Pa]
σ^*_{ss}	reduced solid solution strengthening stress	[Pa]
σ_y	yield strength	[Pa]
τ_l	dislocation line tension	[N]
ω	recovery constant	[Pa]
Ω	atomic volume of the host atom (Al)	1.66×10^{-29} m^3

The explanation of the phenomenology of creep in Al-Mg alloys, which are very synthetically outlined above, and the use of power law to describe related experimental data are generally taken for granted. Yet, the power law is phenomenological in nature, and, for this reason, alternative approaches have been sought to introduce a set of constitutive equations more directly, based on the underlying physics. The recent work by Fernández and González-Doncel [7], which deals with a unified model for the description of creep in Al-Mg alloy, is just one of the many examples. In this line of thought, Sandström proposed another set of equations for fcc metals, which, in his intent, should not require any best-fitting of experimental creep or mechanical data, being based only on a number of physical and microstructural pre-determined parameters [8,9]. The resulting basic model was developed to describe creep phenomena, but has been also successfully applied to high temperature constant strain rate experiments [8]. Since the analysis of the hot-working response of metals is based on compression or torsion constant-strain rate experiments, the use of Sandström's set of equations to predict the material behaviour in this envelope of experimental conditions seems to be a straightforward step. This reasoning has led the authors of this paper to apply the basic model to Al-Mg single-phase alloys, with a major emphasis on the description of the high-temperature conditions typical of hot-working operations.

2. The Model

The model, originally developed for Cu, is based on physically-derived equations, which will be here illustrated for a simple Al alloy, possibly containing elements in solid solution.

Free dislocation density (ρ) [10] is usually related to applied stress by the well-known Taylor equation, written in the form

$$\sigma = \sigma_i + \sigma_{ss} + \sigma_d = \sigma_i + \sigma_{ss} + \alpha m G b \sqrt{\rho} \tag{2}$$

where m is the Taylor factor and $\sigma_d = \alpha m G b \rho^{1/2}$ is the dislocation hardening term. The term σ_i represents the strength of pure annealed metal, that is, the stress that is required to move a dislocation in the absence of other dislocations, while α is a constant. In solid solution alloys, the viscous drag of dislocations is thought to reduce dislocation mobility and control creep response in a wide interval of applied stresses. The term σ_{ss} thus represents the stress required for dislocations to move by viscous drag in the presence of solute atom atmospheres.

The evolution of dislocation density during straining can be expressed as [9]

$$\frac{d\rho}{d\varepsilon} = \frac{m}{bL} - \omega\rho - \frac{2}{\dot{\varepsilon}} M\tau_l\rho^2 \tag{3}$$

where ω is a constant, τ_l is the dislocation line tension ($\tau_l = 0.5Gb^2$), M is the dislocation mobility, and L is the dislocation mean free path, i.e., the distance travelled by a dislocation before it undergoes a reaction, customarily expressed as

$$L = \frac{C_L}{\sqrt{\rho}} \tag{4}$$

C_L being the strain-hardening constant. The first term on the right-hand side of Equation (3) represents the strain hardening effect due to dislocation multiplication, which is more rapid when L, and, consequently, C_L assume low values and/or the dislocation density is high. The second and third terms on the right-hand side of Equation (3) describe the effect of dynamic recovery. Since, at high temperature, the last term of Equation (3), which includes an Arrhenius-type dependence on T, largely predominates on the second term, which is roughly a-thermal, and the main emphasis in this study is on describing high-temperature behaviour ($T > 500$ K), Equation (3) can be simplified to become

$$\frac{d\rho}{d\varepsilon} = \frac{m}{bL} - \frac{2}{\dot{\varepsilon}} M\tau_l\rho^2 \tag{5}$$

The dislocation climb mobility in a pure metal, according to Hirth and Lothe [11], is

$$M_c = \frac{D_{0sd}b}{kT} \exp\left(\frac{\sigma_d b^3}{kT}\right) \exp\left(-\frac{Q_{sd}}{RT}\right) \tag{6}$$

To describe the creep results obtained in the high strain rate regime, which are characterised by higher stress exponents and lower values of the activation energy for creep, the model has been further modified to take into account another mechanism, the glide of dislocations through an obstacle field [12]. Thus, the glide mobility was described by a phenomenological equation in the form

$$M_g \propto \exp\left\{-\frac{Q_g}{RT}\left[1 - \left(\frac{\sigma_d}{R_{\max}}\right)^p\right]^q\right\} \tag{7}$$

where $R_{\max}$, which depends on material structure, is the flow stress that is required to plastically deform the material in the absence of thermal activation and Q_g is the activation energy necessary to overcome the obstacle field. Recent studies [9,13] showed that Equation (7) works very well for pure Cu and Al with $p = 2$ and $q = 1$, with $R_{\max}$ equivalent to the true stress corresponding to the ultimate tensile strength of the material considered, roughly quantified as $1.5R_{uts}$, where R_{uts} is the ultimate tensile strength.

During viscous glide in solid solution alloys, solute atoms have to jump in and out of the atmospheres that spontaneously form around dislocations. Thus, an additional term describing the energy needed to overcome this barrier must be added to the activation energy. This additional term has the form [14]

$$U_{ss} = \frac{\beta R}{bk} \tag{8}$$

with

$$\beta = \frac{1}{3\pi}\frac{(1+v)}{(1-v)}bG\Omega\,\delta_{Mg} \tag{9}$$

where v is the Poisson's ratio (=0.3 in Al), Ω is the average Al-atomic volume, and δ_{Mg} is the volume atomic misfit (details about Ω and δ_{Mg} are given in [14]).

The combination of Equations (5)–(8) [9,13,14] gives the following relationship for climb and glide mobility, to be used for M into Equation (5):

$$M = M_{cg} \cong \frac{D_{0cg}b}{kT} \exp\left[\frac{\sigma_d b^3}{kT}\right] \exp\left\{-\frac{Q_{cg}}{RT}\left[1 - \left(\frac{\sigma_d}{R_{\max}}\right)^2\right]\right\} \exp\left(-\frac{U_{ss}}{RT}\right) \tag{10}$$

where $D_{cg} = D_{0cg}\exp(-Q_{cg}/RT)$ is the appropriate diffusion coefficient.

At steady state, Equation (5), combined with Equations (2) and (4), gives

$$\dot{\varepsilon}_{ss} = \frac{2M_{cg}\tau_l bC_L}{m}\left(\frac{\sigma_d}{\alpha m Gb}\right)^3 \tag{11}$$

The model based on Equations (9) and (10) requires the determination of C_L ($C_L = 86$ in pure Al [13]) and of two of the terms in Equation (2), namely σ_i and σ_{ss}.

In the case of pure Al, the determination of σ_i, which is temperature and strain rate dependent, was based on the assumption that the dislocation density in annealed state (ρ_a) and the σ_i values account for the annealed yield strength of the pure metal [15]. The yield stress is thus given by Equation (2) (which is general, and holds for any single stage of the stress vs. strain curve) [15], where ρ_a is virtually nihil. The traditional way of describing the temperature dependence of yield strength is to assume that, in the creep range, yield strength is proportional to creep strength, whereas below the creep range yield strength is proportional to shear modulus. On these bases, for pure Al, the following expression has been used in [13]

$$\sigma_i = A_y \sqrt{\sigma_{creep} G} \tag{12}$$

where σ_{creep} is the creep stress that corresponds to a given steady-state creep rate. The A_y constant ($A_y = 4.2 \times 10^{-3}$ [13]) was determined so as to obtain a reliable estimate of the yield strength of high-purity Al at room temperature by Equation (2).

An equation for solid solution strengthening stress was given in [14], in the form

$$\sigma_{ss} = \frac{v_d c \beta^2}{b \Omega D_{Mg} kT} I(z_0) \tag{13}$$

where v_d is the dislocation velocity, being c the Mg atomic concentration. The term $D_{Mg} = D_{0Mg} \exp(-Q_{Mg}/RT)$ is the diffusivity of Mg in Al. The term $I(z_0)$ can be calculated by the numerical integration of

$$I(z_0) = \int_1^{z_0} \frac{2\sqrt{2\pi}}{3} z^{-5/2} \exp(z) dz \tag{14}$$

with $z_0 = \beta/bkT$ [14].

3. Description of High Purity Aluminium

Figure 1 plots the steady-state creep rate as a function of stress for Al 99.999% [16] and the model curves given by Equations (10) and (11) with $C_L = 86$, $U_{ss} = 0$, $Q_{cg} = Q_{sd} = 122$ kJ·mol^{-1}, and $D_{cg} = D_{0sd} = 8.34 \times 10^{-6}$ m^2·s^{-1} [13]. The original Figure in [13] was obtained by taking $\omega = 15$; in this study, the value $\omega = 0$ was used, which demonstrates that the simplified model of Equation (5) still works very well. The diffusion coefficient was recalculated in [13] by considering all the data reported in [17]. The basic model gives a very good description of the experimental data for the pure metal, without any need for data fitting, and, for this reason, it is an excellent basis for the implementation of the analysis of Al-Mg alloys.

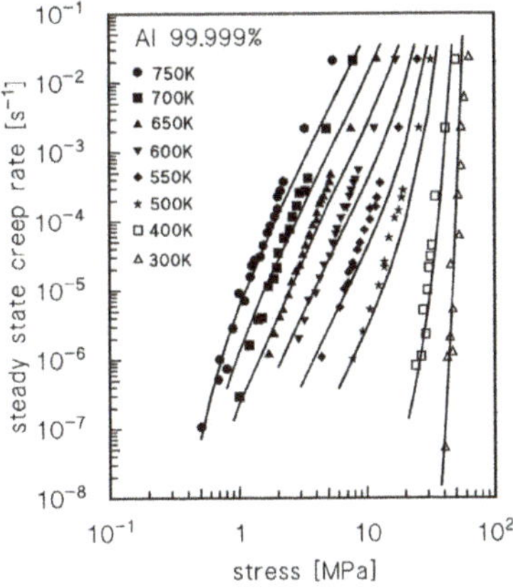

Figure 1. Steady-state creep rate dependence on applied stress for Al 99.999% [16]; the curves were calculated by Equations (10) and (11) with $C_L = 86$, $\omega = \sigma_{ss} = U_{ss} = 0$.

4. Description of High Purity Aluminium-Magnesium Single Phase Alloys

4.1. Diffusion Coefficient

The typical value of the diffusion coefficient parameters for the diffusion of Mg in Al is $D_{0Mg} = 1.24 \times 10^{-4}$ m$^2 \cdot$s^{-1}, $Q_{sd} = 130.5$ kJ$\cdot$mol^{-1} [18]. The accuracy of this estimate was challenged in a recent work [19], which reported a wide collection of literature results on the diffusivity of Mg in Al, which is illustrated in Figure 2. A good fitting is obtained with $Q_{sd} = 119$ kJ$\cdot$mol^{-1} and $D_{0sd} = 1.9 \times 10^{-5}$ m$^2 \cdot$s^{-1}, which actually give a curve very close to the one from [18].

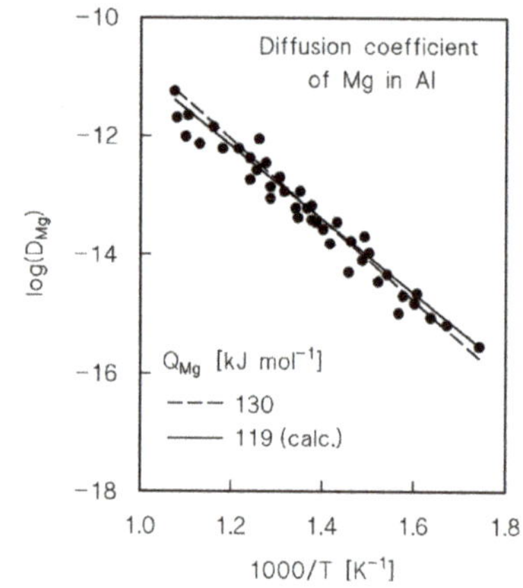

Figure 2. Diffusion coefficient in Al-Mg (set of literature data from [19]). The Figure illustrates the best fitting of the data as well as the curve representing the diffusion coefficient used in the majority of previous studies.

4.2. Drag Stress Calculation and Experimental Datasets on Dislocation Density and Strain Rate

Microstructural data, namely dislocation density variation with applied stress, and previous results on pure Al, will be used to evaluate the accuracy of the model in estimating solid solution stress. Equation (2) can be rewritten as

$$\rho = \left(\frac{\sigma - \sigma_i - \sigma_{ss}}{\alpha m G b} \right)^2 \tag{15}$$

Figure 3, which collects data from different sources [20–22], shows that the experimental value of the dislocation density in Al-Mg is somewhat lower than in pure Al. This behaviour is fully consistent with Equation (15) due to the presence of drag stress. Thus, dislocation density could be properly modelled, as long as the calculated value of drag stress is sufficiently reliable. To proceed in this direction, it is necessary to estimate the dislocation velocity, which can be expressed by combining the well known Orowan equation

$$\dot{\varepsilon} = \rho \, b \, v_d \tag{16}$$

with Equation (11), giving

$$v_d = \frac{C_L}{\alpha \, m^2} M_{cg} b \sigma_d \tag{17}$$

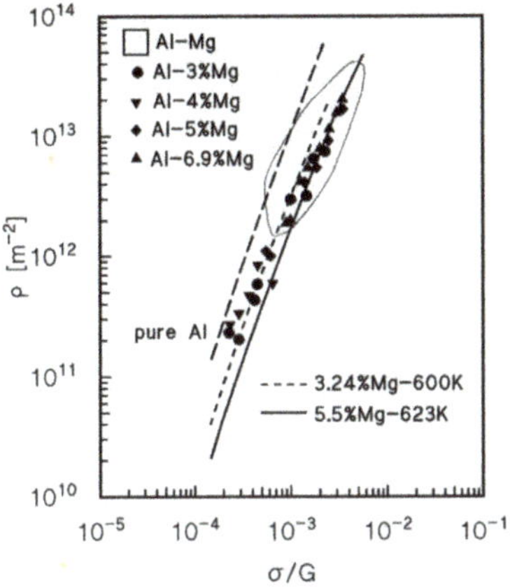

Figure 3. Experimental values of the dislocation density as a function of stress for Al-3%Mg [20–22], Al-4%Mg [21], Al-5%Mg [21], and Al-6.9%Mg [20]. The shaded area, which illustrates the scatter band for Al-Mg alloys, was reported in [22]. The Figure also plots the model curve for pure Al [13] and for Al-Mg alloys (3.24% and 5.5%Mg at 600 and 623 K, respectively).

The combination of Equations (13) and (17) gives

$$\sigma_{ss} = \frac{C_L M_{cg} \sigma_d c \beta^2}{\alpha\, m^2 \Omega D_{Mg} kT} I(z_0) \tag{18}$$

or

$$\sigma_{ss} = \frac{B}{1+B}(\sigma - \sigma_i) \tag{19}$$

where

$$B = \frac{C_L M_{cg} c \beta^2}{\alpha\, m^2 \Omega D_{Mg} kT} I(z_0) \tag{20}$$

The diffusion coefficient for climb and glide D_{cg} can be considered to be equivalent to the self-diffusion coefficient of Al. Once the value of the ultimate tensile strength of the different Al-Mg alloys is quantified (with the above-mentioned assumption, $R_{max} = 1.5 R_{UTS}$), all of the parameters in Equations (10), (12), and (19) are known. The model curves presented in Figure 3 were thus calculated by Equations (15) and (18). Since the agreement between the curves and the experimental data is excellent, a direct and independent confirmation that the estimation of the drag stress is sufficiently reliable to be used in the model for the steady-state creep rate dependence on applied stress is obtained.

Figure 4 shows the sets of experimental creep data used in this study. The first dataset [4] presents the creep rate at an applied strain of 0.2, i.e., reasonably close to the steady state, at a constant temperature, for different Mg contents (Figure 4a). Although being limited to a single temperature, the dataset represents the classical case study, since the three traditional regimes can be easily recognised. Figure 4a clearly shows low stress Regime I, where creep is thought to be climb-controlled (n = 4–5). Above a transition stress σ_c, the rate-controlling mechanism should be viscous glide (n = 3). It is only when the applied stress exceeds a limiting value (σ_{ba}^s, frequently identified with the break-away stress, σ_{ba}) that dislocations are able to break-away from solute atom atmospheres and creep should again become climb-controlled (n = 4–5). The data from [4] are thus well suited to assess the accuracy of the model in describing the effect of different amounts of Mg in solid solution.

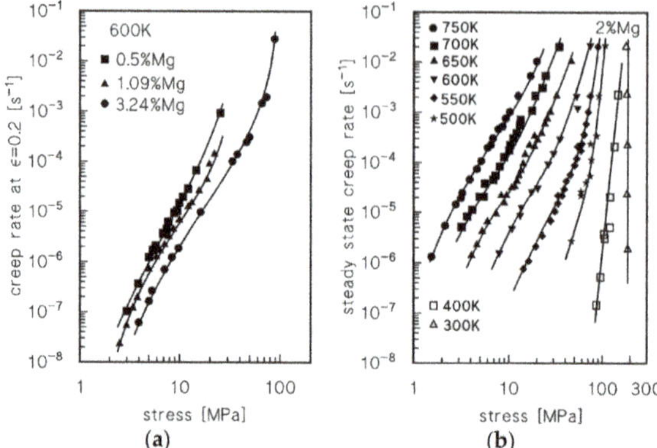

Figure 4. Literature data on the creep of Al-Mg alloys: strain rate at 0.2 strain data from [4] (**a**) and steady-state creep and steady-state flow stress from constant strain rate experiments from [16] (**b**).

The second dataset (Figure 4b), which is related to an Al-2%Mg alloy, was obtained from [16]. This dataset is extremely useful to investigate the accuracy of the model in describing the temperature dependence of the steady-state creep rate.

4.3. Viscous-Glide Controlled Creep: Strain Rate Dependence on Stress and Temperature at $T \geq 523$ K

The first analysis was focused on the data included in Regimes I and II at high temperatures. In the low-intermediate stress regime, solute atmospheres surround dislocations, exerting a drag stress σ_{ss}. Upon substitution of the drag stress into Equations (10) and (11), the model gives the curves that are presented in Figure 5. It is clearly shown that the model gives an accurate description of the experimental data in the low and intermediate stress regimes, although a significant deviation is observed at high stresses. This behaviour could be rationalised by considering that the model can describe the steady-state creep rate with reasonable accuracy as long as $\sigma < \sigma_{ba}^s$. Thus, all of the data points for $\sigma > \sigma_{ba}^s$ represent the steady-state creep rate in conditions, in which the progressive break-away of dislocations results in a reduction of the strengthening effect of solute atom atmospheres, that is, of the σ_{ss} term.

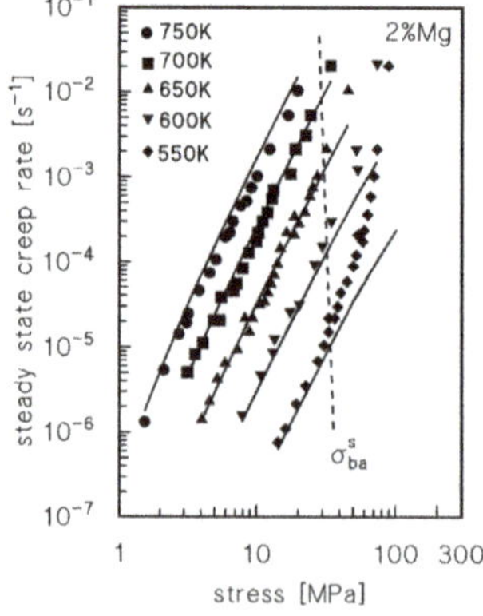

Figure 5. Description of the experimental data for $T > 523$ K from [16] by Equations (10) and (11), where the drag stress σ_{ss} is obtained from Equations (19) and (20).

4.4. Creep Above the Transition for Break-Away of Dislocations from Solute Atom Atmospheres

In the model presented in the previous section, the σ_{ss} term monotonically increases with stress. By contrast, the drag effect should rapidly reduce as dislocations break-away from solute atoms. However, for stresses just above σ_{ba}^s, the solid solution stress cannot be reasonably expected to become abruptly non-existent. This behaviour would indeed result in a step in the strain rate plot that is not evident in Figure 4. Rather, it seems more reasonable to suppose that, as stress increases above σ_{ba}^s, when the dislocation velocity exceeds a limiting value v_{ba}, a progressively increasing fraction of the Mg atoms is expelled from the solute atmospheres. When the stress is sufficiently high that all the dislocations are free from solute atoms, the material behaves essentially like pure Aluminium—the main difference being that here the dislocation should overcome single Mg atoms, and that the alloy still creeps at a lower rate. In parallel, the homogeneous dislocation distribution typical of class A materials, which is associated to high drag stresses, is progressively replaced by a more heterogeneous distribution with substructure formation [1–3].

If such is the situation, the drag stress can no longer monotonically increase, as predicted by Equation (13), but, at a stress correctly identified as break-away stress (σ_{ba}), should present a maximum which should be larger than σ_{ba}^s.

The behaviour described above was here modelled by supposing that the effect of break-away can be described by an equation in the form

$$\sigma_{ss}* = F\,\sigma_{ss} \tag{21}$$

where $0 \le F(\sigma) \le 1$, σ_{ss} being the drag stress calculated by Equation (18) A suitable phenomenological form for F is

$$F = \left[1 + \left(\frac{\sigma}{\sigma_{50}}\right)^{h}\right]^{-1} \tag{22}$$

where σ_{50} is the stress for which the reduced drag stress is a mere half of the computed value of σ_{ss}. For the sake of simplicity, it was here assumed that $\sigma_{50} \cong \sigma_{ba}$. This assumption, in turn, implies that to have a maximum in the drag stress for $\sigma = \sigma_{ba}$, due to the peculiar dependence of σ_{ss} on stress, then the exponent h should range from 2, in the high temperature regime, to 3, at 600 K.

The general expression for the break-away stress was given by Friedel [23]

$$\sigma_{ba} = m\tau_{ba} = m\,A_{ba}\frac{W_m^2 c}{kTb^3} \tag{23}$$

where the maximum interaction energy between solute atoms and an edge dislocation can be expressed as

$$W_m = -\frac{1}{2\pi}\left(\frac{1+\nu}{1-\nu}\right)G|\Delta V_a| \tag{24}$$

being ΔV_a the difference in volume between solute and solvent atoms (that is, with the formalism used in Equation (8), W_m is directly related to β). In Friedel's original formulation $A_{ba} = 1$, giving, for example, a break-away stresses above 400 MPa for 2%Mg at 750 K, a level so high that it is hardly conceivable for this mechanism to play any role in creep. These very high values were considered by some authors [4,24,25] as incompatible with experimental evidence. Their analyses rather suggested that σ_{ba} is one order of magnitude lower, that is, $A_{ba} \cong 0.1$ [24,25] or even $A_{ba} \cong 0.065$ [26,27]: this reduced value of the break-away stress will be here, denoted as σ_{ba}^*. Thus, Equation (23) was here provisionally used with $A_{ba} = 0.065$, to obtain the values of the reduced drag stress σ_{ss}^*. The curves obtained by replacing σ_{ss} with σ_{ss}^* into Equation (11) are presented in Figure 6.

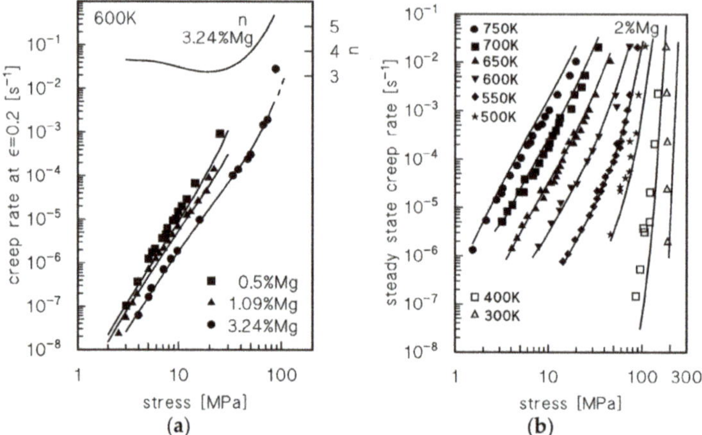

Figure 6. Description of the experimental data from [4] (**a**) and [16] (**b**) by Equations (10) and (11), with σ_{ss}^* from Equations (21)–(24). Figure 6a also shows the variation of $n = \partial \log \dot{\varepsilon} / \partial \log \sigma$ for the alloy with 3.24%Mg.

The model is now able to describe the material behaviour in the whole range of applied stress, and even the complex change in slope n $(=\partial \log \dot{\varepsilon} / \partial \log \sigma)$, usually associated to the transition from Regimes I, II and III (Figure 6a), without requiring specific assumptions on the rate-controlling mechanisms.

Examples of the variation of σ_{ss}^* as a function of temperature and applied stress are presented in Figure 7. The reduction of the drag stress is the effect of a lower amount of Mg in the atmospheres around dislocations: c^* can be identified as the amount of magnesium, still in the solute clouds, which produces a given drag stress σ_{ss}^*. The value of c^*, as calculated by Equation (21), for a single temperature and alloy is presented in Figure 8, which also plots the corresponding σ_{ss}^* and σ_{ss}.

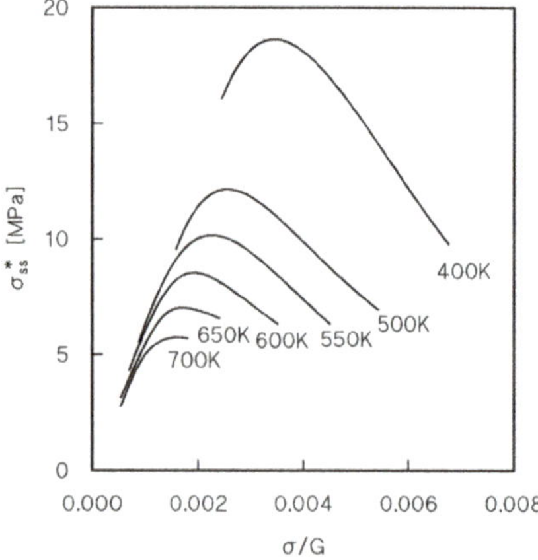

Figure 7. Variation of σ_{ss}^* as a function of temperature and applied stress for Al-2%Mg.

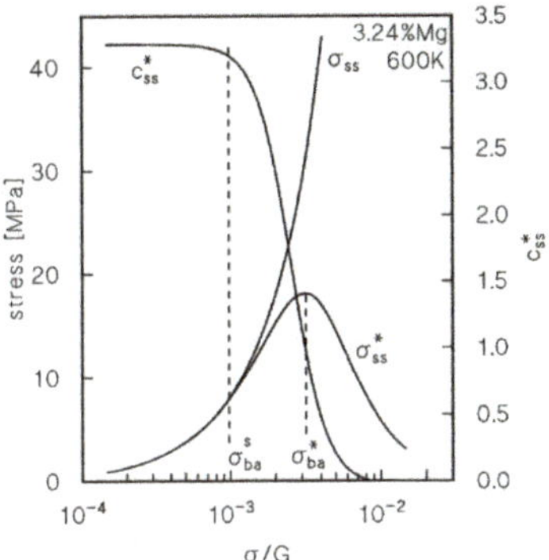

Figure 8. Plots for σ^*_{ss}, c^* and σ_{ss} as a function of modulus compensated stress for 3.24%Mg at 600 K.

4.5. Hot Working as an Extension of Creep: The Model in the High Strain Rate Plasticity Regime

One of the main tasks of this study is to assess whether the Sandström basic model, modified to account for dislocation break-away, is able to also describe high-strain rate/high-temperature data, i.e., the hot working regime. Figures 9–11 illustrate the compression data (stress at 0.3 strain, roughly corresponding to the peak stress of the flow curve) for three different alloys (0.55, 2.2 and 5.5%Mg) [28], and, for comparison purposes, other single creep datasets for 0.48%Mg [4], 2.2%Mg [1,29] and 5.5%Mg [1,30]. The model curves, as calculated by following the same procedure discussed above, i.e., by replacing σ_{ss} with σ^*_{ss}, provide an excellent description of the experimental data (the average error is the stress estimation for a given strain rate is 11, 10, and 7% for the alloys containing 0.55, 2.2, and 5.5%Mg, respectively), with the mere exception of the results for 0.5%Mg at 573 K, where the average error reaches 22% (Figure 7). As a matter of fact, at 573 K and at the highest strain rates at 623 K, the reduced drag stress, as calculated by Equation (21), is extremely low, since the stress is far above σ^*_{ba}. In this condition, isolated atoms can still retard dislocation motion, an effect that has not been considered in the model, leading to the overestimation of the strain rate in the lower temperature regime. With this notable exception, it can thus be concluded that the basic model presented here can be successfully used to predict, with more than reasonable accuracy, the hot workability response of Al-Mg alloys in the temperature regime between 550 and 850 K, without requiring any best fitting of experimental data. Thus, the predictive capability of the model is fully confirmed, suggesting that it could be profitably used in hot working studies.

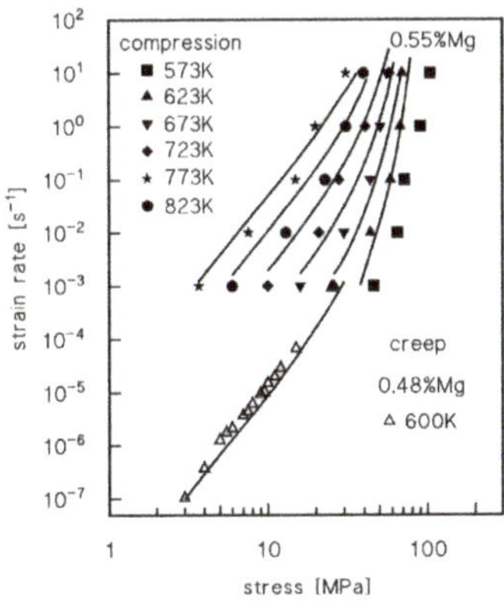

Figure 9. Compression data for 0.55%Mg [28] and creep data for 0.48%Mg [4]. The model curves were calculated by Equations (10)–(11), (21)–(24).

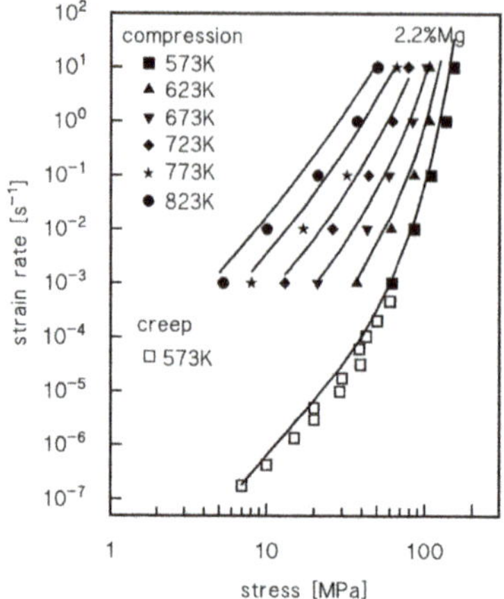

Figure 10. Compression data for 2.2%Mg [28] and creep data from [1,29]. The curves were calculated as in Figure 9.

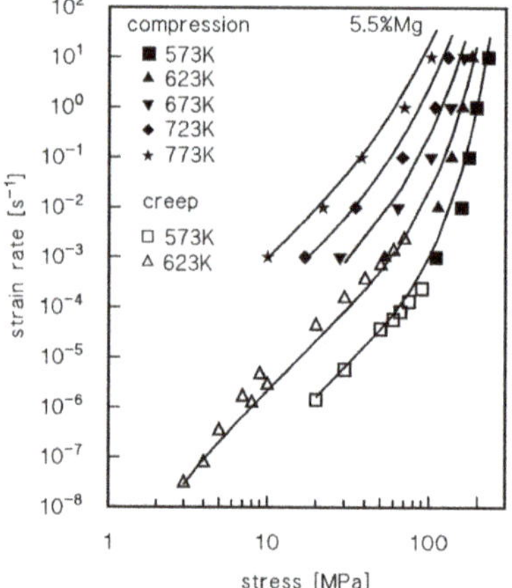

Figure 11. Compression data for 5.5%Mg [28] and creep data from [1,30]. The curves were calculated, as in Figure 9.

5. Conclusions

A unified approach has been presented to describe the high-temperature secondary strain rate dependence on applied stress and temperature for Al-Mg single-phase solid solution alloys in creep and hot working conditions. The physically based constitutive equations proposed contain three important parameters: internal stress, which represents the stress that is required to move a dislocation in the matrix, the strengthening contribution due to solute atom-dislocation interactions (drag stress), and the strain hardening constant C_L, which, in combination with free dislocation density, determines the dislocation mean free path. Once the effect of break-away of dislocations from solute atmospheres has been described by a specific relationship, the model proposed does not require any variation in the constitutive equations to describe the whole experimental range of steady-state creep rate, nor data-fitting, which is a notable advancement over other phenomenological equations.

Acknowledgments: The authors greatly thank Rolf Sandström (Materials Science and Engineering, KTH, Brinellvägen 23, S-10044 Stockholm, Sweden) for stimulating the discussion which provided an invaluable contribution to the work here presented.

Author Contributions: S.S. and C.P. analysed the data; S.S. handled the model; C.P. collected literature data; S.S. wrote the paper.

Conflicts of Interest: The authors declare no conflict of interest.

References

1. Oikawa, H.; Langdon, T.G. The creep characteristics of pure metals and metallic solid solution alloys. In *Creep Behaviour of Crystalline Solids*; Wilshire, R., Evans, R.W., Eds.; Pineridge Press: Swansea, UK, 1985; pp. 33–82. ISBN 9780906674420.
2. Čadek, J. *Creep in Metallic Materials*; Elsevier: Amsterdam, The Netherlands, 1988; pp. 160–175. ISBN 9780444989161.
3. Kassner, M.E.; Pérez-Prado, M.T. *Fundamentals of Creep in Metals and Alloys*; Elsevier: Amsterdam, The Netherlands, 2004; pp. 11–120. ISBN 9780080436371.
4. Oikawa, H.; Honda, K.; Ito, S. Experimental study on the stress range of class I behaviour in the creep of Al-Mg alloys. *Mater. Sci. Eng.* **1984**, *64*, 237–245. [CrossRef]
5. Sato, H.; Maruyama, K.; Oikawa, H. Effects of the third element on creep behaviour of Al-Mg and α-Fe-Be sold solution alloys. *Mater. Sci. Eng. A* **1997**, *234–236*, 1067–1070. [CrossRef]
6. Horita, Z.; Langdon, T.G. High temperature creep of Al-Mg alloys. In *Strength of Metals and Alloys, Proc. ICSMA7*; McQueen, H.J., Bailon, J.P., Dickson, J.I., Jonas, J.J., Akben, M.G., Eds.; Pergamon Press: Oxford, UK, 1986; pp. 797–802. ISBN 0080316409.
7. Fernández, R.; González-Doncel, G. A unified description of solid solution creep strengthening in Al-Mg alloys. *Mater. Sci. Eng. A* **2012**, *550*, 320–324. [CrossRef]
8. Sandström, R. Influence of phosphorus on the tensile stress strain curves in copper. *J. Nucl. Mater.* **2016**, *470*, 290–296. [CrossRef]
9. Sandström, R.; Andersson, H.C.M. Creep in phosphorus alloyed copper during power-law breakdown. *J. Nucl. Mater.* **2008**, *372*, 76–88. [CrossRef]
10. Orlová, A.; Čadek, J. Dislocation structure in the high temperature creep of metals and solid solution alloys: A review. *Mater. Sci. Eng.* **1986**, *77*, 1–18. [CrossRef]
11. Hirth, J.P.; Lothe, J. *Theory of Dislocations*; McGraw-Hill: New York, NY, USA, 1978.
12. Kocks, U.F.; Argon, A.S.; Ashby, M.F. Thermodinamics and kinetics of slip. *Prog. Mater. Sci.* **1979**, *15*, 1–291.
13. Spigarelli, S.; Sandström, R. Basic creep modelling of Aluminium. *Mater. Sci. Eng. A* **2018**, *711*, 343–349. [CrossRef]
14. Korzhavyi, P.A.; Sandström, R. First-principles evaluation of the effect of alloying elements on the lattice parameter of a 23Cr25NiWCuCo austenitic stainless steel to model solid solution hardening contribution to the creep strength. *Mater. Sci. Eng. A* **2015**, *626*, 213–219. [CrossRef]
15. Kassner, M.E. Taylor hardening in five-power-law creep of metals and Class-M alloys. *Acta Mater.* **2004**, *52*, 1–9. [CrossRef]
16. Mecking, H.; Styczynski, A.; Estrin, Y. Steady state and transient plastic flow of Aluminum and Aluminum alloys. In *Strength of Metals and Alloys-ICSMA8*; Kettunen, P.O., Lepistö, T.K., Lehtonen, M.E., Eds.; Pergamon Press: Oxford, UK, 1988; pp. 989–994. ISBN 9781483294155.
17. Zhang, L.; Du, Y.; Chen, O.; Steinbach, I.; Huang, B. Atomic mobilities and diffusivities in the fcc, L1$_2$ and B2 phases of the Ni-Al system. *Int. J. Mater. Res.* **2010**, *101*, 1461–1475. [CrossRef]
18. Rothman, S.J.; Peterson, N.L.; Nowicki, L.J.; Robinson, L.C. Tracer Diffusion of Magnesium in Aluminum Single Crystals. *Phys. Status Solidi (b)* **1974**, *63*, K29–K33. [CrossRef]
19. Zhang, W.; Du, Y.; Zhao, D.; Zhang, L.; Xu, H.; Liu, S.; Li, Y.; Liang, S. Assessment of the atomic mobility in FCC Al-Cu-Mg alloys. *Calphad* **2010**, *34*, 286–293. [CrossRef]
20. Horiuchi, R.; Otsuka, M. Mechanism of high temperature creep of Aluminum-magnesium solid solution alloys. *Trans. Jpn. Inst. Met.* **1972**, *13*, 284–293. [CrossRef]

21. Hayakawa, H.; Nakashima, H.; Yoshinaga, H. Dislocation density and Interaction between dislocations in Al-Mg solution hardened alloys deformed at high temperature. *J. Jpn. Inst. Met.* **1989**, *53*, 1113–1122. [CrossRef]

22. Blum, W.; Zhu, Q.; Merkel, R.; McQueen, H.J. Geometric dynamic recrystallization in hot torsion of Al-5 Mg-0.6 Mn (AA5083). *Mater. Sci. Eng. A* **1996**, *205*, 23–30. [CrossRef]

23. Friedel, J. *Dislocations*; Pergamon Press: Oxford, UK, 1964; ISBN 9781483135922.

24. Yavari, P.; Langdon, T.G. An examination of the breakdown in creep by viscous glide in solid solution alloys at high stress level. *Acta Metall.* **1982**, *30*, 2181–2196. [CrossRef]

25. Endo, T.; Shimada, T.; Langdon, T.G. The deviation from creep by viscous glide in solid solution alloys at high stresses-characteristics of the dragging stress. *Acta Metall.* **1984**, *32*, 1991–1999. [CrossRef]

26. Northwood, D.O.; Moerner, L.; Smith, I.O. The Breakdown in Creep by Viscous Glide in Al-Mg Solid Solution Alloys at High Stress Levels. *Phys. Status Solidi (a)* **1984**, *84*, 509–515. [CrossRef]

27. Kim, H.K. Low-temperature creep behavior of ultrafine-grained 5083 Al alloy processed by equal-channel angular pressing. *J. Mech. Sci. Technol.* **2010**, *24*, 2075–2081. [CrossRef]

28. Prasad, Y.V.R.K.; Sasidhara, S. *Hot Working Guide—A Compendium of Processing Maps*; ASM International: Almere, The Netherlands, 1997; pp. 98–100. ISBN 9781627080910.

29. Oikawa, H.; Sugawara, K.; Karashima, S. Creep Behavior of Al-2.2 at% Mg Alloy at 573 K. *Trans. Jpn. Inst. Met.* **1978**, *19*, 611–616. [CrossRef]

30. Pahutová, M.; Čadek, J. On Two Types of Creep Behaviour of F.C.C. Solid Solution Alloys. *Phys. Status Solidi (a)* **1979**, *56*, 305–313.

 *metals*

Article

High-Temperature Behavior of High-Pressure Diecast Alloys Based on the Al-Si-Cu System: The Role Played by Chemical Composition

Elisabetta Gariboldi [1,*], Jannis Nicolas Lemke [1], Ludovica Rovatti [1], Oksana Baer [2], Giulio Timelli [3] and Franco Bonollo [3]

[1] Dipartimento di Meccanica, Politecnico di Milano Via La Masa 1, 20156 Milano, Italy; jannisnicolas.lemke@polimi.it (J.N.L.); ludovica.rovatti@polimi.it (L.R.)

[2] Laboratory for Machine Tools and Production Engineering (WZL), RWTH Aachen University, 52074 Aachen, Germany; O.Baer@wzl.rwth-aachen.de

[3] Dipartimento di Tecnica di Gestione di Sistemi Industriali, Università di Padova, Stradella S. Nicola, 3-36100 Vicenza, Italy; timelli@gest.unipd.it (G.T.); bonollo@gest.unipd.it (F.B.)

* Correspondence: elisabetta.gariboldi@polimi.it; Tel.: +39-02-2399-8224

Received: 1 April 2018; Accepted: 9 May 2018; Published: 11 May 2018

Abstract: Al-Si-Cu foundry alloys are widely applied in the form of high-pressure diecast components. They feature hypo- or nearly eutectic compositions, such as AlSi9Cu3(Fe), AlSi11Cu2(Fe), and AlSi12Cu1(Fe) alloys, which are used in the present study. Diecast specimens, with a thickness of 3 mm, were used for tension tests. The short-term mechanical behavior was characterized at temperatures from 25 up to 450 °C. At temperatures above 200 °C, the tensile strength properties (YS and UTS) of the investigated alloys were severely affected by temperature, and less by chemical differences. Material hardness and ductility indexes better highlighted the differences in the mechanical behavior of these age-hardenable alloys and allowed us to relate them to the microstructure and its changes that took place at test temperatures. Thermodynamic calculations were found to be useful tools to predict phases formed during solidification, as well as those related to precipitation strengthening. By means of the performed comprehensive material characterization, deeper knowledge of the microstructural changes of Al-Si-Cu foundry alloys during short-term mechanical behavior was obtained. The gained knowledge can be used as input data for constitutive modeling of the investigated alloys.

Keywords: high temperature; tensile properties; microstructural changes; AlSi9Cu3(Fe); AlSi11Cu2(Fe); AlSi12Cu1(Fe)

1. Introduction

Al alloys based on the Al-Si-Cu system, specifically of hypo- or nearly eutectic composition, are widely applied to manufacture parts by high-pressure die casting processes (HPDC). They are frequently employed for the production of automotive components where a high strength-to-weight ratio is of great appeal. Even if their chemical composition allows an age-hardening response, and thus the mechanical properties can be improved by suitable heat treatment [1–4], Al-Si-Cu-based alloys are mostly applied in the as-diecast condition.

The high-temperature mechanical behavior of Al-Si-based casting alloys has mainly been investigated for piston alloys [5,6], of nearly eutectic or hypereutectic composition. Only recently has attention been focused on hypoeutectic alloys [7,8]. As a matter of fact, during their lifecycle, structural parts made of these alloys are held at moderate or high temperatures (for example in parts close to heat sources where the high thermal conductivity of Al alloys is appreciated) and the knowledge of high-temperature tensile properties in their actual temper condition is of interest. While many experimental studies have been devoted to the thermal stability of Al-Si-Cu casting alloys and to their direct effects on hardness and room-temperature mechanical properties [7,8], less attention has been paid to the temperature dependence of stress-strain curves and the tensile properties in a wide temperature range [8]. Small efforts have also been dedicated to describe the effects of microstructural stability on the mechanical properties after long-term exposure to elevated temperatures, for instance by means of long-term experimental creep tests [9] or tensile testing in a wide range of strain rates [8].

Further, to the authors' knowledge, no direct comparison of the high-temperature behavior of different Al-Si-Cu casting alloys is currently available in the literature, as a function of chemical composition and corresponding microstructural features. As a matter of fact, the mechanical response of Al-Si casting alloys is highly process- and geometry-dependent since foundry defects, as well as process-related microstructural features, can significantly affect properties such as yield strength and ductility [1,2].

The present paper aimed to evaluate the short-term high-temperature behavior of three widely diffused Al-Si-Cu alloys for high-pressure diecasting, which are often proposed as alternative alloys for moderate/high-temperature service. They were then analyzed taking into account the effects of chemical composition and of the correlated microstructural features. For this reason, experimental tests were performed on specimens produced by an optimized and strictly controlled high-pressure die casting process, leading to sound parts (with a very limited amount of pores) with proper microstructure.

2. Materials and Methods

2.1. Alloys and Specimen Manufacturing

Three secondary AlSiCu(Fe) alloys were selected within the standard production of an industrial foundry, according to EN 1706 standard [10], in order to investigate a set of widely diffused alloys with different amounts of Si and Cu, considered as the main elements that can vary the amount of eutectic and secondary precipitates. The selected alloys were EN AC 46000-AlSi9Cu3(Fe), EN AC 46100-AlSi11Cu2(Fe), and EN AC 47100-AlSi12Cu1(Fe), whose actual chemical compositions as well as compositional limits are given in Table 1. The chemistries of the alloys are similar to that of the A380.0 (UNS A13800), A383.0 (UNS A03830), and A413.0 (UNS A04130) alloys, respectively [11]. For the sake of simplicity, the three materials will be hereafter referred as alloys A, B, and C, as shown in Table 1.

The specimens for the tensile tests were high-pressure diecast using a cold chamber die casting machine. Specifically, the alloys were molten in a 300-kg furnace and held at 690 °C for 30 min prior to being diecast. The cavity was specifically designed to produce testing specimens and simple shapes for experimental purposes, with a cast part mass of about 0.9 kg. Both the die design and the process parameters aimed to minimize the porosity level and to obtain a controlled microstructure. Oil circulation channels in the die were used to stabilize the temperature (at ~230 °C). The main process parameters were: in-gate speed of about 45 m/s, filling time of about 10 ms, and intensification pressure at 40 MPa. Additional details of the casting procedure, die design, microstructural features, defect content [12–14], and mechanical properties at room temperature [13] have already been described and discussed in previous works and in CEN T/R 16748 [15].

Table 1. Experimental chemical compositions (mass %) of the investigated materials and compositional limits according to EN 1706 standard [10].

	Alloy		Si	Fe	Cu	Mn	Mg	Cr	Ni	Zn	Pb	Sn	Ti	Al
A	EN AC 46000-AlSi9Cu3Fe	actual	8.227	0.799	2.825	0.261	0.252	0.083	0.081	0.895	0.083	0.026	0.041	bal
		limits	8.0–11.0	<1.3	2.0–4.0	<0.55	0.05–0.55	<0.15	<0.55	<1.20	<0.35	0.15	<0.25	bal
B	EN AC 46100-AlSi11Cu2(Fe)	actual	10.895	0.889	1.746	0.219	0.224	0.082	0.084	1.274	0.089	0.029	0.047	bal
		limits	10.0–12.0	<1.1	1.5–2.5	<0.55	<0.30	<0.15	<0.45	<1.70	<0.25	0.15	<0.25	bal
C	EN AC 47100-AlSi12Cu1(Fe)	actual	10.510	0.721	0.941	0.232	0.242	0.045	0.080	0.354	0.055	0.025	0.038	bal
		limits	10.5–13.5	<1.3	0.7–1.2	<0.55	<0.35	<0.10	<0.30	<0.55	<0.20	0.10	<0.20	bal

Specimens used for tensile tests are shown in Figure 1. They were flat bars with nominal thickness of 3 mm. The length and width of the gauge length were 35 and 10 mm, respectively. Outside the gauge length, within a distance of 12.5 mm, the width of the specimen gradually increased to the 20 mm of the gripping ends. Two holes, 6 mm in diameter, spaced 70 mm apart, were drilled to insert steel pins fixing the specimens in the loading train. No other machining operation was performed on the specimens. A representative macrograph of a transverse section of a specimen gauge length evidences the low defect amount of the cast parts (see Figure 2) as well as the extension of the skin layer compared to the overall specimen cross-section. The tensile behavior to be derived from these specimens is thus reasonably closer to that of the chill layer than that of the inner regions characterized by a coarser microstructure. Specimens were tested in T1 condition, after a stabilization period following the casting process.

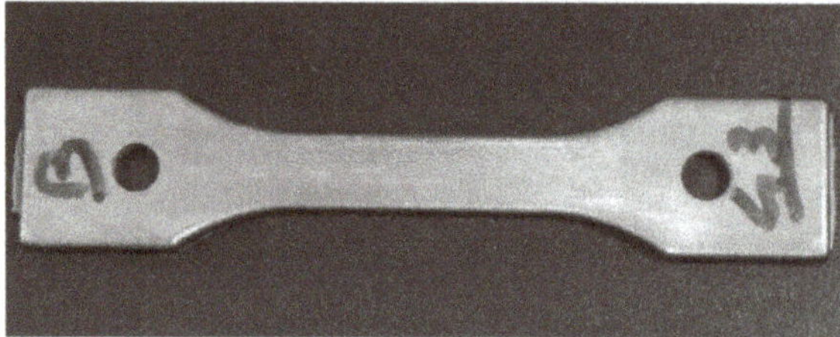

Figure 1. View of a specimen used for the experimental tensile tests.

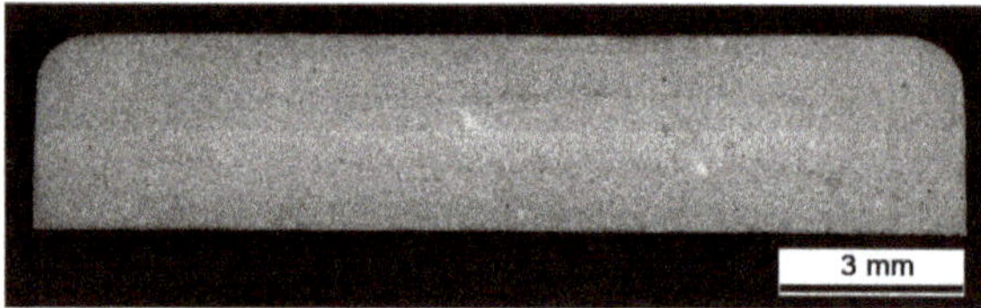

Figure 2. Transversal section of a specimen (alloy A) after metallographic polishing and etching to reveal casting defects and metallurgical features.

2.2. Tensile Testing

Tensile testing was carried out in an electromechanical CSR-30 machine (CERMAC, Pozzo d'Adda, 20060 Pozzo d'Adda, Italy), specifically designed for high-temperature testing, equipped with a three-zone vertical furnace, at temperatures in the range of 25–450 °C with 50 °C temperature steps. The temperature range extended from the conventional application of the alloys, including moderate- and high-temperature service, to the maximum investigated temperature, 450 °C, which is not far from the temperatures of the solution treatment to which these alloys could be subjected and at which strain could arise from stresses induced by the same component weight, from thermal gradients and, locally, by the presence of pressurized pores [15]. Tests were performed according to CEN ISO 6892-2 standard [16]. The test temperature was measured by three S-type thermocouples directly placed on the specimens, the signal of which was also used to control the three-zone vertical furnace used for the test.

Attention was paid to avoid excessive holding time in the furnace before testing, in order to limit microstructural changes. The heating time was set at approximately 30–40 min (including 10 min holding at the test temperature, as required by the reference standard) and it was shortened for the tests at the highest temperatures. These times were the minimum possible to avoid overshooting, i.e., to exceed the set temperature during the heating stage of the test, a risk that is higher at moderate than at high set temperatures. Tensile testing was carried out in displacement control. The displacement rate was fixed at 0.012 mms^{-1} for all testing temperatures, with the target strain rate in the plastic range of

$2.5 \times 10^{-5}\ \mathrm{s}^{-1}$, corresponding to range 2 in CEN ISO 6892-2 standard for high-temperature tensile tests [16]. Specimen elongation was measured by fixing the extensometric system to the loading pins and reading the relative displacement of pins by Linear Variable Differential Transducers. A reference length close to 60 mm was calculated as suggested by the above standard and used in order to estimate the strain values. Elongation (A%) and reduction of area (RA%) at rupture were estimated on the basis of changes in the specimen gauge length and in the cross-section. The area of the latter was approximated to remain rectangular at the end of the tests. At high temperatures, the material ductility reached very high values and tests were interrupted at about 35% strain, when the extensometric system limit was reached.

2.3. Analyses on Tested Specimens

Specimens tested at room temperature (RT), 350, and 450 °C were then cut longitudinally, polished, and etched with Keller's reagent to observe microstructural features with the special aim of distinguishing the features between the most stressed and strained regions, close to final fracture, and those in gripping ends where no load was applied. The metallographic samples were then observed by means of optical microscope (OM, Aristomet, Leitz, Wetzlar, Germany) and scanning electron microscope (SEM, EVO50, Carl Zeiss AG, Oberkochen, Germany). In the latter case, energy-dispersive X-ray microanalyses (EDX, OXFORD Instruments, Adingdon, UK) were also performed, using OXFORD INCA software for microanalyses. The size of the α-Al grains was estimated with optical microscopy at low magnification, covering an area of 0.01 mm^2 and analyzing three distinct regions of the sample. SEM fractographic analyses were carried out combining secondary electrons and backscattered electrons probes to highlight the correlation between morphological features and intermetallic particles.

Vickers hardness tests were performed on the above samples in the regions of the gripping ends. Test were done according to ISO 6507-1:2005 standard on a VMHT30 microindenter (Leica, Wien, Austria), with a 19.6-N load applied to the indenter for 15 s. Five indentations were made for each experimental condition. These tests aimed to evaluate the effects brought about by aging for about 20–30 min (corresponding to the time spent at temperature during holding and testing).

3. Results

3.1. Tensile Behaviour

The tensile behavior of the investigated alloys was significantly affected by the testing temperature, as clearly shown in Figures 3 and 4.

At RT, the 0.2% offset yield stress (YS) and ultimate tensile strength (UTS) of the alloys fell in the 155–175 MPa and 270–290 MPa ranges, respectively. The properties of alloys A and B were close to the average values derived from previous tests at room temperature on sets of at least seven specimens carried out on the same production batches of specimens [12]. In the optimized casting process conditions, maximum deviations of ±15 MPa, ±3 MPa, and $\pm0.2\%$ from the average UTS, YS, and elongation at rupture (A%) were observed from the abovementioned study [12,17]. In a related paper [13], it was demonstrated that the stress-strain response of the alloy in the uniform plastic zone is very close for specimens of the same alloy/process up to their final fracture. This latter occurred earlier in specimens containing more severe defects. This fact introduced a relatively wider scatter in the UTS data compared to that of the YS data. The limited specimens' availability prevented in the present case the repetition of tests and data scatter cannot be presented directly. Due to the progressively increasing ductility of the materials with temperature, it can be reasonably expected that the experimental scatter of tensile properties reduces as test temperature is increased.

At RT, the UTS of alloys A and B were comparable, while the lowest values were reached for alloy C. In the RT-150 °C range, data points in Figure 3a,b show a plateau or a slight increase at about 100 °C for YS and/or UTS. A first drop in tensile properties can be observed in the intermediate temperature range (150–250 °C), where alloy C mostly differentiates with respect to the other two materials in terms of UTS and displays a particularly stable resistance to plastic deformation. Above 250 °C, both strength parameters decrease gradually with increasing temperature, with the highest temperature-effect on UTS. In the same temperature range, the YS and UTS are very close for the investigated alloys; the first becomes lower than 100 MPa at about 300 °C, while UTS decreases below 100 MPa only above 320 °C. The two strength parameters become comparable only above 350 °C and they decrease to about 10 MPa as the test temperature is increased to 450 °C. The YS and UTS data shown in Figure 3 are fitted by third-order polynomials. These fitting curves could be used as a simple empirical model to describe the strength experienced by the alloys in the temperature range up to 450 °C.

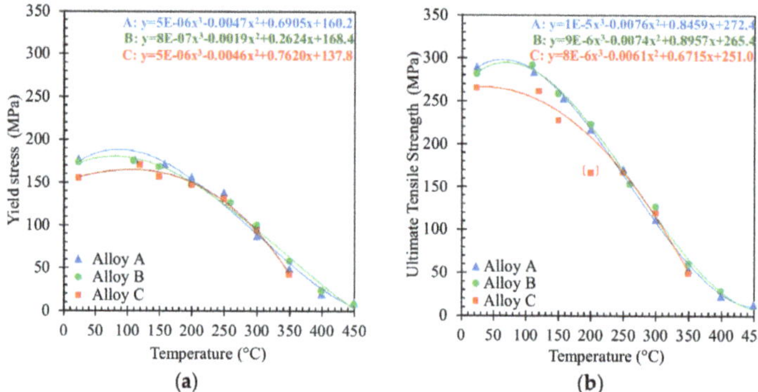

Figure 3. (**a**) Yield stress (YS) and (**b**) ultimate tensile strength (UTS) plotted as functions of test temperatures for the investigated alloys. Experimental data were fitted by third-order polynomials.

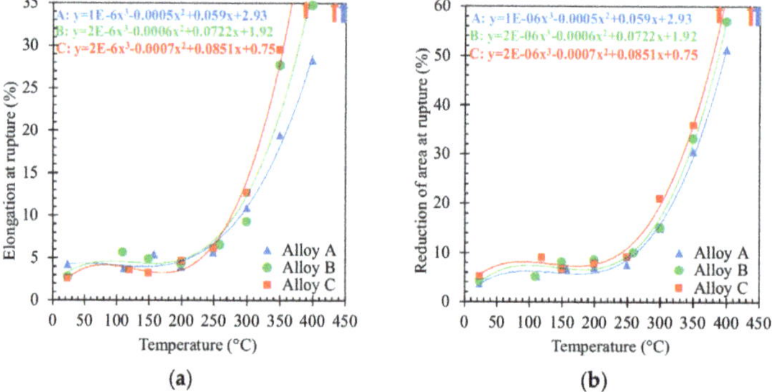

Figure 4. (**a**) Elongation at rupture and (**b**) reduction of area at rupture plotted as functions of test temperatures for the investigated alloys. Experimental data were fitted by third-order polynomials.

Test temperature also significantly affected the ductility parameters, as illustrated in Figure 4a,b for A% and RA%, respectively. At temperatures below 200 °C, A% remains at about 5% and RA% is slightly higher. Both ductility indexes do not show significant differences between the investigated alloys (up to 200 °C). Differences could be partly hindered by experimental scatter involved in measurements of small elongation of the broken gauge lengths. Above 250 °C, the ductility rapidly increases and above 450 °C it exceeds 35% for all of the experimental alloys. In the range of 250–450 °C, alloys C and A display the highest and lowest increase of ductility, respectively.

As conducted for the strength properties, a third-order polynomial fitting was adopted for the ductility as well. At high temperatures, the alloys can be easily ranked in terms of ductility, where the fitting lines could be actually used to describe changes of both the ductility parameters induced by temperature.

The mechanical behavior of the alloy can be further described by their flow curves. The terms of true stress vs. true plastic strain correlation was modeled by means of the Ramberg-Osgood relationship between true plastic strain $\varepsilon_{p,t}$ and true stress σ_t [18]:

$$\sigma_t = K_T \times (\varepsilon_{p,t}) n_T \tag{1}$$

where the constants K_T (strength coefficient) and n_T (strain hardening coefficient) are derived for each test by fitting experimental data in the engineering stress range from YS to UTS. The temperature dependence of these material parameters is shown in Figure 5. Different from the trends observed at low temperatures for YS and UTS, the strength coefficient monotonically decreases as the test temperature increases. Apart from room temperature, where alloy A displays the lowest K_T value, in the intermediate temperature range (110–250 °C) the lowest values are calculated for alloy C.

As far as the strain hardening is concerned, the alloy chemistry seems to play a clear role only at room temperature. The strain hardening coefficient reduces as the temperature increases up to about 350 °C, where the trend changes.

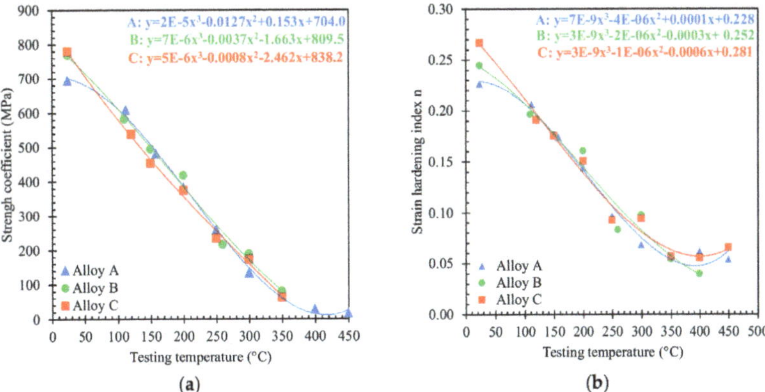

Figure 5. (a) Strength coefficient K_T and (b) strain hardening coefficient n_T plotted as functions of test temperatures for the investigated alloys. Experimental data were fitted by third-order polynomials.

3.2. Hardness Testing

The microhardness evolution as a function of holding temperature in unloaded regions of tested specimens is presented in Figure 6. A similar trend is displayed by the three materials as the test temperature rises. In the 100–150 °C range a slight decrease of hardness can be noticed with respect to room temperature. This behavior is compatible both to relieving effects of residual stresses induced by the manufacturing process [19] and to the loss of solute atoms and relative strengthening from the matrix, not fully balanced by particle strengthening.

The second feature of the common trend of the three alloys is a hardness peak between 200 and 250 °C followed by the third feature measured above 250 °C, i.e., a monotonic decrease. The differences between the three alloys are more evident than those displayed by the strength properties YS and UTS. In the whole investigated temperature range, alloys A and C were clearly the hardest and softest materials, respectively. Further, material softening is anticipated in alloy C, for which the onset of peak appears to be shifted at higher temperatures. The hardness peaks are reasonably in the 250–300 °C range, with peak hardness and temperature increasing with the Cu content of the alloys.

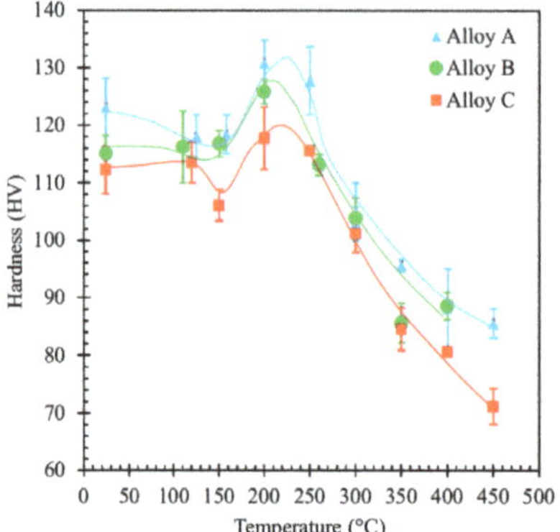

Figure 6. Microhardness of the investigated alloys measured at room temperature in unloaded zones of the specimen after tensile testing at different temperatures.

3.3. Microstructural Observations

The microstructure of the alloys is reported in in Figure 7. It is made of relatively small globular-dendritic grains of aluminum solid solution (Al_{ss} phase, light gray in micrographs in Figure 7) surrounded by the eutectic structure (grey in Figure 7) characterized by fine Si particles. A relatively thick sound layer is found starting from the surface of castings while trapped air/gas porosity is concentrated in the central part of the specimens (Figure 2). The volume fraction occupied by Al_{ss} grains varied within different regions of the samples, being 62 ± 5% in alloy A and 57 ± 6% in both alloys B and C. The size of Al_{ss} grains varied slightly for the three materials. Details for alloys A and B are described in Reference [12].

Particles of the primary Fe-containing α-phase, usually referred to as sludge (pointed out by the arrows in Figure 7), are observed in all of the diecast alloys, although with different features. They are coarse in alloy A (typically in the range of 15–20 μm), and their average size decreases in alloy B and then further in alloy C. Alloy B is also characterized by the lowest volume fraction of these sludge particles.

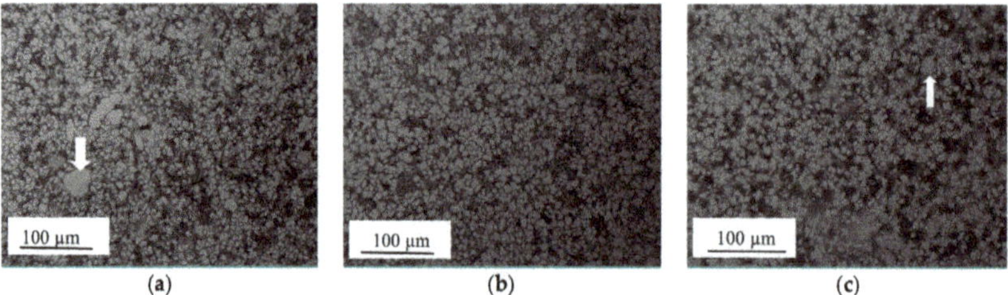

(a)   (b)   (c)

Figure 7. Microstructures of the investigated alloys in the as–diecast condition: (a) alloys A; (b) B; and (c) C.

The OM observations of the specimens after testing at room temperature and 350 °C are presented in Figure 8. At room temperature, the fracture surface lays within the relatively thin eutectic structure, at the interface between the eutectic and a slightly deformed Al_{ss} phase. This also intercepts coarse sludge particles, which also are fractured even when their position is far from the fracture surface. In specimens tested at higher temperatures, the microstructural rearrangement corresponding to a larger plastic strain was more evident. The Al_{ss} grain deformation increased from alloy A to alloy C. At this temperature, the fracture path remained within the eutectic structure. SEM observations revealed the presence of different intermetallic particles, on which EDX point microanalyses were performed. Representative microanalyses are here proposed in Table 2. It is worth mentioning that under the standard conditions for SEM microanalyses of Al-Si alloys, the depth of the interaction volume between the electron beam and the analyzed particles is about 1.5 μm (neglecting differences among alloying elements) [20] and a similar size can be considered on the interaction surface. Consequently, the results of the point microanalyses of the intermetallic particles give an intermediate composition between the matrix and the particles rather than the composition of the latter.

Table 2. EDX (energy-dispersive X-ray) microanalyses (in mass %) of points highlighted in the micrographs of Figure 9.

Alloy	Figure	Point	Si	Fe	Cu	Mn	Mg	Cr	Ni	Al
A	9c	A	4.06	1.59	5.50	-	-	-		Bal
A	9c	B	8.57	16.07	4.94	4.85	-	1.20		Bal
A	9c	C	35.81	2.90	22.04	-	1.18		0.81	Bal
B	9f	A	8.54	5.39	13.27	-	-	-	-	Bal
B	9f	B	8.82	-	24.95	-	-	-	-	Bal
B	9f	C	4.35	-	25.4	-	-	-	2.18	Bal
C	9g	A	11.59	12.14	0.61	4.85	-	1.17	-	Bal

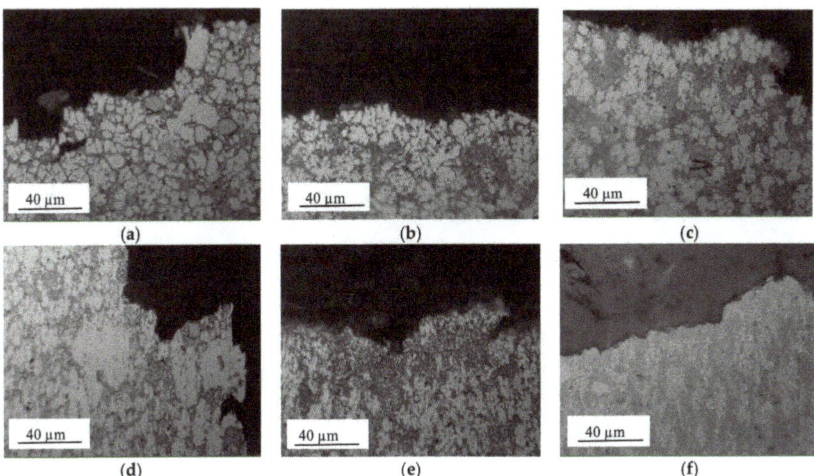

Figure 8. Representative metallographic sections close to the fracture surface of specimens tested at (**a–c**) room temperature and (**d–f**) 350 °C. (**a,d**) alloy A; (**b,e**) alloy B; (**c,f**) alloy C.

In alloy A (Figure 9a–e), the coarse polygonal sludge particles have a chemical analysis roughly corresponding to Al-18Fe-10Si-8.0Mn-4Cr, with a comparatively smaller amount of Cu. In this alloy, sludge particles are surrounded by an elongated Al_2Cu phase (point C in Figure 9c) and seldom by finer Fe-rich particles containing the same alloying elements. Cu-containing particles are mainly of the Al_2Cu phase, but occasionally Mg or Ni can be detected in them. Cu-rich particles appear either in globular or elongated morphologies, in the latter case underlying grains and surrounding the smallest, secondary α-particles (as in point B in Figure 9c). In the same alloy A, the amount of the brightest Cu-rich particles tends to reduce as the test temperature is increased, particularly in the temperature range from 350 to 450 °C. At this last temperature, high-magnification images revealed the presence of globular eutectic Si particles (Figure 9e).

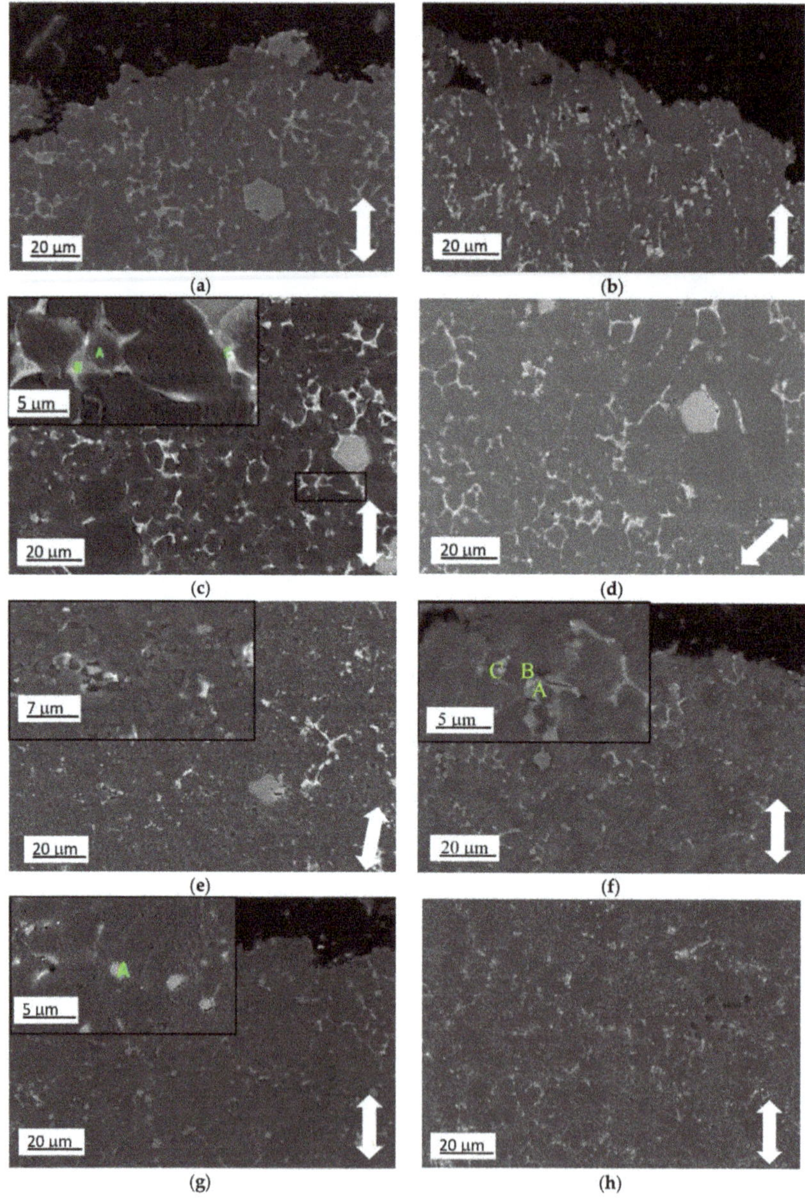

Figure 9. Representative SEM-BSE (backscattered electrons) micrographs of the longitudinal metallographic sections of tensile tested specimens taken close to the final fracture section or 5 mm away from it. (**a–e**): alloy A tested at 25 °C (**a,c**), 350 °C (**b,d**), and (**e**) 450 °C; (**f**) alloy B tested at 25 °C; (**g**) alloy C tested at 25 °C; (**h**) alloy C tested at 450 °C. White arrows display the loading direction. Cu-rich phase particles are the brightest, α-particles have an intermediate gray color.

As the amount of Cu decreases from alloy A to alloy C, Cu-containing particles, mainly surrounding homogeneous grains, can be observed in lower amount and size (compare Figure 9a with

Figure 9f,g). As far as Fe-containing intermetallic particles are concerned, in addition to the coarse α-particles previously described, smaller particles with similar a morphology, possibly a secondary α-family, was also observed in alloys (points A in Figure 9f,g), with their average size being lower than that observed in alloy A.

SEM analyses of tested samples confirm that at room temperature the fracture of all of the alloys occurred in the interdendritic regions where, in addition to Al_{ss} and eutectic Si, other coarse and brittle intermetallic phases were located (Figure 9a,f,g). Micrographs taken using a backscattered electrons (BSE) probe revealed the presence of early decohesions at the matrix-particle interfaces.

At 350 °C, the greater ductility of the matrix promoted the reorientation of elongated particles along the loading direction. Strain also caused further fragmentation of interdendritic particles, specifically in the case of coarse particles. The result was that at this temperature, the alloys displayed higher strains and concurrently withstood higher amounts of damage (i.e., decohesions) before the final fracture (compare Figure 9a,b).

Specimens tested at the highest temperature exceeded strains of 35% without reaching the final fracture. However, they displayed some evidence of microstructural damage in their gauge length, as shown in Figure 9e,h, which refer to alloys A and C, respectively. The increased matrix ductility with respect to 350 °C, the presence of spheroidized Si particles, and the reduction of the amount and size of the Cu-containing particles prevented early fracture, which occurred in any case at or within intermetallic particles, starting with the coarser ones (Figure 9e). The microstructural damage observed in alloy C is less marked than that in alloy A, as a result of the lower amount and smaller size of particles from which damage starts. Fracture surface analyses, here reported in Figure 10, confirmed that in all cases fracture was of a ductile type, with dimples of the Al_{ss} phase often formed at intermetallic particles (brighter in BSE fractographs) as well as at the eutectic silicon. The increased local deformation of the matrix at high temperatures can be observed by comparing Figure 10a,b for the alloy B, while the lower ductility of alloy A and the presence in it of a higher amount of coarser particles can be observed in Figure 10c.

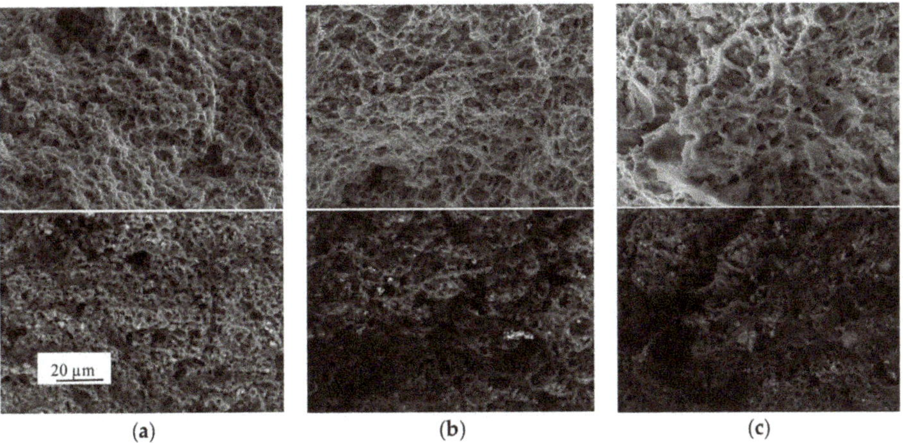

(a) (b) (c)

Figure 10. Representative fractographs (all at the same magnification) of tensile tested specimens: (a) alloy B, 25 °C; (b) alloy B, 350 °C; (c) alloy A, 350 °C. For each alloy the top fractograph was taken by a secondary electrons probe, while the bottom fractography employed backscattered electrons to highlight the presence of intermetallic phases.

4. Discussion

At temperatures above 200 °C, the tensile strength (YS and UTS) of the investigated alloys were severely affected by temperature, and less by chemical differences. These, on the other hand, had clear effects on ductility (Figure 4), which was particularly high in alloy C above 250 °C. The greatest ductility of this alloy was also demonstrated by the stress-strain curve, which displayed the relative maximum below 150 °C. Microstructural observations suggested a damage mechanism where coarse intermetallic particles (including primary α-phase), eutectic, as well as the deformability of Al_{ss} grains play an important role, with clear differences in the investigated alloys according to the testing temperature. The deformability of Al_{ss} grains was particularly sensitive to the age-hardening response of the alloys, as deduced from hardness of the tested specimens, which differentiate the alloys' behavior more than the tensile strength parameters.

The effects of alloy chemistry on the short-term high-temperature properties of Al-Si-Cu alloys should thus be mainly attributed to coarse intermetallic phases resulting from alloy solidification, eutectic, as well as the amount of intragranular strengthening particles, which mainly precipitate during high temperature exposure prior or during testing. Due to the relatively wide and partly overlapping elemental ranges according to EN 1706 standard (see Table 1), the actual rather than the nominal alloy chemistry has to be taken into account to verify compositional effects. In the present case, the Si content, known to affect the amount of the most significant eutectic phase, was slightly higher in the alloy nominally containing 11 mass % Si than in that with a nominal content of 12 mass %. On the other hand, the amount of Cu, related to the possibility to age-harden alloys, decreased from alloy A to C, according to their nominal composition.

Considering the Zn content, due to the low amount of Mg in the alloys (0.22–0.25 mass %), it could not induce age-hardening effects, but it remained in the Al-rich solid solution (Al_{ss} phase) [1,21]. The beneficial, even if slight, effect of Zn on the tensile properties at RT decreased from alloys B to A and C. Further, strengthening by Zn was not significantly affected by exposure at higher temperatures, where Zn solubility in Al_{ss} was greater than that at RT. The role of this element will thus not be discussed further in this paper.

The same decreasing trend from alloys B to A and C could be observed for iron content, although in a quite narrow range of 0.721–0.889 mass %. Iron is known to govern material ductility by affecting the amount of Fe-containing particles, as well as their shape and size. Minor elements such as Mn, Cr, and Ni can also play an important role in the amount, morphology, and distribution of intermetallic particles, specifically in many Fe-containing phases reported in the literature [21–25]. The composition and presence of Fe-containing phases will be discussed later, as related to thermodynamic simulation results. Ni, which gives the possibility to form some intermetallic phases in the alloys of interest (see the review work by Rana et al. [22]) was found in a relatively similar amount in all of the investigated alloys and reasonably plays no significant role in differentiating the alloy microstructures and properties. Thus, the effect of Ni will also not be further taken into account. Similarly, the effects of Ti, known as a refining element, and of Sn, whose addition is reported to improve alloy fluidity but also induce soldering effects [1], will not be considered hereafter.

4.1. Thermodynamic Simulations and Microstructures/Properties at Room Temperature

Thermodynamic simulations were carried out to verify the effect of the alloy composition on phases formed during solidification and on those related to precipitation strengthening during the following exposure at high temperatures before and during tensile testing. The alloy composition was simplified, neglecting the presence of Ti, Sn, Pb, and other minor elements and impurities, as previously discussed.

The solidification was simulated under Scheil-Gulliver (SG) assumptions (no back-diffusion in solid phases) and assuming intermetallic phases of fixed chemical composition. In Al-Si-Mg alloys, the SG approach has already been proved by Dorè et al. [26] to be very close to the solidification path under more complex assumptions in the case of solidification times of 10 s, which are higher than the solidification time employed in the present case. The same SG approach has been often applied for other alloys since it is able to account for a wider solidification range and a higher number of intermetallic phases generally observed in casting alloys with respect to the simplified solidification description in equilibrium conditions [27,28].

A summary of the main interest results for the present study is plotted in Figure 11. They will be discussed below, comparing them, whenever possible, to the experimental results and/or to literature data. The liquidus temperature (T_{liq}) decreased progressively from alloy A to alloy C, and the solidification temperature range reduced correspondingly. By following the solidification sequence, the first stage led to the formation of the primary α-phase in all cases, the so-called sludge, a common feature of the solidification sequence of the three alloys. Primary α formed at highest temperature range in alloy A (629–593 °C), and in the lowest and shortest range for alloy C (611–585 °C).

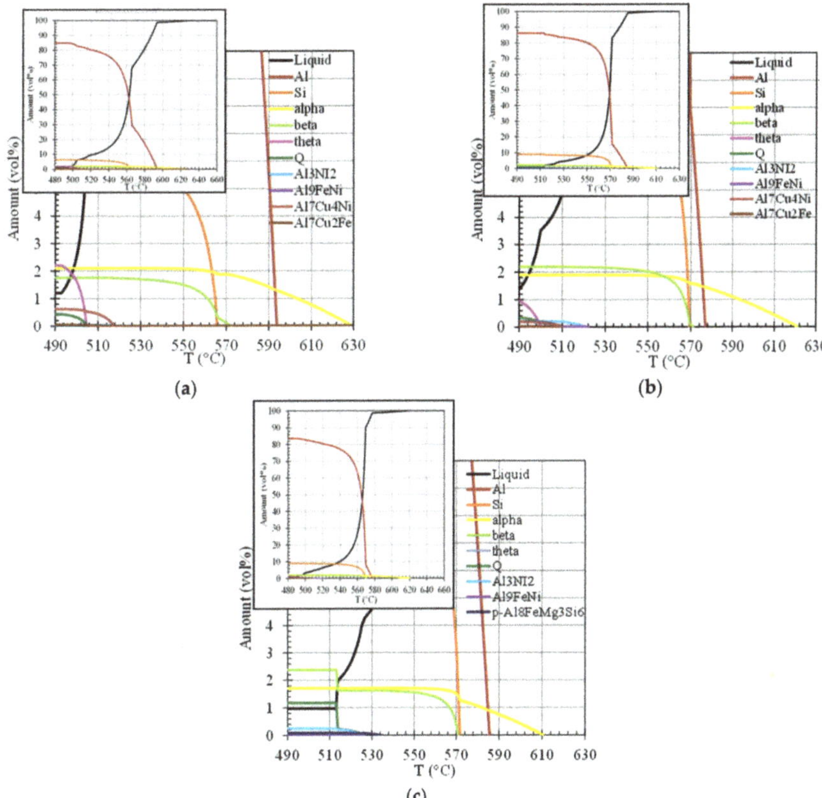

Figure 11. Modeled evolution of the amount of phases during non-equilibrium solidification. (**a**) Alloy A; (**b**) alloy B; and (**c**) alloy C. For each alloy the macroscopic changes related to liquid and Al$_{ss}$ are visible in the top-left-side of the plot, whose y-scale is extended to 100 vol %.

In the literature, the presence of sludge is often related to the sludge factor (SF), an Fe equivalent calculated as a weighted sum of Fe, Mn, and Cr, whose weight factors are 1, 2, and 3, respectively, as in Reference [29]. The SF of the present materials is 1.57 for both alloys A and B, and 1.32 for alloy C. Further, experimental observations by Ferraro et al. [25] showed that the size of sludge particles increases as the Fe and Mn contents increase, corresponding to the increase of the temperature for the onset of solidification, as observed experimentally for Fe by Shabestari [29] and from computed phase diagrams for Mn by Cao [30]. Thus, on the basis of SF and temperature ranges for the first stage of solidification, alloys A and B should be characterized by the highest amount of the primary sludge particles, slightly coarser in alloy A, while alloy C should be characterized by a minor amount of primary α-particles—both features confirmed here by microstructural observations.

A second step of solidification can be considered to last up to the onset of eutectic formation, predicted at 566, 569, and 572 °C for alloys A, B, and C, respectively. Most of this temperature range is characterized by the transformation of liquid into Al_{ss} and α-phase. At the end of this solidification stage, the amount of Al, in the form of homogeneous Al_{ss} grains, is nearly 30, 8.7, and 15.5 vol % for alloys A, B, and C, respectively, thus with a significant difference for the two alloys with close Si content, which is different from present experimental observations. Experimental results also led to higher amounts of Al_{ss}, partly probably due to difficulties in attributing Al_{ss} to homogeneous grains rather to eutectic, partly due to the shifting of the eutectic composition to higher Si contents caused by alloy compression acting during solidification, as explained by Kaufman et al. [27] and neglected by the applied thermodynamic model.

The predicted amounts of α-phase formed during this second solidification step were 0.60, 0.14, and 0.38 vol % for alloys A, B, and C, respectively. The morphology of this phase cannot be defined by the present approach and part of it could add to primary α-particles formed during the earlier stage. At the lowest temperatures of this range, the model predicts the concurrent formation of relatively low or no amounts of pro-eutectic β-Al_5FeSi phase; 0.31, 0.07, and 0 vol % for alloys A, B, and C, respectively.

The third and last stage of solidification can be considered to be characterized by the concurrent formation of silicon and intermetallic phases together with the Al_{ss}, forming binary and/or more complex eutectics. The predicted amount of the Si phase at the end of solidification was close to 9 vol % for alloys B and C, and less than 7 vol % in low-Si alloy A, mostly formed at the beginning of this solidification stage. The Al_{ss} homogeneous grains and Si-containing eutectic structure will not be discussed in detail here, except to mention that minor differences in the morphology of the Si phase could be predicted as a result of similar temperatures and cooling rates.

During the first part of this stage of solidification, the amount of α-phase further increased and stabilized, while β also appeared in alloy C. Primary α-particles with an average size decreasing from alloy A to C were experimentally observed according to their predicted nucleation and formation temperature ranges. Additionally, many α-pro-eutectic particles of a rather polyhedral morphology were experimentally found in all of the alloys.

The observed amount of primary and secondary α followed the trend suggested by thermodynamic simulations, confirming at the same time other results reported in the literature. On the other hand, β-particles were detected only in exceptional cases. In addition to the possible differences of modeled and experimental solidification conditions, the predicted amount of the above Fe-containing particles could be related to simplified assumptions made about their chemistry. Their composition has been widely described in the literature [25,28–34], specifically for α-phase, which can be written in the most general form as $Al_x(Fe,Mn,Cr,Cu)_ySi_z$. In the thermodynamic software, the α-phase is modeled as stoichiometric $Al_{15}(Fe,Mn,Cr,Cu)_{3.6}Si_2$, while the β-phase is modeled as Al_5Fe_2Si. During solidification, the modeled composition of the α-phase change, as shown in Figure 12 for alloy A, with progressive relative enrichment in Fe and Cu. Correspondingly, the predicted ratio between the sum of Fe and its substitutional elements and Si (all in mass %) increased from 3.45 to 3.63. This ratio is well within the range of 3–4.5 for Fe/Si suggested for the α-phase in Al-Si-Fe alloys by Ferdian [31], who also suggested the narrower and lower range of 1.8–2 for the β-phase. The actual composition of Fe-containing particles differed from predictions: EDX analyses of these particles confirmed the presence of substitutional elements for Fe in all particles, with (Fe + Mn + Cr + Cu)/Si ratios in the range of 2.8–3.2 for coarse sludge particles (Figure 9a,c) and lower (2.2–2.4) for the small particles (A in Figure 9f). This confirms that the difficulties in correct predictions of the type of Fe-containing particles in the investigated secondary alloys are at least partially related to simplified assumptions made about their chemistry. Even if the actual type and amount of Fe-containing phases cannot be carefully predicted, their total amount and main features, such as the coarseness of primary sludge particles in alloy A, which proved to be detrimental for its ductility, can nevertheless be derived from the thermodynamic models.

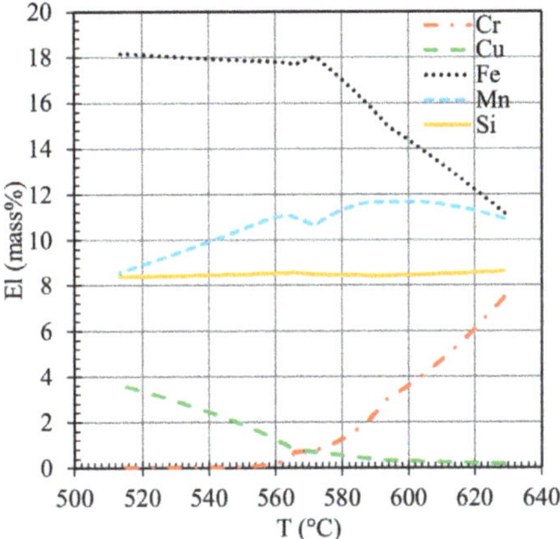

Figure 12. Predicted changes in the chemical composition of the α-phase (Al could be derived as a complement of 100%) according to Scheil-Gulliver conditions for alloy A.

Other phases formed in a last stage of the same third solidification stage, i.e., below ~530 °C. According to SG assumptions, the formation of the θ-Al$_2$Cu phase is possible for alloys A and B (2.16 and 1.03%, respectively), while in equilibrium conditions it occurs in all of the alloys only after the completion of solidification. Non-equilibrium solidification leads also to the presence of a quaternary Q-Al$_5$Cu$_2$Mg$_8$Si$_6$ phase, which is predicted in equilibrium (1.7%) at the end of solidification only for alloy C. Experimental observations confirmed the presence of coarse Cu-containing particles, whose overall amount decreased with the amount of Cu. In alloys A and B, they were θ-Al$_2$Cu particles, which were mixed with Q phase particles in alloy C.

Other intermetallic containing transition metals such as Ni, Cu, or Fe were found in minor amounts. No Zn-containing phase was predicted to form.

During the manufacturing process of the investigated materials, the solidification stage was followed by fast cooling up to the extraction of the thin HPDC specimens from the die, followed by slower air cooling to room temperature. According to literature [9], the cooling stage does not significantly alter the amount of phases.

Thus, at the end of solidification, the predicted microstructure in the alloys included Al$_{ss}$ grains, silicon, α, β, and other intermetallic phases, such as Al$_2$Cu, Q, and possibly minor amounts of other intermetallics. The amounts of phases different from Si and Al$_{ss}$ predicted at room temperature were 7.1, 5.9, and 3.8 vol % for alloys A, B, and C, respectively. The overall amount of these phases, containing high atomic number elements, qualitatively observed in SEM-BSE micrographs of Figure 9, confirmed their higher amount in alloy A, as well as the lowest amount in alloy C, as predicted. X-ray diffraction analyses as a means to verify the presence of phases in minor amounts in AlSi9Cu3Fe had been previously tested by some of the authors of the present paper [35], but no other phases were identified in addition to Al$_{ss}$, Si, α, and θ-Al$_2$Cu. Therefore, the results would not reasonably supply different results for the other two investigated alloys.

The fast cooling during and after solidification also affected the composition of Al$_{ss}$, which at room temperature is a supersaturated solid solution. The degree of supersaturation was greater in the alloy containing the highest amount of Cu at the end of solidification [3,36]. The time spent to conduct the manufacturing process and mechanical testing of the cast specimens was sufficiently long to cause natural aging in these age-hardenable alloys. A careful prediction of the microstructural changes taking place within Al_{ss} grains is not easy. Neglecting the amount of Cu inside coarse particles, that of intragranular nanometric particles formed during this stabilization period, could be roughly correlated to the equilibrium amount of the stable phases belonging to the actual precipitation sequence: θ-Al_2Cu or Q-$Al_5Cu_2Mg_8Si_6$. The amounts of the latter phase calculated in equilibrium at room temperature were in the range 0.71–0.81 vol %, while those of θ-Al_2Cu were 4.3, 2.3, and less than 1% for alloys A, C, and B, respectively. Considering the amount of coarse particles in the above phases stable at RT, the expected response to natural aging is strongest in alloy A and mildest in alloy C.

A direct confirmation of the presence of the strengthening particles of nanometric size by TEM observations was outside the scope of the present work. Nevertheless, the higher RT yield strength of the alloy with the highest Cu, notwithstanding its lowest amount of eutectic structure, can be considered as an indirect proof of this.

In alloy A, the beneficial effect of the highest volume fraction of fine strengthening particles, the size of which can be assumed to be similar in the three naturally aged alloys, prevailed over other microstructural effects, leading to the highest strength at RT. The alloy hardness was also improved by the relevant amount of the coarse and hard intermetallic phases formed during solidification. Even if no critical plate-like or acicular morphology was noticed for these phases, the combination of their higher amount and bigger size is responsible for the lowest ductility of alloy A, for which stress-strain curves do not display the relative maximum below 200 °C.

4.2. High-Temperature Mechanical Behaviour and Microstructural Features

When addressing the high-temperature behavior of the investigated alloys, the considerations of the phases formed during solidification and stabilization at room temperature should be coupled to the microstructural modifications taking place during heating and holding time prior or during to the test to explain the high-temperature tension tests results and the hardness of the pulled specimens. These microstructural modifications will be discussed here in the three temperature ranges suggested by hardness test results; RT–150 °C, 150–250 °C, and 250–450 °C.

As the test temperature increased up to 150 °C, the tensile properties showed a plateau or a slight increase in YS and/or UTS followed by slight decrease at the highest temperature, with a concurrent slight increase in elongation and decrease of hardness that, at 150 °C, was more pronounced for alloy C. This behavior, recently reported by Czerwinski [37] for isochronal tests on A380 alloy, is compatible with stress relief occurring at these low temperatures. While heating up to these temperatures, no drastic evolution of intragranular precipitates in the previously stabilized condition can be reasonably expected, nor compositional change in intermetallic phases containing low-diffusivity elements. In this temperature range, the combination of strength and ductility properties does not fundamentally change with respect to that at room temperature. The most brittle alloy A reached UTS values comparable to those of alloy B, characterized by a lower YS. Opposite trends, in terms of hardness, strength, and ductility, were noticed for alloy C, characterized by a low amount of Cu and a Si content close to that of alloy B, but also by the shortest solidification range, for which microstructural investigations confirmed the predicted low amount of coarse Fe- and Cu-containing intermetallic particles. At the microscopical level, the fracture mode that can be derived from Figures 8–10 was of ductile type in all the of alloys/temperatures, with dimples in many cases located at intermetallic phases, the coarser of which (sludge particles) often fractured. A progressive increase of the ductility of these alloys with temperature can be correlated to the microstructural changes taking place within the matrix: the formation of intermetallic particles and then their coarsening/dissolution, as well as changes in the amount of the intermetallic phases.

In the second temperature range, i.e., from about 150 to 250 °C, the importance of Cu as an element affecting age-hardenability became evident, specifically in the hardness of pulled specimens, and partly on the yield strength that, at 250 °C, did not follow the simple trend suggested by the third-order polynomial fitting lines in Figure 3a. The strengthening effects of precipitates formed in this temperature range and the concurrent reduction of the matrix ductility were hindered in UTS and ductility indexes, and most significantly affected by test temperature. The effect on the strain-hardening index was even more critical to detect since the formation of precipitates reduces strain hardening, similar to an increased test temperature.

As the test temperature increased, the amount of Al_2Cu in equilibrium progressively reduced as the solvus temperature was approached and the strengthening phase changes and coarsening reduced their strengthening effect. The peak of hardness values, originated by an optimal amount and size distribution of strengthening particle, was higher as the amount of Cu increased, but also broader due to the higher solvus temperature for the Cu-containing phases. The initial microstructural condition of the alloys as well as the combination of heating/holding time and temperature affected the hardness peak and the mechanical properties of each alloy. In the present case, the initial alloy microstructure and precautions during the experimental testing reasonably led to strengthening effects far from the highest possible for Cu-containing alloys, such as those attainable by a suitable combined solution treatment and artificial aging [3].

The steep reduction of hardness above the peak value and the increase in ductility can be considered as transition to the third and higher temperature range (250–450 °C). Here, the strengthening effects of fine intragranular precipitates based on the θ-Al_2Cu sequence was rapidly lost due to the abovementioned fine particle coarsening and dissolution. Particles of the Q precipitation sequence behaved similarly, but with a higher solvus temperature, as observed by Bassani et al. [38]. As a matter of fact, the calculated solvus temperature for θ-Al_2Cu and Q for alloy A were 409 and 477 °C, respectively, while in equilibrium conditions no θ-Al_2Cu was predicted above about 225 °C in alloys B and C. The combination of a higher amount of fine precipitates possible in alloy A and higher solvus temperatures corresponded to a higher hardness peak temperature, as well as a more gradual hardness reduction and ductility increase above the peak. The presence of coarse Al_2Cu particles was also affected by temperature and reduced progressively as the test temperature increased. Their dissolution was similarly observed above 250 °C by Zamami et al. [8] for the same alloy grade of alloy A. At the highest temperatures, Fe-containing particles can modify, too. Each particle group has its own dissolution kinetics, related to their initial size and to the alloy chemistry, which defines their equilibrium amount at different temperatures. Since these coarse particles are related to the fracture mechanism, the higher amount of stable coarser sludge particles and pro-eutectic Fe-containing intermetallic and coarse Al_2Cu particles kept the ductility of alloy A at the lowest levels (as confirmed by fractographic and microstructural analyses), even in the high temperature range, in agreement with the findings reported by Samuel et al. [39]. The part of the thermal cycle anticipating the tensile test, specifically the holding time spent at high investigated temperatures, could alter the amount and distribution of coarse particles, thus affecting the alloy ductility but with reasonably low effects on the strength properties of the alloys.

Lastly, it is worth mentioning that creep effects can play a role on the mechanical behavior of Al-Si alloys as the temperature is increased above 100 °C, resulting in a progressively more significant strain-rate dependence of the tensile properties of the alloys, as experimentally proven recently above 200 °C [8]. In the temperature range where the precipitation of strengthening phases can take place, strain rate effects are particularly difficult to predict.

5. Conclusions

- The short-term tensile behavior of the three investigated Al-Si-Cu alloys showed a clear decrease of tensile properties above 200 °C, with a steep decrease occurring at testing temperatures between 250 and 350 °C, corresponding to a significant increase in ductility.
- The tensile behavior of alloys at high temperatures can be described by K_T, n_T parameters, for which simple correlations with temperature have been proposed.

- Material hardness and ductility indexes better evidence the differences in the mechanical behavior of the different alloys and can be related to the microstructural changes taking place at the test temperatures.

- Microstructural damage mechanisms are mainly correlated, at all of the investigated temperatures, to the amount and size of secondary phases and coarse intermetallic particles. The AlSi12Cu1(Fe) alloy, characterized by a lower amount of these particles, displayed the highest ductility indexes, particularly above 300 °C, when the highest ductility of its Al_{ss} grains combined to the lowest amount of coarse intermetallic phases characterizing this alloy.

- The hardness of specimens pulled at 150–250 °C suggests an age-hardening strengthening effect, even in the alloy stabilized at room temperature for a long time. The strengthening effect increased with the actual Cu content of the alloy, corresponding to a higher amount of fine particles formed within the Al_{ss} phase. Since their evolution is highly affected by the thermal cycle of the alloy, the actual peak temperature and hardness as well as other mechanical properties in the investigated temperature range are influenced by the thermal history of the material before the tensile test.

- The presence of coarse intermetallic phases in different alloys, related to their ductility, as well as the relevance of age-hardening effects by θ-Al_2Cu and Q-$Al_5Cu_2Mg_8Si_6$ precipitation sequences, related to their strength at intermediate temperatures, can be derived from the results of thermodynamic simulations.

Author Contributions: All the authors actively participated to the research: F.B. and G.T. worked on the material selection, and the optimization of specimens design and the manufacturing process; E.G., O.B., and J.N.L. organized, carried out, and elaborated the results of high-temperature tensile tests and their post-test microstructural analyses; L.R. performed the microhardness and microstructural observations; G.T. performed thermodynamic calculations; E.G. and G.T. discussed mechanical-microstructural correlations; E.G. wrote the draft and all the authors revised and improved it.

Acknowledgments: The authors would like to thank Raffineria Metalli Capra for the supplying of alloys and SAEN for the high-pressure diecasting of specimens. The process development to produce the investigated material was carried out within the NADIA Project, supported by the European Union (FP6-2006, grant agreement No. 026563-2).

Conflicts of Interest: The authors declare no conflict of interest.

References

1. Kaufman, J.G.; Rooy, E.L. *Aluminium Alloy Castings: Properties, Processes and Applications*; ASM International: Materials Park, OH, USA, 2004; pp. 61–68, ISBN 0-87170-803-5.

2. Sjolander, E. Heat Treatment of Al-Si-Cu-Mg Casting Alloys. Ph.D. Thesis, Jönköping University, Jönköping, Sweden, 2011.

3. Lumley, R.N.; ODonnell, R.G.; Gunasegaram, D.R.; Givord, M. Heat treatment of high-pressure die castings. *Metall. Mater. Trans.* **2007**, *38*, 2564–2574. [CrossRef]

4. Hurtalova, L.; Tillova, E.; Chalupova, M. The structure analysis of secondary (recycled) AlSi9Cu3 cast alloy with and without heat treatment. *Eng. Trans.* **2013**, *61*, 197–218.

5. Jeong, C.Y. Effect of alloying elements on high temperature mechanical properties for piston alloy. *Mater. Trans.* **2012**, *53*, 234–239. [CrossRef]

6. Molina, R.; Amalberto, P.; Rosso, M. Mechanical characterization of aluminium alloys for high temperature applications Part1: Al-Si-Cu alloys. *Metall. Sci. Technol.* **2012**, *29*, 5–11.

7. Ferguson, J.B.; Lopez, H.F.; Cho, K.; Kim, C.S. Temperature effects on the tensile properties of precipitation-hardened Al-Mg-Cu-Si alloys. *Metals* **2016**, *6*, 43. [CrossRef]

8. Zamani, M.; Seifeddine, S.; Jarfors, A.E.W. High temperature tensile deformation behaviour and failure mechanisms of an Al–Si–Cu–Mg cast alloy—The microstructural scale effect. *Mater. Des.* **2015**, *86*, 361–370. [CrossRef]

9. Yousefi, M.; Dehnavi, M.; Miresmaeili1, S.M. Microstructure and impression creep characteristics of Al-9Si-xCu aluminium alloys. *Metall. Mater. Eng.* **2015**, *21*, 115–125.

10. *EN 1706 Standard, Aluminium and Aluminium Alloys. Castings. Chemical Composition and Mechanical Properties*; CEN–Comité Européen de Normalisation: Brussel, Belgium, 2007.

11. *ASTM B85/B85M—14 Standard Specification for Aluminum-Alloy Die Castings*; ASTM International: West Conshohocken, PA, USA, 2014.

12. Timelli, G.; Ferraro, S.; Groselle, F.; Bonollo, F.; Voltazza, F.; Capra, L. Caratterizzazione meccanica e microstrutturale di leghe di alluminio pressocolate. *Metall. Ital.* **2011**, *1*, 5–16.

13. Timelli, G.; Bonollo, F. Quality mapping of aluminum alloy die castings. *Metall. Sci. Technol.* **2008**, *26*, 2–8.

14. Ferrero, S.; Timelli, G. Influence of sludge particles on the tensile properties of die cast secondary aluminium alloys. *Metall. Mater. Trans. B* **2015**, *46*, 1022–1034. [CrossRef]

15. Ozhoga-Maslovskaja, O.; Gariboldi, E.; Lemke, J.N. Conditions for blister formation during thermal cycles of Al-Si-Cu-Fe alloys for high pressure diecasting. *Mater. Des.* **2016**, *92*, 151–159. [CrossRef]

16. *ISO 6892-2:2011. Metallic Materials—Tensile Testing—Part 2: Method of Test at Elevated Temperature*; ISO: Geneva, Switzerland, 2011.

17. *CEN T/R 16748:2014 Aluminium and Aluminium Alloys. Mechanical Potential of Al-Si Alloys for High Pressure, Low Pressure and Gravity Die Casting*; CEN–Comité Européen de Normalisation: Brussel, Belgium, 2014.

18. Ramberg, W.; Osgood, W.R. *Description of Stress–Strain Currves by Three Parameters*; Technical Note No. 902; National Advisory Committee for Aeronautics: Washington, DC, USA, 1943.

19. Wiesner, D.J.; Watkins, T.R.; Ely, T.M.; Spooner, S.; Hubbard, C.R.; Williams, J.C. Residual stress measurements of cast aluminum engine blocks using diffraction. *JCPDS-Int. Centre Diffr. Data Adv. X-ray Anal.* **2005**, *48*, 136–142.

20. Rossi, J.L.; Pilkington, R.; Trumper, R.L. Graphical Manipulation of EDS Data Obtained by SEM. In Proceedings of the 13 International Conference UMST X Ray Optics and Microanalysis 1992, Manchester, UK, 31 August–4 September 1992; pp. 275–278.

21. Hatch, J.E. *Aluminium: Properties and Physical Metallurgy*, 10th ed.; ASM International: Materials Park, OH, USA, 2005; p. 238, ISBN 978-0871701763.

22. Rana, R.S.; Purohit, R.; Das, S. Reviews on the influences of alloying elements on the microstructure and mechanical properties of aluminum alloys and aluminum alloy composites. *Int. J. Sci. Res. Publ.* **2012**, *2*, 1–7.

23. Belov, N.A.; Aksenov, A.A.; Eskin, D.G. *Iron in Aluminium Alloys—Impurity and Alloying Element*; Taylor and Francis: London, UK, 2002; pp. 78–80, ISBN 9780415273527.

24. Kaufmann, H.; Fragner, W.; Uggowitzer, P.J. Influence of variations in alloy composition on castability and process stability. Part 1: Gravity and pressure casting. *Int. J. Cast Met. Res.* **2005**, *18*, 273–278. [CrossRef]

25. Ferraro, S.; Fabrizi, A.; Timelli, G. Evolution of sludge particles in secondary die-cast aluminum alloys as function of Fe, Mn and Cr contents. *Mater. Chem. Phys.* **2015**, *153*, 168–179. [CrossRef]

26. Dorè, X.; Combeau, X.; Rappaz, M. Modelling of microsegregation in ternary alloys: Application to the solidification of Al-Mg-Si. *Acta Mater.* **2000**, *48*, 3951–3962. [CrossRef]

27. Kaufmann, H.; Uggowitzer, P.J. *Metallurgy and Processing of High Integrity Light Metal Pressure Castings*; Schiele & Schon: Berlin, Germany, 2007; pp. 7–12, 187–191, ISBN 9783794907540.

28. Zoqui, E.J. Alloys for semisolid processing. In *Comprehensive Materials Processing, Volume 5—Casting, Semi-Solid Forming and Hot Metal Forming*; Hashmi, S., Ed.; Elsevier: New York, NY, USA, 2014; pp. 163–188, ISBN 9780080965338.

29. Shabestari, S.G. The effect of iron and manganese on the formation of intermetallic compounds in aluminum–silicon alloys. *Mater. Sci. Eng. A* **2004**, *383*, 289–298. [CrossRef]

30. Cao, X. Effect of iron and manganese contents on convection-free precipitation and sedimentation of primary α-Al(FeMn)Si phase in liquid Al-11.5Si-0.4Mg alloy. *J. Mater. Sci.* **2004**, *39*, 2303–2314. [CrossRef]

31. Ferdian, M.D. Effet de la Vitesse de Refroidissment sur la Taille des Grains, la Modification Eutectique et la Precipitation D'intermetalliques Riches en fer dans des Alliages Al-Si Hypotectiques. Ph.D. Thesis, Universitè de Tolouse, Toulouse, France, 2014.

32. Karl, M.V. A crystallographic identification of intermetallic phases in Al–Si alloys. *Mater. Lett.* **2005**, *59*, 2271–2276. [CrossRef]

33. Taylor, J.A. Iron-containing intermetallic phases in Al-Si based casting alloys. *Procedia Mater. Sci.* **2012**, *1*, 19–33. [CrossRef]

34. Ji, S.; Yang, W.; Gao, F.; Watson, D.; Fan, Z. Effect of iron on the microstructure and mechanical property of Al–Mg–Si–Mn and Al–Mg–Si diecast alloys. *Mater. Sci. Eng. A* **2013**, *564*, 30–139. [CrossRef]

35. Timelli, G.; Fabrizi, A. The effects of microstructure heterogeneities and casting defects on the mechanical properties of high-pressure die-cast AlSi9Cu3(Fe) Alloys. *Metall. Mater. Trans. A* **2014**, *45*, 5486–5498. [CrossRef]

36. Mohamed, A.M.A.; Samuel, A.M.; Samuel, F.H.; Doty, H.W. Influence of additives on the microstructure and tensile properties of near-eutectic Al–10.8% Si cast alloy. *Mater. Des.* **2009**, *30*, 3943–3957. [CrossRef]

37. Czerwinski, F.; Shaha, S.K.; Kasprzak, W.; Friedman, J.; Chen, D.L. Aging characteristics of the Al-Si-Cu-Mg cast alloy modified with transition metals Zr, V and Ti. *IOP Conf. Ser. Mater. Sci. Eng.* **2016**, *117*, 012031. [CrossRef]

38. Bassani, P.; Gariboldi, E.; Ripamonti, D. Thermal analysis of Al–Cu–Mg–Si alloy with Ag/Zr additions. *J. Therm. Anal. Calorim.* **2008**, *91*, 29–35. [CrossRef]

39. Samuel, A.M.; Gauthier, J.; Samuel, F.H. Microstructural Aspects of the Dissolution and Melting of Al$_2$Cu Phase in Al-Si Alloys during Solution Heat Treatment. *Metall. Mater. Trans. A* **1996**, *27*, 1785–1798. [CrossRef]

Article

Hot Deformation Behavior of a Spray-Deposited Al-8.31Zn-2.07Mg-2.46Cu-0.12Zr Alloy

Xiaofei Sheng [1,2], **Qian Lei** [1,3,*], **Zhu Xiao** [1] **and Mingpu Wang** [1]

[1] School of Materials Science and Engineering, Central South University, Changsha 410083, China; auden1@126.com (X.S.); xiaozhumse@163.com (Z.X.); wangmp@csu.edu.cn (M.W.)
[2] School of Materials Science and Engineering, Hubei University of Automotive Technology, Shiyan 442002, China
[3] Department of Materials Science and Engineering, College of Engineering, University of Michigan, Ann Arbor, MI 48109, USA
* Correspondence: qianlei@umich.edu; Tel.: +1-734-763-5282

Received: 20 June 2017; Accepted: 1 August 2017; Published: 4 August 2017

Abstract: Metallic materials have a significant number of applications, among which Al alloys have drawn people's attention due to their low density and high strength. High-strength Al-based alloys, such as 7XXX Al alloys, contain many alloying elements and with high concentration, whose microstructures present casting voids, segregation, dendrites, etc. In this work, a spray deposition method was employed to fabricate an Al-8.31Zn-2.07Mg-2.46Cu-0.12Zr (wt %) alloy with fine structure. The hot deformation behavior of the studied alloy was investigated using a Gleeble 1500 thermal simulator and electron microscopes. The microstructure evolution, variation in the properties, and precipitation behavior were systematically investigated to explore a short process producing an alloy with high property values. The results revealed that the $MgZn_2$ particles were detected from inside the grain and grain boundary, while some Al_3Zr particles were inside the grain. An Arrhenius equation was employed to describe the relationship between the flow stress and the strain rate, and the established constitutive equation was that: $\dot{\varepsilon} = [\sinh(0.017\sigma)]^{4.049} \exp[19.14 - (129.9/RT)]$. An appropriate hot extrusion temperature was determined to be 460 °C. Hot deformation (460 °C by 60%) + age treatment (120 °C) was optimized to shorten the processing method for the as-spray-deposited alloy, after which considerable properties were approached. The high strength was mainly attributed to the grain boundary strengthening and the precipitation strengthening from the nanoscale $MgZn_2$ and Al_3Zr precipitates.

Keywords: alloy; 7XXX Al alloy; spray deposited; hot deformation behavior; precipitation

---

1. Introduction

Aluminum alloys of the 7XXX series (Al-Zn-Mg-Cu) have a large number of applications in automobile and aviation industries because of their considerable combination of properties, e.g., high strength, low density, outstanding workability, etc. [1–4]. Their excellent performance attracts much interest from the manufacturing industry [5,6]. High strength makes these products strong, and good ductility allows them to be formed and deformed easily. Meanwhile, high fatigue strength and lifetime are related to the strength and ductility. Strength and ductility are equally important in the damage tolerance and durability [7,8]. Precipitation hardening plays an important role in strengthening the 7XXX aluminum alloys [9]. Increasing the additions of Zn and Mg will enhance the volume fraction of precipitates and improve the mechanical properties [10]. However, a large amount of Zn and Mg results in coarse grains and segregation in ingots and results in poor ductility and low fracture strength [11]. Moreover, the 7XXX alloys with high Zn and Mg content were easy to failure in service. An important and effective way to approach the superior properties is by refining the structure, within which thermal-mechanical heat treatment and severe plastic deformation have been developed in the past decade. Severe plastic deformation techniques, such as equal-channel accumulative roll bonding [12], angular pressing [13], high-pressure torsion, and friction stir processing [14], are effective approaches to refine the microstructure. However, the above techniques yield small sizes and quantities in the laboratory, but are not suitable for industrialized production [15]. Fortunately, spray deposition technology has been used to manufacture high-performance 7XXX series aluminum alloys in the past [16]. Spray forming combines the advantages of rapid solidification, high solute content, and homogeneous microstructure containing fine grains, avoiding the macro segregation. Thus, it is thought that the spray deposition technique could be used to enhance the mechanical properties of aluminum alloys. The additions of Mg and Zn enhanced the mechanical properties because of the η'-$MgZn_2$ precipitates after the aging treatment [17,18]. The addition of other alloying elements were also effective to enhance the properties. The additions and Zr and Sc could be precipitated as $Al_3(Sc, Zr)$ precipitates during aging treatment, and these precipitates are functionally important in pinning the dislocation and preventing grain boundary migration. As a result, the yield strength was enhanced [19,20].

Therefore, an Al-8.31Zn-2.07Mg-2.46Cu-0.12Zr alloy (wt %) was designed, fabricated by the spray deposition solidification method, and investigated to develop a novel high-strength high-ductility aluminum alloy. Previous research on spray forming Al-Zn-Mg-Cu alloy only focused on the microstructure and properties of the spray-deposited ingots [21,22], and the thermal-mechanical treatment that was an important and efficient method to improve the properties. In previous decades, most researchers have focused on the melting, casting, homogenization-processed solid solution treatment, aging treatment, and so forth. Precipitation behavior, the crystal orientation relationship between matrix and precipitates, homogenization behavior, recovery, crystallization behavior, and cold working hardening have been reported in the past decades. However, hot deformation and shortened processing have not been reported yet.

In this work, an Al-8.31Zn-2.07Mg-2.46Cu-0.12Zr alloy was prepared by the spray deposition method. Subsequently, the hot deformation behavior of the above alloy ingot was investigated systematically. The constitutive equation, hot working map, microstructure evolution, precipitation of the compressed samples, and strength mechanism was explored in this work. For the industry baseline, a shorter processing was developed on the basis of the hot compressed result.

2. Materials and Methods

The composition of the studied alloy was Al-8.31Zn-2.07Mg-2.46Cu-0.12Zr (wt %). The ingots were prepared using an SFZD-5000 type environmental chamber at Haoran Co., Ltd., Jiangsu, China [23]. Then the spray-deposited ingot was cut into small cylindrical samples with a dimension of 10 mm in diameter and 15 mm in height. Hot deformation experiments were carried out on the Gleeble-1500 thermal simulator (Dynamic Systems Inc., Poestenkill, NY, USA); the strain rates were 0.001 s^{-1}, 0.01 s^{-1}, 0.1 s^{-1}, and 1 s^{-1}, and the deformation temperatures were 340 °C, 380 °C, 420 °C, and 460 °C, respectively. The total deformation for each sample was 60%. Hardness values were measured with an HV-5 hardness tester (Laizhou Huayin Testing Instrument Co., Ltd., Laizhou, China), with a load of 2 kg and a load time of 15 s, each average hardness was taken from at least seven measurements. The X-ray diffraction analyses on polished discs samples were performed using a D8 Discover X-ray diffractometer (Bruker AXS Inc., Karlsruhe, Germany), with a scanning step of 8°/min, a Cu-Kα radiation graphite monochromator filtering, a tube voltage of 40 kV, a tube current of 250 mA. The X-ray diffraction (XRD) result was analyzed with an MDI Jade software (5.0, MDI, Livermore, CA, USA). Phase identification was conducted on an FEI Quanta 650 scanning electron microscopy (SEM) (FEI Company, Hillsboro, OR, USA) with an electron back-scattered diffraction (EBSD) detector. The differential scanning calorimetry (DSC) analysis test was conducted on SDT-Q600 differential thermal equipment (Ta Instruments, New Castle, DE, USA). The sample was heated in an argon gas environment, the heating rate is 10 °C/min, the heating temperature region is 25–680 °C, the reference sample is Al_2O_3. Optical microscopy was performed on a Leica DM microscope (Leica Microsystems, Wetzlar, Germany), and samples were etched by Keller's reagent after grinding and polishing. The transmission electron microscopy (TEM) samples were prepared by dual-jet electropolishing technique with a solution containing 30% nitric acid and 70% methanol at $-30\,^\circ$C. Transmission electron microscopy (TEM) observations were carried out on a TECNAI G2 F20 transmission electron microscope (FEI Company, Hillsboro, OR, USA) with an operation voltage of 200 kV.

3. Results

3.1. Initial Microstructure of As-Spray-Deposited Samples

Figure 1 shows back-scattered electros images (BSE) and XRD of the as-spray-deposited Al-Zn-Mg-Cu-Zr alloy. The average grain size was about 50 ± 10 µm after measurements (more than 60 grains were counted and measured from BSE SEM images). The XRD result and index revealed that some $MgZn_2$ intermetallic compounds were formed in the as-spray-deposited samples (Figure 1c). Figure 2 shows the EBSD phase identify results of the as-spray-deposited alloy. $MgZn_2$ and Al_3Zr phases were detected from the EBSD results. The $MgZn_2$ particles were distributed in the grain and on the grain boundary, while the Al_3Zr particles distributed inside the grain at the nanoscale. The white contrast particles in Figures 1b and 2a were determined to be $MgZn_2$. For the as-spray-deposited sample, there are no dendrites and segregation due to rapid solidification, as shown in Figures 1 and 2.

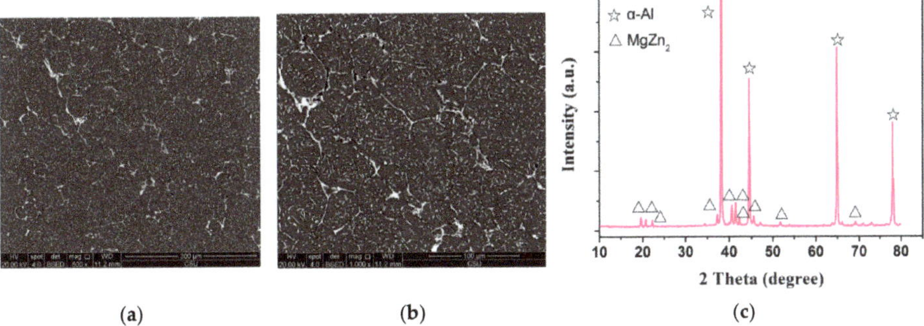

Figure 1. Back-scattered electrons images (**a**,**b**) and XRD (X-ray diffraction) pattern (**c**) of as-spray-deposited Al-8.31Zn-2.07Mg-2.46Cu-0.12Zr alloy.

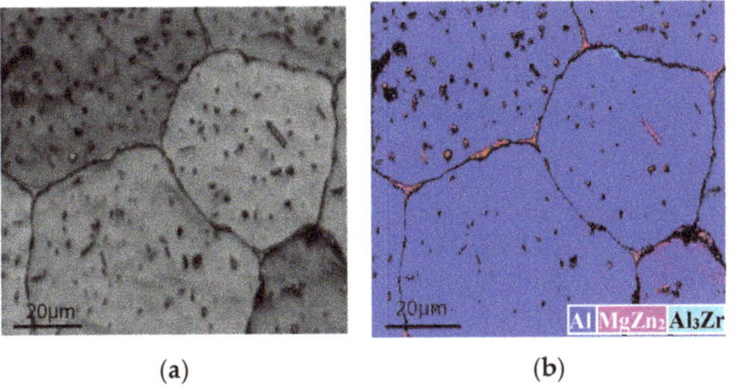

Figure 2. The SEM image and EBSD phase map of as-spray-deposited Al-8.31Zn-2.07Mg-2.46Cu-0.12Zr alloy: (**a**) secondary electron image; (**b**) EBSD phase maps showing the detected phases and their distribution.

3.2. Compressive Stress-Strain Curves

Hot-deformation is an effective way to refine the structure, to eliminate the defects, and to shape the structural parts. Figure 3 shows the true stress-true strain of the as-spray-deposited samples after hot-compressive deformation at 340–460 °C with strain rates of 0.001–1 s^{-1}. As the samples were hot compressed at 340 °C and 380 °C, the true-stress was rapidly increased with the true-strain at the early deformation process, and then slowly increased with the strain after a true strain of around 5% due to the balance between hardening and recovery softening. As the samples were hot compressed at 420 °C and 460 °C, the true-stress rapidly increased to a mountain peak and then decreased, followed by slowly increasing with the strain when the strain rate was less than 0.1 s^{-1}. The decline in true-stress contributed to the softening from the grains' subdivision and the coarsening of initial grains. The true-stress increased with the strain rate when the sample was hot-compressed at a certain temperature, while the true-stress decreased with the deformation temperature.

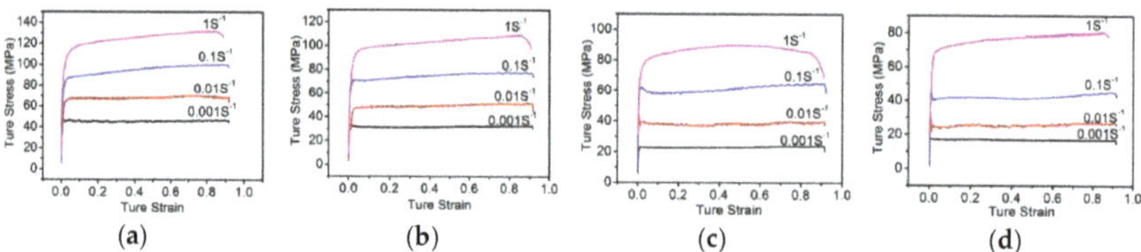

Figure 3. True stress-true strain curves of the specimens hot compressed at temperatures of 340 °C (**a**); 380 °C (**b**); 420 °C (**c**); and 460 °C; and (**d**) with different strain rates (0.001–1 s^{-1}).

3.3. Microstructure Evolution

EBSD orientation mapping is an advanced technology to present the microstructure of the deformed samples, especially for aluminum alloys. Figure 4 shows the EBSD orientation maps of the samples after hot-compression at different temperatures with strain rates of 1 s^{-1} and 0.001 s^{-1}. The red lines in Figure 4 presented large angle grain boundaries (larger than 15 degrees), and green lines showed small angle grain boundaries (less than 15 degrees). The fraction of small angle grain boundaries decreased with the deformation temperature when the strain rate was 0.001 s^{-1}, while remaining constant at a strain rate of 1 s^{-1}, even if the deformation temperature increased. The grain size was increased when the deformation temperature was increased from 340 °C to 460 °C for the strain rate of 0.001 s^{-1}. Figure 5 shows the area fraction of the grain size in the compressed samples, on the basis of the EBSD mapping results from Figure 4. The grain size was 40–65 μm when the strain rate was 0.001 s^{-1}, while it was 0–30 μm when the strain rate was 1 s^{-1}. Figure 6 shows SEM images of samples hot-compressed at 340–460 °C with different deformation rates of 1 s^{-1} and 0.001 s^{-1}. As compared with the as-spray-deposited microstructure (Figure 1b), the fraction of MgZn$_2$ phase decreased when the sample was hot compressed at elevated temperatures. The fraction of the MgZn$_2$ phase decreased with the hot deformation temperature. The fraction of the MgZn$_2$ in the samples compressed at a certain temperature decreased with the decrease of the strain rate (from 1 s^{-1} to 0.001 s^{-1}).

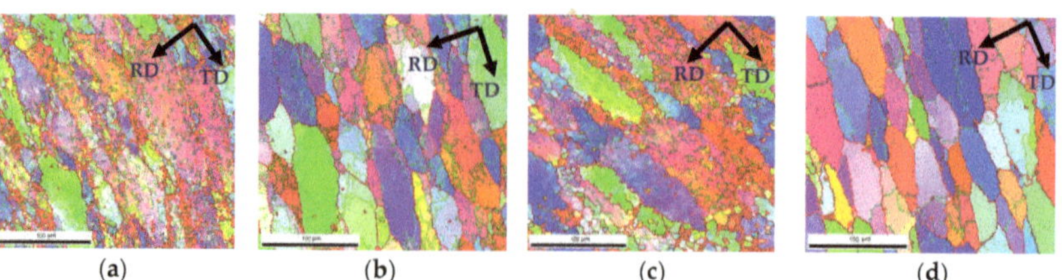

Figure 4. *Cont.*

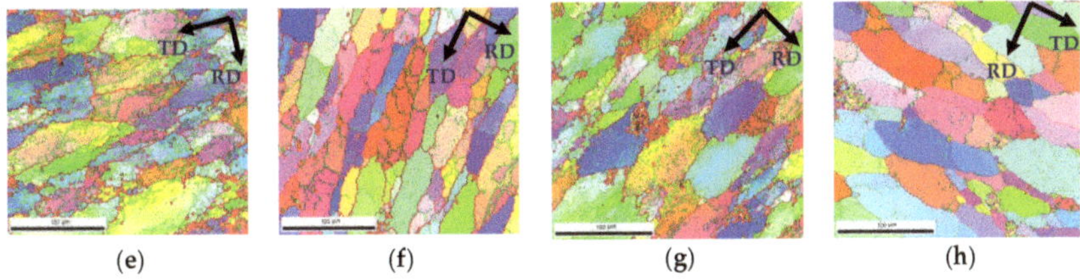

Figure 4. Typical EBSD orientation maps: (**a**) 340 °C, 1 s^{-1}; (**b**) 340 °C, 0.001 s^{-1}; (**c**) 380 °C, 1 s^{-1}; (**d**) 380 °C, 0.001 s^{-1}; (**e**) 420 °C, 1 s^{-1}; (**f**) 420 °C, 0.001 s^{-1}; (**g**) 460 °C, 1 s^{-1}; and (**h**) 460 °C, 0.001 s^{-1}.

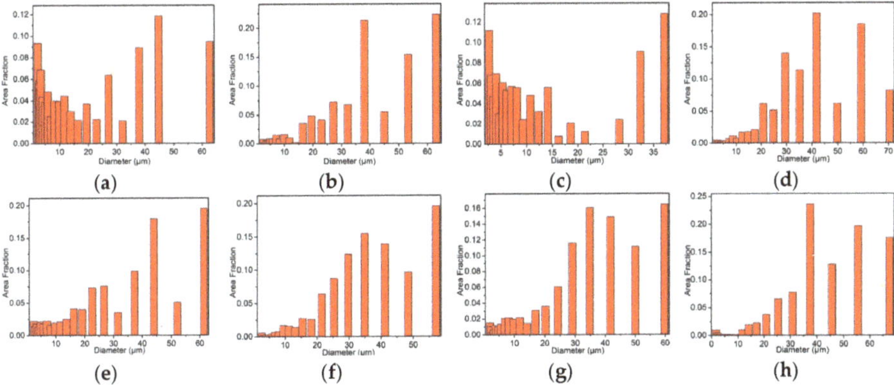

Figure 5. Variations of gain size in the samples after hot-compressing with different conditions: (**a**) 340 °C, 1 s^{-1}; (**b**) 340 °C, 0.001 s^{-1}; (**c**) 380 °C, 1 s^{-1}; (**d**) 380 °C, 0.001 s^{-1}; (**e**) 420 °C, 1 s^{-1}; (**f**) 420 °C, 0.001 s^{-1}; (**g**) 460 °C, 1 s^{-1}; and (**h**) 460 °C, 0.001 s^{-1}.

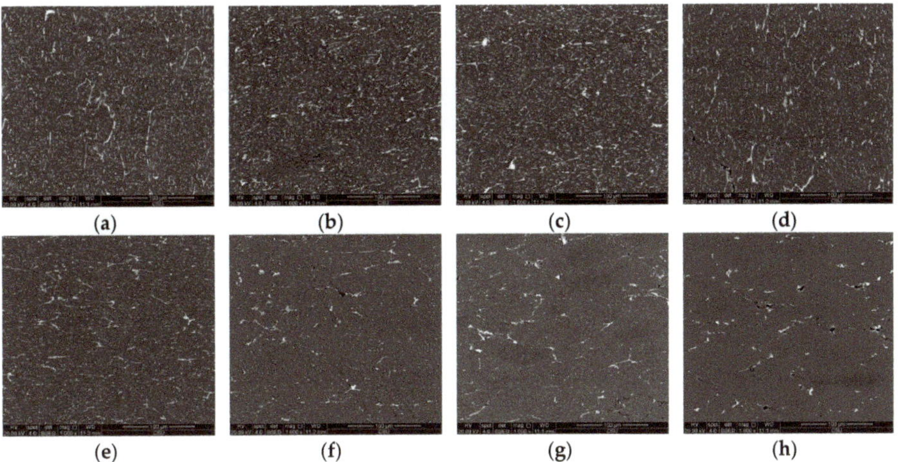

Figure 6. Back-scattered electron SEM images: (**a**) 340 °C, 1 s^{-1}; (**b**) 340 °C, 0.001 s^{-1}; (**c**) 380 °C, 1 s^{-1}; (**d**) 380 °C, 0.001 s^{-1}; (**e**) 420 °C, 1 s^{-1}; (**f**) 420 °C, 0.001 s^{-1}; (**g**) 460 °C, 1 s^{-1}; and (**h**) 460 °C, 0.001 s^{-1}.

3.4. Hardness of the Compressive Samples after Age Treatments

Figure 7 gives the hardness variations of the hot compressed samples aged at 120 °C for different durations. For the samples after being hot-compressed at strain rates of 1 s^{-1} and 0.001 s^{-1}, the curves present the same trend. The initial hardness of compressed samples increased with the hot deformation temperature. The hardness of the hot-compressed samples increased with the aging time, and almost kept constant as they were aging at certain temperatures for 15 h.

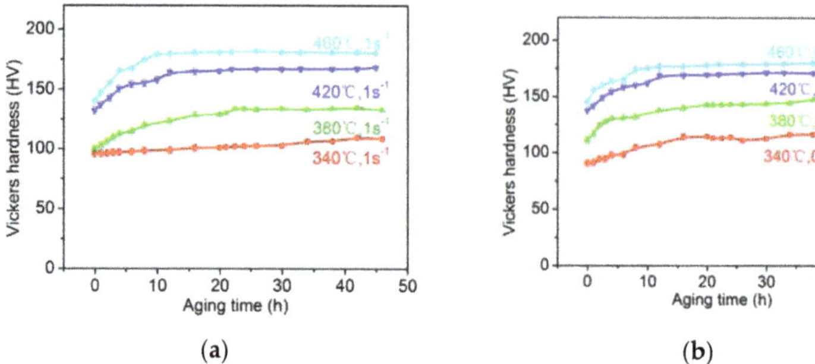

Figure 7. Hardness variation of the hot-compressed samples with an aging time at 120 °C. The samples were compressed with strain rates of: (**a**) 1 s^{-1}; and (**b**) 0.001 s^{-1}.

4. Discussion

4.1. Constitutive Equation

The constitutive relationships were used to describe the hot deformation behavior. Some models are employed to reveal the hot deformation behavior, in which one of the most extended versions of the hot working constitutive equation is as follows [24]:

$$\dot{\varepsilon} = AF(\sigma)\exp[-Q/(RT)], \tag{1}$$

where $\dot{\varepsilon}$ is the strain rate; A is a material constant which is independent of temperature; σ is peak true stress; Q is the activation energy; R is the gas constant (8.31); and T is the absolute temperature (K).

At low stress lever ($\alpha\sigma < 0.8$):

$$F(\sigma) = \exp(\beta\sigma), \tag{2}$$

At high stress lever ($\alpha\sigma > 1.2$):

$$F(\sigma) = \sigma^n, \tag{3}$$

β and n are the material constants, $\alpha = \beta/n$. Substituting Equations (2) and (3) into Equation (1):

$$\dot{\varepsilon} = A_1\exp(-Q/RT)\exp(\beta\sigma), \tag{4}$$

$$\dot{\varepsilon} = A_2\exp(-Q/RT)\sigma^n, \tag{5}$$

where parameters B_1 and B_2 are set to $B_1 = A_1\exp(Q/RT)$ and $B_2 = A_2\exp(Q/RT)$; when the temperature is fixed, B_1 and B_2 are temperature constants. Taking the logarithm of both sides of Equations (4) and (5), respectively, gives:

$$\ln\dot{\varepsilon} = \ln B_1 + \beta\sigma, \tag{6}$$

$$\ln\dot{\varepsilon} = \ln B_2 + n\ln\sigma, \tag{7}$$

The values of n and β are from the slopes of the lines in $\ln \dot{\varepsilon}$ and σ, $\ln \dot{\varepsilon}$ and $\ln \sigma$, and plots, respectively. Figure 8 plotted them, the slope value of β is 0.09491, and the slope value n is 5.512, $\alpha = \beta/n = 0.01722$.

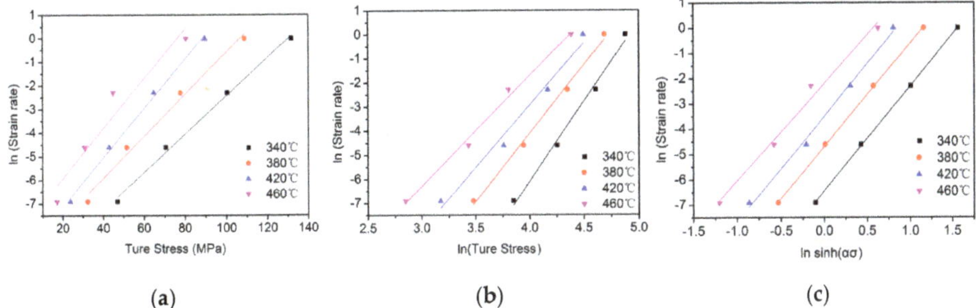

Figure 8. (a) $\ln \dot{\varepsilon}$-σ; (b) $\ln \dot{\varepsilon}$-$\ln\sigma$; (c) $\ln \dot{\varepsilon}$-$\ln[\sinh(\alpha\sigma)]$.

From Figure 8, it can be seen that the logarithm of the true stress and the logarithm of the strain rate met a linear relationship. In addition, flow stress, $F(\sigma)$, could be described by Equation (8):

$$F(\sigma) = [\sinh(\alpha\sigma)]^x \tag{8}$$

Substituting Equation (8) into Equation (1):

$$\dot{\varepsilon} = A[\sinh(\alpha\sigma)]^x \exp[-Q/(RT)] = [\sinh(\alpha\sigma)]^x \exp[\ln A - Q/(RT)]. \tag{9}$$

Taking the logarithm of both sides of Equation (9):

$$\ln \dot{\varepsilon} = \ln A + x \ln[\sinh(\alpha\sigma)] - Q/(RT). \tag{10}$$

The value of x can be obtained from the slope of the lines in $\ln \dot{\varepsilon}$ and $\ln[\sinh(\alpha\sigma)]$, shown in Figure 7c, and $x = 4.053$.

Taking the Zener-Hollomon parameter [25]:

$$Z = A[\sinh(\alpha\sigma)]^x = \dot{\varepsilon}\ \exp(Q/RT). \tag{11}$$

Taking the logarithm of both sides of Equation (11):

$$\ln[\sinh(\alpha\sigma)] = (1/x)\cdot(\ln \dot{\varepsilon} - \ln A) + (Q/1000xR)\cdot(1000/T). \tag{12}$$

$x = 4.053$. $\ln[\sinh(\alpha\sigma)] - 1000/T$ could be imitated. From Figure 9a, $P = Q/1000xR = 3.856$, and $Q = 1000PxR = 129.9$ kJ/mol. The activation energy of a studied material derived from Equation (11) was used as an indicator of the degree of difficulty for the material in the hot deformation process. It provides a guideline to optimize the hot working process, and also furnishes additional information on the microstructure and flow stress evolution in successive deformation processes.

According to Equation (11):

$$Z = A[\sinh(\alpha\sigma)]^x, \tag{13}$$

Taking the logarithm of both sides of Equation (13):

$$\ln Z = \ln A + x \ln[\sinh(\alpha\sigma)], \tag{14}$$

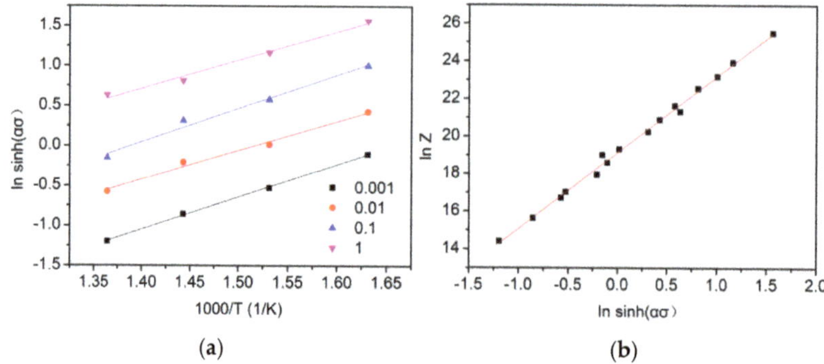

Figure 9. $\ln[\sinh(\alpha\sigma)] - 1000/T$ (**a**); $\ln Z - \ln[\sinh(\alpha\sigma)]$ (**b**).

Then, $Z = \dot{\varepsilon}\exp(Q/RT)$, R is the gas constant, and Q is 129.9 kJ/mol. $\ln Z$ could be calculated from the slope of the fitting line in Figure 9b. $\ln A$ is 19.14, x is 4.045, which is very close to n (4.053). Here, taking the average value, $x = 4.049$.

Substituting the values of A, Q, x, α into Equation (11), the flow stress constitutive equation for alloy can be expressed by the Arrhenius equation:

$$\dot{\varepsilon} = [\sinh\alpha\sigma]^{x}\exp[\ln A - (Q/RT)], \tag{15}$$

$$\dot{\varepsilon} = [\sinh(0.017\sigma)]^{4.049}\exp[19.14 - (129.9/RT)]. \tag{16}$$

4.2. Hot Deformation Behavior

In the first deformation, the true-stress increased rapidly (Figure 3). When the deformation temperature was constant, the stress peaks increased with the strain rate. When the deformation temperature is 340–460 °C and the strain rate is 0.001–0.1 s^{-1}, the steady-state flow stress characteristic was the main phenomenon. At a certain temperature and a fixed strain rate, the performance of the dynamic recovery characteristics occurred when the true-strain exceeded a certain value (around 5%). With increasing the true-strain, the internal space of the material also increased the dislocation density, which was hardening the material, while dynamic recovery occurred; finally, the hardening and softening approached a balance. Thus, the curve for the true-stress and the true strain presents a horizontal line. The peak stress increases with both the decrease of the deformation temperature and the growth of the strain rate.

Figure 10a shows the differential scanning calorimetry (DSC) result of the as-spray-deposited specimen. There were two endothermic peaks in the DSC curve. Combined with XRD analysis result (Figure 1c), the endothermic peak at 474.2 °C is corresponding to the MgZn$_2$ phase redissolution, the other one is corresponding to the melting point of the alloy matrix. Thus, the hot deformation temperature should be lower than 474.2 °C. The temperature of the sample will increase when it was deformed at elevated temperature. An extrusion test was carried out at 470 °C for the as-spray-deposited alloy with an extrusion ratio of 6.25. The microstructure of the hot extruded sample is shown in Figure 10b, which shows a significant over-burning and widening of the grain boundaries. Thus, the hot deformation temperature should be less than 470 °C. According to Figure 6d, 460 °C was determined as the hot deformation temperature for the studied alloy.

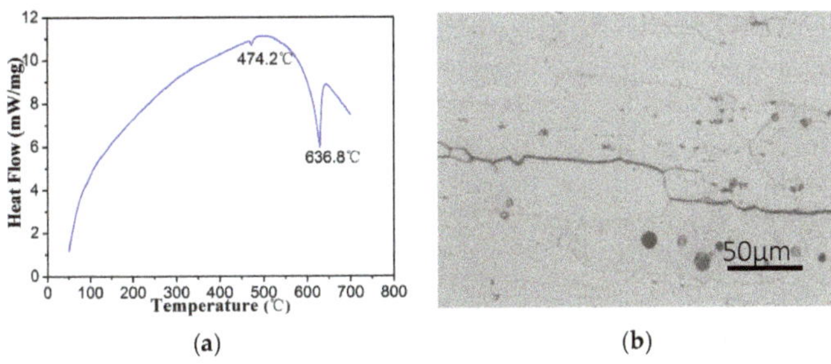

(a)   (b)

Figure 10. (**a**) The DSC results of the as-spray-deposited specimen; and (**b**) the microstructure of the sample after hot extrusion at 470 °C with an extrusion ratio of 6.2.

4.3. Shortening the Production Process

Homogenization was employed in the as-cast Al-Zn-Mg alloys (7XXX serious alloy) to eliminate the serious dendritic segregation before hot deformation in the previous investigations [26,27], for serious dendritic segregation. In this work, through spray deposition, the high composition Al-Zn-Mg alloys (7XXX serious alloy) were made without serious dendritic segregation. Moreover, the result of the test of hot deformation behavior means that here homogenization is not required for the spray-deposited Al-8.31Zn-2.07Mg-2.46Cu-0.12Zr alloy. For the Al-Zn-Mg-Cu-Zr system alloy, a very complex thermal-mechanical treatment path has been set up in a previous work (path I) [28]. On the basis of the above investigation, a simpler production processing method may be set up for the studied alloy; here, three other paths (paths I, II, III) referred to the previous work [28–30]. The four paths were listed as following: (I) as-deposited + homogenization treatment (450 °C/24 h + 460 °C/2 h) + hot deformation (420 °C by 6.25) + solid solution treatment (470 °C/1 h) + age treatment (120 °C) [28]; (II) as-spray-deposited + homogenization treatment (450 °C/24 h + 460 °C/2 h) + hot deformation (420 °C by 6.25) + solid solution treatment (470 °C/1 h) + age treatment (120 °C) [29]; (III) as-spray-deposited + hot deformation (420 °C by 6.25) + solid solution treatment (470 °C/1 h) + age treatment (120 °C) [30]; (IV) as-spray-deposited + hot deformation (460 °C by 60%) + age treatment (120 °C) (this work). Figure 11 shows the hardness variations as a function of the aging time. The hardness increased with the age time, especially at the early state. The peak hardness values were 210 HV, 208 HV, 205 HV, and 180 HV when the spray deposited samples were treated with path I, path II, path III, and path IV, respectively. For the path IV, the stages of homogenization treatment and solid solution treatment were eliminated, resulting in a lower energy cost, and the hardness is only slightly less than those in other spray-deposited samples. Thus, a shortened processing method of hot deformation (460 °C by 60%) + age treatment (120 °C) was developed for the as-spray-deposited Al-8.31Zn-2.07Mg-2.46Cu-0.12Zr alloy in this work.

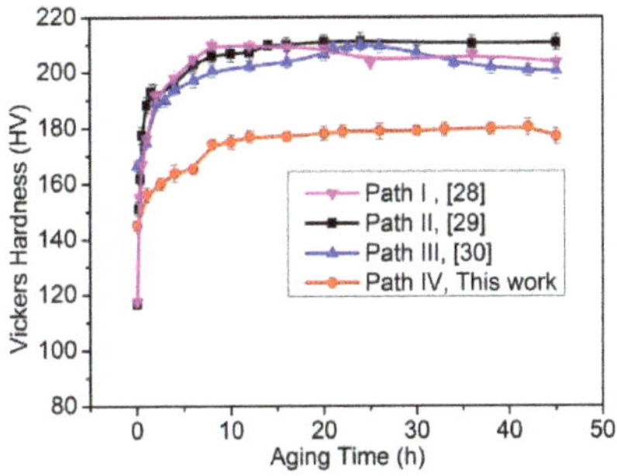

Figure 11. Hardness variations with the aging time with the different paths.

4.4. Strengthening Mechanism

Figure 12 shows the bright-field (BF) TEM images in addition to selected-area diffraction patterns (SADPs) of the spray-deposited alloy treated with the path IV (hot deformation at 460 °C by 60% + aging at 120 °C for 10 h). The as-spray deposited alloy were hot compressed by 60% at 460 °C and followed by aging at 120 °C for 10 h; nano-scale precipitates were captured in the BF TEM photographs (Figure 12a). A few diffraction spots from the precipitates were observed from SADPs (Figure 12b). According to the crystal geometry and crystal diffraction, indexing of the SADPs showed that the rod-like precipitates were hexagonal η'-MgZn$_2$ phase particles with parameters of a = 0.496 nm and c = 1.402 nm. There is a coherence relationship between Al and η'-MgZn$_2$: $(001)_{\eta'}$ // $(110)_{Al}$ and $[100]_{\eta'}$ // $[001]_{Al}$ [31]. Moreover, some nano-scale Al$_3$Zr precipitates were detected in the BF images, as marked by the arrows. Precipitation strengthening is the most effective strengthening mechanism in the studied alloy.

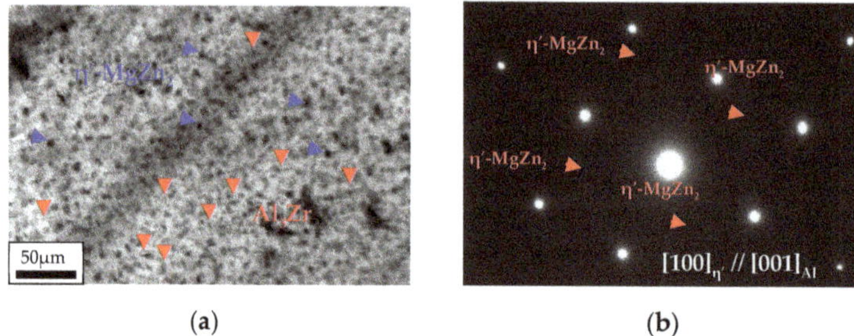

(**a**)　　　　　　　　　　(**b**)

Figure 12. BF TEM image (**a**) of the specimen after treat with path V (hot deformation at 460 °C by 60% + aging at 120 °C for 10 h) and the corresponding SADPs (**b**).

The high strength of the Al-Zn-Mg-Cu-Zr alloy was mainly attributed to precipitation strengthening and grain boundary strengthening.

(I) Precipitation strengthening. An increase in yield stress caused by dislocation bypassing η'-MgZn$_2$ particles in the Orowan process is given by [32–34]:

$$\Delta\sigma_{Orowan} = \frac{0.81 MGb}{2\pi(1-v)^{1/2}} \frac{\ln(d_p/b)}{(\lambda - d_p)},\tag{17}$$

$$\lambda = \frac{1}{2}d_P\sqrt{\frac{3\pi}{2f_v}},\tag{18}$$

where M is the Taylor factor, G is the shear modulus, b is the Burgers vector, v is the Poisson's ratio, λ is the space between particles, d_p is the average diameter of particles; and f_v is the volume fraction of η'-MgZn$_2$ particles.

(II) Grain boundary strengthening. As shown in the EBSD images, when the studied alloy was hot-compressed to a reduction of 60% in thickness. For path IV, samples have some grains with a dimension of 40–60 µm. These grain boundaries hinder the dislocation motion, leading to a remarkable increase in yield strength [34]. An increase in yield stress due to the grain boundary can be estimated using the Hall-Petch equation [35]:

$$\Delta\sigma_{H\text{-}P} = kd^{-\frac{1}{2}},\tag{19}$$

where k is a constant (3.5–4.5 MPa·mm$^{-1/2}$ for Al alloys [36]), d is the diameter of grain size. The smaller the grain size, the more numerous the boundaries, and the higher the strength of the studied alloy.

5. Conclusions

In summary, an Al-8.31Zn-2.07Mg-2.46Cu-0.12Zr alloy was synthesized using the spray deposition method. In the as-spray-deposited sample, the $MgZn_2$ particles were detected from inside the grain and grain boundary, while some nanoscale Al_3Zr particles were inside the grain. Hot compression experiments showed that the true-stress increased with the true-strain. The Arrhenius equation was employed to describe the relationship between peaks of the flow stress and the strain rate. The activation energy Q of the studied alloy was 129.9 kJ/mol, and the constitutive equation was established: $\dot{\varepsilon} = [\sinh(0.017\sigma)]^{4.049} \exp[19.14 - (129.9/RT)]$. The appropriate hot deformation temperature was determined to be 460 °C. A short processing baseline has been optimized: as-spray-deposited + hot deformation (460 °C by 60%) + age treatment (at 120 °C), by which considerable properties were approached for the studied alloy. The high strength of the studied alloy was mainly attributed to the grain boundary strengthening from small grains with tens of microns, and the precipitation strengthening from nanoscale $MgZn_2$ and Al_3Zr precipitates.

Acknowledgments: The authors thank to the financial support by the National Key Technology R and D Program (2014BAC03B08), the National Key Research and Development Program of China (2016YFB0301300), the start-up funding from the Central South University, the Fundamental Research Funds for the Central Universities of Central South University (2014ZZTS018), and the Hunan Provincial Innovation Foundation for Postgraduate (CX2014B048).

Author Contributions: Xiaofei Sheng and Qian Lei conceived and designed the experiments; Xiaofei Sheng performed the experiments; Xiaofei Sheng and Qian Lei analyzed the data; Mingpu Wang and Zhu Xiao contributed materials and discussions; and Xiaofei Sheng and Qian Lei wrote the manuscript.

Conflicts of Interest: The authors declare no conflict of interest.

References

1. Ferguson, J.B.; Schultz, B.F.; Mantas, J.C.; Shokouhi, H.; Rohatgi, P.K. Effect of Cu, Zn, and Mg concentration on heat treating behavior of squeeze cast Al-(10 to 12)Zn-(3.0 to 3.4)Mg-(0.8 to 1)Cu. *Metals* **2014**, *4*, 314–321. [CrossRef]

2. Xu, Y.Q.; Zhan, L.H. Effect of creep aging process on microstructures and properties of the retrogressed Al-Zn-Mg-Cu alloy. *Metals* **2016**, *6*, 189. [CrossRef]

3. Zhou, M.; Lin, Y.C.; Deng, J.; Jiang, Y.Q. Hot tensile deformation behaviors and constitutive model of an Al-Zn-Mg-Cu alloy. *Mater. Des.* **2014**, *59*, 141–150. [CrossRef]

4. Ma, S.M.; Sun, Y.H.; Wang, H.Y.; Lü, X.S.; Qian, M.; Ma, Y.L.; Zhang, C.; Liu, B.C. Effect of a minor Sr modifier on the microstructures and mechanical properties of 7075 T6 Al alloys. *Metals* **2017**, *7*, 13. [CrossRef]

5. Zhao, H.L.; Yao, D.M.; Qiu, F.; Xia, Y.M.; Jiang, Q.C. High strength and good ductility of casting Al-Cu alloy modified by Pr_xO_y and La_xO_y. *J. Alloys Compd.* **2011**, *509*, 43–46. [CrossRef]

6. Lang, Y.J.; Zhou, G.X.; Hou, L.G.; Zhang, J.S.; Zhuang, L.Z. Significantly enhanced the ductility of the fine-grained Al-Zn-Mg-Cu alloy by strain-induced precipitation. *Mater. Des.* **2015**, *88*, 625–631. [CrossRef]

7. Yuan, W.H.; Zhang, J.; Zhang, C.C.; Chen, Z.H. Processing of ultra-high strength SiCp/Al-Zn-Mg-Cu composites. *J. Mater. Process. Technol.* **2009**, *209*, 3251–3255. [CrossRef]

8. Krasilnikov, N.A.; Sharafutdiniv, A. High strength and ductility of nanostructured Al-based alloy, prepared by high-pressure technique. *Mater. Sci. Eng. A* **2007**, *463*, 74–77. [CrossRef]

9. Marlaud, T.; Deschamps, A.; Bley, F.; Lefebvre, W.; Baroux, B. Influence of alloy composition and heat treatment on precipitate composition in Al-Zn-Mg-Cu alloys. *Acta Mater.* **2010**, *58*, 248–260. [CrossRef]

10. Sharma, M.M.; Amateau, M.F.; Eden, T.J. Aging response of Al-Zn-Mg-Cu spray formed alloys and their metal matrix composites. *Mater. Sci. Eng. A* **2006**, *424*, 87–96. [CrossRef]

11. Dong, J.; Cui, J.Z.; Yu, F.X.; Zhao, Z.H.; Zhuo, Y.B. A new way to cast high-alloyed Al-Zn-Mg-Cu-Zr for super-high strength and toughness. *J. Mater. Process. Technol.* **2006**, *171*, 399–404. [CrossRef]

12. Roy, S.B.Y.; Nataraj, B.R.; Suwas, S.; Kumar, S.; Chattopadhyay, K. Accumulative roll bonding of aluminum alloys 2219/5086 laminates: Microstructural evolution and tensile properties. *Mater. Des.* **2012**, *36*, 529–539. [CrossRef]

13. Shaeri, M.H.; Salehi, M.T.; Seyyedein, S.H.; Abutalebi, M.R.; Park, J.K. Microstructure and mechanical properties of Al-7075 alloy processed by equal channel angular pressing combined with aging treatment. *Mater. Des.* **2014**, *57*, 250–257. [CrossRef]

14. Sun, N.; Apelian, D. Friction stir processing of aluminum cast alloys for high performance applications. *J. Miner. Metal. Mater. Soc.* **2011**, *63*, 44–50. [CrossRef]

15. Naeem, H.T.; Mohammed, K.S.; Ahmad, K.R. Effect of friction stir processing on the microstructure and hardness of an aluminum–zinc–magnesium–copper alloy with nickel additives. *Strength Plast.* **2015**, *116*, 1035–1046. [CrossRef]

16. Sharma, M.M.; Amateau, M.F.; Eden, T.J. Hardening mechanisms of spray formed Al-Zn-Mg-Cu alloys with scandium and other elemental additions. *J. Alloys Compd.* **2006**, *416*, 135–142. [CrossRef]

17. Dorward, R.C.; Beerntsen, D.J. Grain structure and quench-rate effects on strength and toughness of AA7050 Al-Zn-Mg-Cu-Zr alloy plate. *Metall. Mater. Trans. A* **1995**, *26*, 2481–2484. [CrossRef]

18. Sharma, M.M.; Amateau, M.F.; Eden, T.J. Mesoscopic structure control of spray formed high strength Al-Zn-Mg-Cu alloys. *Acta Mater.* **2005**, *53*, 2919–2924. [CrossRef]

19. Jia, Y.D.; Cao, F.Y.; Ning, Z.L.; Guo, S.; Ma, P.; Sun, J.F. Influence of second phases on mechanical properties of spray-deposited Al-Zn-Mg-Cu alloy. *Mater. Des.* **2012**, *40*, 536–540. [CrossRef]

20. Liu, S.D.; Zhang, X.M.; Chen, X.M.; You, J.H.; Zhang, X.Y. Effect of Zr content on quench sensitivity of AlZnMgCu alloys. *Trans. Nonferr. Met. Soc. China* **2007**, *17*, 787–792. [CrossRef]

21. Cai, Y.H.; Liang, R.G.; Su, Z.P.; Zhang, J.S. Microstructure of spray formed Al-Zn-Mg-Cu alloy with Mn addition. *Trans. Nonferr. Met. Soc. China* **2011**, *21*, 9–14. [CrossRef]

22. Li, F.X.; Liu, Y.Z.; Jiang, Y.; Luo, X. Effect of processing parameters on the relative density of spray rolling 7050 aluminum alloy strip. *Int. J. Adv. Manuf. Technol.* **2013**, *67*, 2771–2778. [CrossRef]

23. Yu, H.C.; Wang, M.P.; Sheng, X.F.; Li, Z.; Chen, L.B.; Lei, Q.; Chen, C.; Jia, Y.L.; Xiao, Z.; Chen, W.; et al. Microstructure and tensile properties of large-size 7055 aluminum billets fabricated by spray forming rapid solidification technology. *J. Alloys Compd.* **2013**, *578*, 208–214. [CrossRef]

24. Sellars, C.M.; McTegart, W.J. On the mechanism of hot deformation. *Acta Metall.* **1966**, *14*, 1136–1138. [CrossRef]

25. Medina, S.F.; Hernandez, C.A. General expression of the Zener-Hollomon parameter as a function of the chemical composition of low alloy and microalloyed steels. *Acta Mater.* **1996**, *44*, 137–148. [CrossRef]

26. Kolb, G.K.H.; Scheiber, S.; Antrekowitsch, H.; Uggowitzer, P.J.U.; Pöschmann, D.; Pogatscher, S. Differential scanning calorimetry and thermodynamic predictions—A comparative study of Al-Zn-Mg-Cu alloys. *Metals* **2016**, *6*, 180. [CrossRef]

27. Wu, H.; Wen, S.P.; Huang, H.; Gao, K.Y.; Wu, L.X.; Wang, W.; Nie, Z.R. Hot deformation behavior and processing map of a new type Al-Zn-Mg-Er-Zr alloy. *J. Alloys Compd.* **2016**, *685*, 869–880. [CrossRef]

28. Wang, Z.A.; Wang, M.P.; Yang, W.C.; Zhang, Q. Ageing precipitation and hardening behavior of 1973 high strength and high toughness aluminum alloy. *Chin. J. Nonferr. Met.* **2011**, *21*, 522–528.

29. Yu, H.C. Microstructure and Mechanical Properties of Spray-Deposition 7055 High Strength Aluminum Alloys. Ph.D. Thesis, Central South University, Changsha, China, May 2015. (In Chinese)

30. Liu, B.; Wang, M.P.; Lei, Q.; Daun, Y.L.; Liu, L.X.; Yu, H.C. Microstructure and properties of Al-Zn-Mg-Cu-Zr alloy prepared by spray deposition method. *Chin. J. Nonferr. Met.* **2015**, *25*, 1773–1780.

31. Totten, G.E.; MacKenzie, D.S. *Handbook of Aluminum*; Marcel Dekker, Inc.: New York, NY, USA, 2003; Volume 1, pp. 287–288, ISBN:0824704940.

32. Mabuchi, M.; Higashi, K. Strengthening mechanism of Mg-Si alloys. *Acta Mater.* **1996**, *44*, 4611–4618. [CrossRef]

33. Lee, J.S.; Jung, J.Y.; Lee, E.K.; Park, W.J.; Ahn, S.; Nack, J.K. Microstructure and properties of titanium boride dispersed Cu alloys fabricated by spray forming. *Mater. Sci. Eng. A* **2000**, *277*, 274–283. [CrossRef]

34. Zhou, Y.J.; Song, K.X.; Xing, J.D.; Zhang, Y.M. Precipitation behavior and properties of aged Cu-0.23Be-0.84Co alloy. *J. Alloys Compd.* **2016**, *658*, 920–930. [CrossRef]

35. Yang, G.; Li, Z.; Yuan, Y.; Qian, L. Microstructure, mechanical properties and electrical conductivity of Cu-0.3Mg-0.05Ce alloy processed by equal channel angular pressing and subsequent annealing. *J. Alloys Compd.* **2015**, *640*, 347–354. [CrossRef]

36. Zheng, Z.Q. *Fundamentals of Materials Science*; Central South University Press: Changsha, China, 2005; pp. 302–306. ISBN 978-7-5487-0948-0. (In Chinese)

Article

Effect of Fe-Content on the Mechanical Properties of Recycled Al Alloys during Hot Compression

Hongzhou Lu [1], Zeran Hou [2], Mingtu Ma [3] and Guimin Lu [1,*]

[1] School of Resources and Environmental Engineering, East China University of Science and Technology, Meilong Road 130, Xuhui District, Shanghai 200237, China; hongzhoulu@foxmail.com

[2] State Key Laboratory of Rolling & Automation, Northeastern University, Box 105, No. 11, Lane 3, Wenhua Road, Heping District, Shenyang 110819, China; houzrneu@163.com

[3] China Automotive Engineering Research Institute, No. 9 Jinyu Ave., New North Zone, Chongqing 401122, China; Mingtuma@126.com

* Correspondence: gmlu@ecust.edu.cn; Tel.: +86-021-6425-2065

Received: 10 June 2017; Accepted: 7 July 2017; Published: 10 July 2017

Abstract: It is unavoidable that Fe impurities will be mixed into Al alloys during recycling of automotive aluminum parts, and the Fe content has a significant effect on the mechanical properties of the recycled Al alloys. In this work, hot compression tests of two Fe-containing Al alloys were carried out at elevated temperatures within a wide strain rate range from $0.01\,\mathrm{s}^{-1}$ to $10\,\mathrm{s}^{-1}$. The effect of Fe content on the peak stress of the stress vs. strain curves, strain rate sensitivity and activation energy for dynamic recrystallization are analyzed. Results show that the recycled Al alloy containing 0.5 wt % Fe exhibits higher peak stresses and larger activation energy than the recycled Al alloy containing 0.1 wt % Fe, which results from the fact that there are more dispersed AlMgFeSi and/or AlFeSi precipitates in the recycled Al alloy containing 0.5 wt % Fe as confirmed by SEM observation and energy spectrum analysis. It is also shown that the Fe content has little effect on the strain rate sensitivity of the recycled Al alloys.

Keywords: Al alloy; remanufacturing; hot rolling; activation energy

---

1. Introduction

Energy saving and emission reduction are the motivations of remanufacturing Al alloys recycled from automotive parts. The energy consumption of remanufacturing recycled Al alloys is only about 2.8 kWh/kg [1], compared to 45 kWh/kg in producing Al alloys from pure Al. Additionally, the emission of CO_2 in remanufacturing of recycled Al alloy is only 5% of that when using pure Al to produce Al alloys. As reported by Das [2], there are four challenges in remanufacturing recycled automotive Al alloys: (1) improvement of the recycling efficiency of Al alloy parts; (2) wide application of automatic shredding and sorting technologies; (3) development of new Al alloys which can be directly remanufactured from the recycled automotive Al alloys; and (4) investigation of the effect of impurity elements (e.g., Fe) on the Al alloy properties; where aspect (3) is the most challenging. During the recycling and remanufacturing of automotive Al alloys, it is unavoidable that a certain amount of Fe will be mixed into the original Al alloys [3]. Das et al. [4] and Adam et al. [5] reported that the Fe content in Al alloy sheets remanufactured from recycled Al alloys can reach up to 1.0 wt %.

Fe has been regarded as harmful to the mechanical properties of Al alloys, especially to casting Al alloys. This is mainly due to the fact that the solubility of Fe in Al is very low and Fe exists as brittle AlFe phase or AlFeSi phase in Al alloys. Nevertheless, Fe is a natural impurity to Al alloys, since Fe can be mixed in to Al alloys during electrolytic Al production, metallurgy melting, casting, alloying processes etc. [6]. During automotive recycling, steel parts may be mixed into Al alloys due to non-prompt sorting, which could be the main reason contributing to rich Fe in recycled Al alloys.

At present, many efforts have been made to control or eliminate the impact of Fe-rich phases in Al alloys [7,8]. For example, physical methods, e.g., precipitation method, dilution method, filtration method, centrifugal removal method, electromagnetic removal method etc., have been applied to decrease the Fe content in recycled Al alloys during melting process; other possible approaches are to change the forms of Fe-rich phases in Al alloys to decrease the formation of acicular or lath Fe-rich phases, or to improve the morphologies of Fe-rich phases in Al alloys to reduce the harmful effects of Fe impurities. As an extensive process that is mainly employed in the recycling of vehicles, it is difficult to further reduce the Fe content in remanufactured Al alloys. In addition, the current Al alloy purification methods are difficult to apply in industry. Matsubar et al. [9] decreased the Fe content from 2.07 wt % to 0.27 wt % in Al alloys by using the centrifugal removal method in the laboratory. However, this method is complicated and difficult to control. The electromagnetic removal method was also applied in decreasing the Fe content from 1.13 wt % to 0.41 wt % in Al alloys [10], but it exhibited low cost performance.

The effect of Fe impurities on the deformation behavior as well as mechanical properties of recycled Al alloys will be studied in this work. Hot compression tests at various strain rates and elevated temperatures are performed to acquire stress vs. strain curves and therefore analyze the effect of Fe content on strain rate sensitivity and activation energy of the recycled Al alloys, which are helpful to design rolling process for recycled Al alloys containing Fe.

2. Materials and Methods

Table 1 lists the chemical compositions of the two recycled Al alloys containing 0.1 wt % Fe and 0.5 wt % Fe, respectively. Casting ingots were prepared first for the two recycled Al alloys. After homogenizing treatment, cylindrical specimens with a diameter of 10 mm and a length of 15 mm were machined from each ingot. For all specimens, a groove with a depth of 0.2 mm and a diameter of 9 mm was machined at each end to contain a certain of lubricant composed of 75% graphite, 20% oil and 5% trimethyl nitrate, which minimized the friction effect on measuring flow stress during hot compression. The detailed geometries of the specimen are illustrated in Figure 1a.

Table 1. The chemical composition of the recycled Al alloys with different contents of Fe (unit in wt %).

Si	Mg	Fe	Mn	Al
1.05	0.54	0.10	0.069	Balance
0.97	0.56	0.50	0.071	Balance

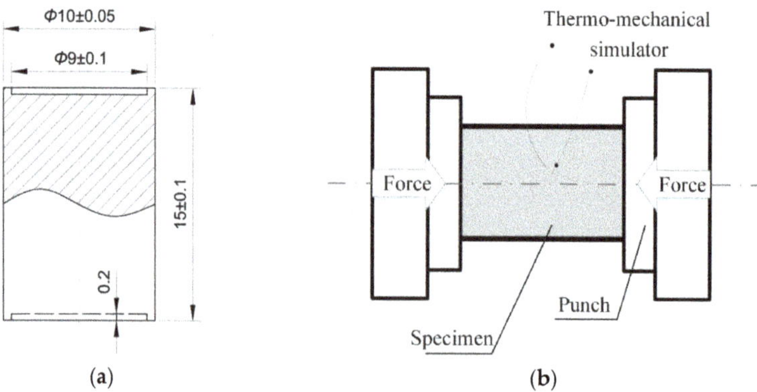

Figure 1. (a) Geometries of the cylindrical specimen for hot compression tests (units in mm) and (b) and illustration of hot compression in the MMS300 thermo-mechanical simulator.

Hot compression tests were conducted on a MMS300 thermo-mechanical simulator (Northeastern University, Shenyang, China), which is capable of automatically recording compressive true stress, true strain, etc. Temperatures were measured by a thermocouple as shown in Figure 1b and controlled by a feed-back program. Specimens were heated by applying direct current, namely, resistance heating, where the heating rate was set at 10 K/s. Each specimen was soaked at the preset deformation temperature for 3 min to make the temperature of the tested specimen uniform. The deformation temperatures were 673 K, 723 K, 773 K and 823 K. At each temperature, four groups of specimens were tested at nominal strain rates ($\dot{\varepsilon}_N$) of 0.01 s^{-1}, 0.1 s^{-1}, 1 s^{-1} and 10 s^{-1}. Each test was stopped when the specimen was compressed to a strain level of 0.6, and then the specimen was water quenched to room temperature. The temperature history of each specimen is plotted in Figure 2.

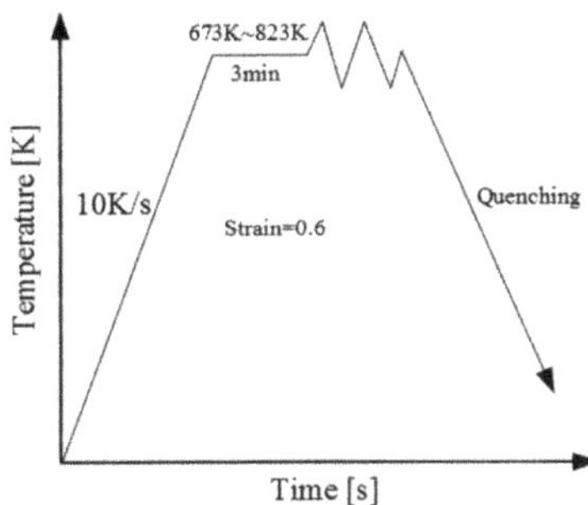

Figure 2. A schematic diagram of the temperature history in hot compression test.

3. Results

3.1. Flow Curves

The compressive true stress vs. true strain curves of the 0.1 wt % Fe and 0.5 wt % Fe containing Al alloys are presented in Figures 3 and 4, respectively. For both Al alloys, most curves indicate that typical dynamic recrystallization, especially discontinuous dynamic recrystallization (DDRX), occurred during hot compression, which is evident from work hardening occurred at the initial deformation stage firstly, followed by a peak stress, then the flow stress decreases due to recrystallization. After recrystallization, the newly formed grain work hardens, and recrystallization occurs again, which is repeated as deformation proceeds and results in wavy true stress vs. true strain curves in later stages. Beyond the peak stress, the overall trend is that the flow stress decreases as the true strain increases, which could result from dynamic recovery, as dynamic recovery is always regarded to be accompanied by dynamic recrystallization during hot deformation of metals. Both dynamic recrystallization and dynamic recovery are softening mechanisms during hot deformation. At the larger $\dot{\varepsilon}_N$, i.e., 10 s^{-1} (Figures 3d and 4d), the phenomenon of DDRX, especially at lower temperatures (i.e., 673 K and 723 K), is not as obvious as that observed from the true stress vs. true strain curves tested under other conditions, and continuous dynamic recrystallization (CDRX) and dynamic recovery may dominate the softening mechanism in those specimens. This may be due to the fact that time was limited for DDRX during hot compression at $\dot{\varepsilon}_N$ = 10 s^{-1}, e.g., there was only 0.06 s when the specimen was deformed to a strain of 0.6, in particular, at lower deformation temperatures. A similar phenomenon was noted by Haghdadi et al. [11] when they investigated the DRX phenomenon in the ferrite phase of austenite/ferrite steel.

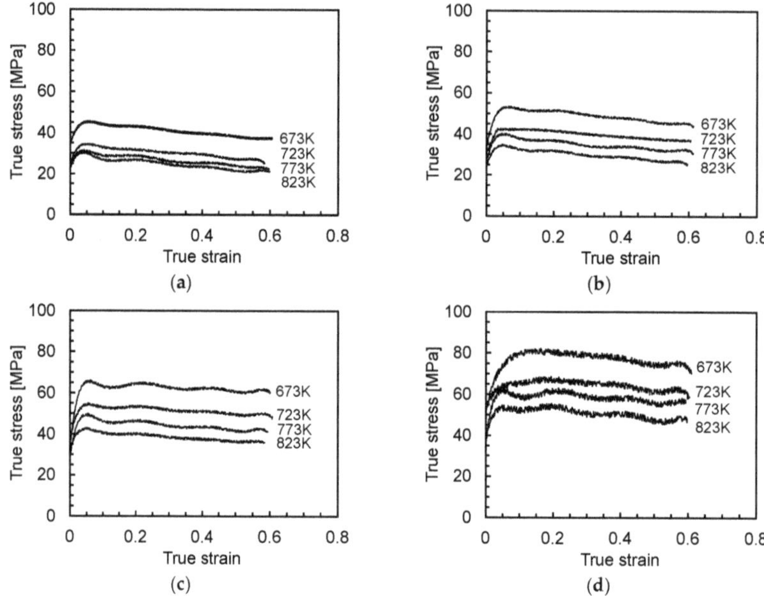

Figure 3. Compressive true stress vs. true strain curves of the recycled Al alloy containing 0.1 wt % Fe at nominal strain rates (**a**) $\dot{\varepsilon}_N = 0.01\ \mathrm{s}^{-1}$; (**b**) $\dot{\varepsilon}_N = 0.1\ \mathrm{s}^{-1}$; (**c**) $\dot{\varepsilon}_N = 1\ \mathrm{s}^{-1}$; and (**d**) $\dot{\varepsilon}_N = 10\ \mathrm{s}^{-1}$.

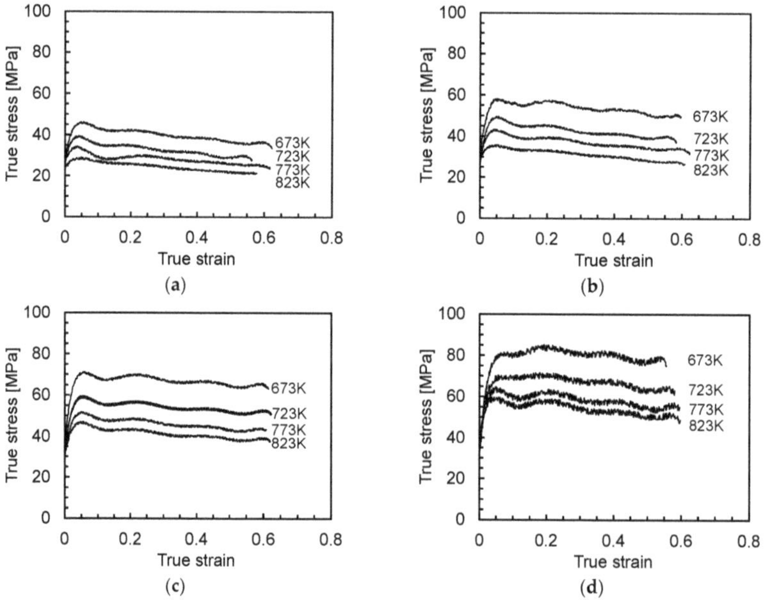

Figure 4. Compressive true stress vs. true strain curves of the recycled Al alloy containing 0.5 wt % Fe at nominal strain rates (**a**) $\dot{\varepsilon}_N = 0.01\ \mathrm{s}^{-1}$; (**b**) $\dot{\varepsilon}_N = 0.1\ \mathrm{s}^{-1}$; (**c**) $\dot{\varepsilon}_N = 1\ \mathrm{s}^{-1}$; and (**d**) $\dot{\varepsilon}_N = 10\ \mathrm{s}^{-1}$.

3.2. Strain Rate Sensitivity

The peak stresses (σ_P) of the Fe-containing recycled Al alloys are summarized in Figure 5a, where the data of the 0.1 wt % Fe containing Al alloy and the 0.5 wt % Fe containing Al alloy are indicated by solid and dashed curves, respectively. It is observed that the Al alloy containing 0.5 wt % Fe exhibited greater σ_P than that containing 0.1 wt % Fe at all testing conditions except the test at 823 K and 0.01 s^{-1}. At all tested temperatures, σ_P of both Fe-containing Al alloys increases with increasing strain rate; with increasing temperature, σ_P decreases. The approach following Min et al. [12] and Sung et al. [13], as expressed by Equation (1), was used to evaluate the strain rate sensitivity of σ_P.

$$M = \frac{\Delta \ln(\sigma_P)}{\Delta \ln(\dot{\varepsilon}_N)} \tag{1}$$

where M is the strain rate sensitivity exponent. A difference method was here applied to calculate the average M at each temperature, $\overline{M}(T)$, by considering the σ_P values at four strain rates from 0.01 s^{-1} to 10 s^{-1}.

$$\overline{M}(T) = \frac{\ln\left(\sigma_{P(10)}\right) + \ln(\sigma_{P(1)}) + \ln(\sigma_{P(0.1)}) + \ln(\sigma_{P(0.01)})}{2\ln(100)}, \tag{2}$$

where $\sigma_{P(0.01)}$, $\sigma_{P(0.1)}$, $\sigma_{P(1)}$ and $\sigma_{P(10)}$ are the peak stresses at strain rates of 0.01 s^{-1}, 0.1 s^{-1}, 1 s^{-1} and 10 s^{-1}, respectively. $\overline{M}$ values of both Al alloys are presented in Figure 5b as a function of temperature. Although the 0.5 wt % Fe containing Al alloy exhibits greater σ_P, Fe content does not have an obvious effect on $\overline{M}$. Both Al alloys exhibit nearly identical $\overline{M}$ at 673 K. For the 0.1 wt % Fe containing Al alloy, $\overline{M}$ increases from 8.68×10^{-2} s^{-2} to 1.01×10^{-1} s^{-2} when T increases from 673 K to 723 K; beyond 723 K, $\overline{M}$ decreases as T increases. M of the 0.5 wt % Fe containing Al alloy exhibits a different behavior as the temperature increases, e.g., $\overline{M}$ changes little as T increases from 673 K to 773 K, and increases to 1.03×10^{-1} s^{-2} at $T = 823$ K.

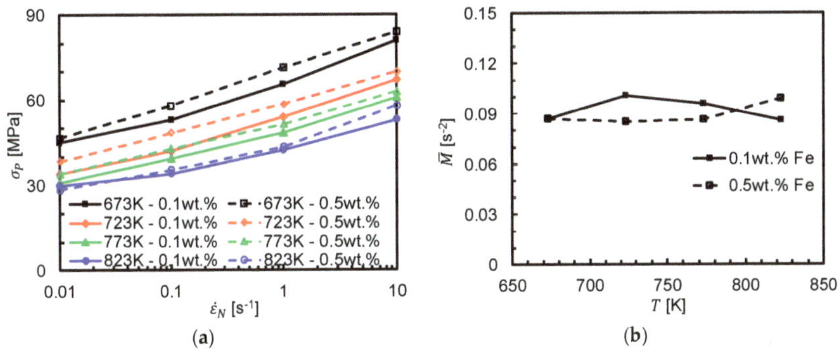

Figure 5. (a) Peak stress (σ_P) vs. nominal strain rate ($\dot{\varepsilon}_N$) curves and (b) the average strain rate sensitivity exponent ($\overline{M}$, Equation (2)) of both recycled Al alloys.

3.3. Activation Energy for Dynamic Recrystallization and Peak Stress Model

The peak stress is of significance to the design of rolling processes and predicting the maximum rolling forces. The peak stress (σ_P) is the result of dynamic balance between work hardening and softening mechanisms (e.g., dynamic recrystallization) of the material during hot deformation. Dynamic recrystallization is regarded to be related to the activation energy (Q). In this section, Q will be calculated for the Fe-containing recycled Al alloys, and a model will be deduced to predict σ_P as a function of strain rate ($\dot{\varepsilon}_N$) and deformation temperature (T).

3.3.1. Governing Equations

According to Jonas et al. [14], Huang et al. [15], and Shi et al. [16], the strain rate ($\dot{\varepsilon}$) can be expressed as a function of flow stress (σ) by Equation (3)

$$\dot{\varepsilon} = A_1 \sigma^m, \tag{3}$$

or by Equation (4).

$$\dot{\varepsilon} = A_2 \exp(\beta\sigma), \tag{4}$$

where A_1, m, A_2, and β are material parameters non-related to temperature. To describe the relationship between $\dot{\varepsilon}$ and σ in a wider temperature range, Zener et al. [17] proposed a temperature-compensated strain rate factor as there is a thermal activation process during hot deformation, i.e., the Zener-Hollomon parameter Z, which is expressed as

$$Z = \dot{\varepsilon} \exp(Q/RT), \tag{5}$$

where A and n are material parameters, and R is the air constant (8.314 J·mol^{-1}·K^{-1}). In addition, Jonas et al. [14] proposed a hyperbolic sine function, Equation (6), which includes Q and T compared to Equations (3) and (4), to describe the relationship between $\dot{\varepsilon}$ and σ.

$$\dot{\varepsilon} \exp(Q/RT) = A[\sinh(\alpha\sigma)]^n, \tag{6}$$

where

$$\alpha = \beta/m. \tag{7}$$

As a result, the flow stress (σ) can be expressed as a function of strain rate ($\dot{\varepsilon}$) and temperature (T) by Equation (8)

$$\sigma = \frac{1}{\alpha}\sinh^{-1}\left[\frac{\dot{\varepsilon}}{A}\exp(Q/RT)\right]^{\frac{1}{n}}. \tag{8}$$

3.3.2. Identification of Parameters

As the peak stress (σ_P) is an important parameter to the rolling process, σ and $\dot{\varepsilon}$ in Equation (8) will be replaced by σ_P and $\dot{\varepsilon}_N$, respectively. Therefore, σ_P is expressed as a function of $\dot{\varepsilon}_N$ and T by following Equation (8). The material parameters in Equation (8) will be identified for the investigated Fe-containing recycled Al alloys.

According to Jonas et al. [14], Huang et al. [15], and Shi et al. [16], it is assumed that the relationship between σ_P and $\dot{\varepsilon}_N$ at temperatures 673–723 K and 773–823 K follows Equations (4) and (3), respectively. After applying logarithmic function to both sides of Equations (3) and (4), we have

$$\ln\dot{\varepsilon}_N = \ln A_1 + m\ln\sigma_P, \tag{9}$$

$$\ln\dot{\varepsilon}_N = \ln A_2 + \beta\sigma_P. \tag{10}$$

By performing linear regression analysis, β and m are calculated as the average slopes of the $\ln\dot{\varepsilon}_N$ vs. $\ln\sigma_P$ data at 673–723 K and 773–823 K as presented in Figures S1 and S5, respectively. Then α is calculated by Equation (7). By applying logarithm function to both sides of Equation (6), we have

$$\ln\dot{\varepsilon}_N = n\ln[\sinh(\alpha\sigma_P)] + \ln A - Q/RT, \tag{11}$$

or

$$\ln[\sinh(\alpha\sigma_P)] = \frac{Q}{nR}T + \frac{\ln\dot{\varepsilon}_N - \ln A}{n}. \tag{12}$$

Then n and Q/nR can be obtained as the average slope from linear regression analysis on the $\ln \dot{\varepsilon}_N$ vs. $\ln[\sin h(\alpha\sigma_P)]$ and $\ln[\sin h(\alpha\sigma_P)]$ vs. $1/T$ data at the four temperatures, respectively, which refer to Figures S2, S6, S3 and S7. Consequently, the activation energy Q can be solved. By noting Equation (5), Equation (11) can be rewritten as

$$\ln[\sin h(\alpha\sigma_P)] = \frac{1}{n}\ln Z - \frac{1}{n}\ln A. \tag{13}$$

By substituting the known α, n, Q, $\dot{\varepsilon}_N$, T and σ_P data at all test conditions into Equation (13), we can plot $\ln[\sin h(\alpha\sigma_P)]$ vs. $\ln Z$ as Figures S4 and S8. Apparently, $\ln[\sin h(\alpha\sigma_P)]$ has a linear relationship with $\ln Z$, and the slope of the fitted line is very close to $1/n$. Thus, A can be solved from the intercept, i.e., $-\frac{1}{n}\ln A$.

To this end, not only the activation energies for dynamic recrystallization of the Fe-containing recycled Al alloys are calculated, but also all material parameters in the model predicting σ_P (refer to Equations (14) and (15)) are identified, which are listed in Table 2.

$$\sigma_P|_{0.1Fe} = 56.14 \sin h^{-1}\left[2.817 \times 10^{-10}\dot{\varepsilon}_N \exp\left(\frac{1.690 \times 10^4}{T}\right)\right]^{0.1142}, \tag{14}$$

$$\sigma_P|_{0.5Fe} = 46.95 \sin h^{-1}\left[4.161 \times 10^{-10}\dot{\varepsilon}_N \exp\left(\frac{1.858 \times 10^4}{T}\right)\right]^{0.1245}. \tag{15}$$

Based on Equations (14) and (15), the predicted peak stresses ($\sigma_{P(pre)}$) and experimentally measured peak stresses ($\sigma_{P(exp)}$) are listed and compared in Table 3. It is seen that the maximum error (δ) defined by $\frac{\sigma_{P(pre)} - \sigma_{P(exp)}}{\sigma_{P(exp)}} \times 100\%$ occurs at $\dot{\varepsilon}_N = 0.01$ s^{-1} and $T = 823$ K, which is -9.87%, while most of the errors are smaller than 5%.

Table 2. Material parameters of the Fe-containing recycled Al alloys.

Al Alloys	m	β	α	n	Q [kJ mol^{-1}]	A
0.1 Fe [1]	11.02	0.1963	0.01781	8.760	140.5	3.550×10^9
0.5 Fe [1]	9.452	0.2013	0.02130	8.030	154.5	2.569×10^9

[1] "0.1 Fe" and "0.5 Fe" indicate the recycled Al alloys containing 0.1 wt % Fe and 0.5 wt % Fe, respectively.

Table 3. Comparison between the predicted peak stresses ($\sigma_{P(pre)}$) and experimental peak stresses ($\sigma_{P(exp)}$) at the investigated strain rates ($\dot{\varepsilon}_N$) of the two Fe-containing Al alloys.

Alloys	$\dot{\varepsilon}_N$ [s^{-1}]	T [K]	$\sigma_{P(exp)}$ [MPa]	$\sigma_{P(pre)}$ [MPa]	δ [%]
		673	45.0	43.0	−4.38%
	0.01	723	34.0	36.3	6.69%
		773	31.0	31.1	0.26%
		823	30.0	27.0	−9.87%
		673	53.0	53.3	0.48%
	0.1	723	42.0	45.4	8.13%
		773	39.5	39.2	−0.64%
0.1 Fe		823	34.0	34.4	1.05%
		673	65.5	64.8	−1.11%
	1	723	54.0	56.0	3.66%
		773	48.5	48.9	0.80%
		823	42.5	43.2	1.53%
		673	81.0	77.3	−4.53%
	10	723	67.0	67.8	1.16%
		773	61.0	59.9	−1.80%
		823	53.0	53.4	0.75%

Table 3. *Cont.*

Alloys	$\dot{\varepsilon}_N$ [s^{-1}]	T [K]	$\sigma_{P(exp)}$ [MPa]	$\sigma_{P(pre)}$ [MPa]	δ [%]
		673	46.5	47.8	2.90%
	0.01	723	38.5	39.8	3.31%
		773	34.0	33.6	−1.29%
		823	28.5	28.7	0.86%
		673	58.0	58.8	1.37%
	0.1	723	48.5	49.6	2.33%
		773	43.0	42.4	−1.46%
0.5 Fe		823	35.5	36.6	3.13%
		673	71.5	70.8	−0.94%
	1	723	58.5	60.8	3.90%
		773	51.5	52.6	2.16%
		823	43.5	46.0	5.65%
		673	84.0	83.7	−0.39%
	10	723	70.0	73.0	4.25%
		773	63.0	64.1	1.71%
		823	58.0	56.7	−2.28%

4. Discussion

Although the effects of temperature and strain rate on the flow stress of commercialized Al alloys have been extensively reported, the effect of Fe content on hot deformation behavior has rarely been investigated. From Figure 5a, increasing the Fe content in the recycled Al alloy results in an increase of the peak stress, which suggests that the recycled Al alloy containing higher Fe content is more difficult to deform. Nevertheless, the strain rate sensitivity is nearly independent on the Fe content, as compared in Figure 5b.

Before reaching the peak stress, the flow stress increases rapidly first, and then more slowly as the strain increases, which is different from the continuous work hardening of most metals deformed at room temperature. In this stage, dislocation accumulation is the main hardening mechanism and the main softening mechanism is dynamic recovery, which is due to dislocation annihilation and/or dislocation rearrangement. During the competition of work hardening and softening, dislocation cells and walls were formed. When the flow stress reached the peak stress, the grains were elongated and recrystallization occurred. The grains nucleated preferably at the original grain boundaries, and the newly formed sub-grain boundaries evident by dislocation cells/walls. The recycled Al alloy containing 0.5 wt % Fe requires larger activation energy for dynamic recrystallization of the alloy with a lower Fe content, which may be due to the fact that the Al alloy containing a higher Fe content exhibits larger flow stress. This result is consistent with the report from Jeniski et al. [18]: Fe-containing dispersed precipitates can increase the resistance for dislocation movements and dynamic recrystallization. It is noted that the activation energy of recycled Al alloy containing 0.1 wt % Fe is very close to the activation energy for bulk self-diffusion of pure aluminum, 142 kJ·mol^{-1}. However, the activation energies of both investigated Al alloys are much lower than those of both AA2026 [19] and AA7150 [20], but close to that of cast A356 Al alloy [21], during hot compression. Figure 6 presents the SEM images of the as-cast recycled Al alloys. There are two types of precipitates in both alloys: rod precipitates (denoted by solid arrows) and round precipitates (denoted by hollow arrows). According to energy spectrum analysis, the rod precipitates are either AlMgFeSi or AlFeSi phases, while the round precipitates are mainly Mg_2Si phase. By applying the Image-Pro plus 6.0 software, the statistical analysis results of the precipitates having a size range from 1 μm to 200 μm shown in Figure 6 are listed in Table 4. Note that the precipitates smaller than 1 μm are ignored here. It is seen that the maximum diameter and average diameter of the precipitates in both recycled Al alloys are similar, while the number density and the area density of the precipitates in the 0.5 wt % Fe containing Al alloy are two times that in the 0.1 wt % Fe containing Al alloy. As there are more dispersed precipitates in the 0.5 wt % Fe containing Al alloy and the deformation of the dispersed precipitates is also dependent on the Fe content, the resistance of dynamic recrystallization is larger, and the required activation energy is higher in the 0.5 wt % Fe containing Al alloy than in the 0.1 wt % Fe containing Al alloy.

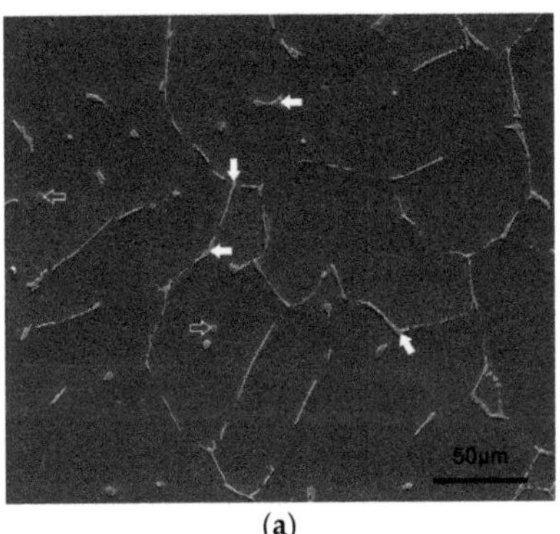

(a)

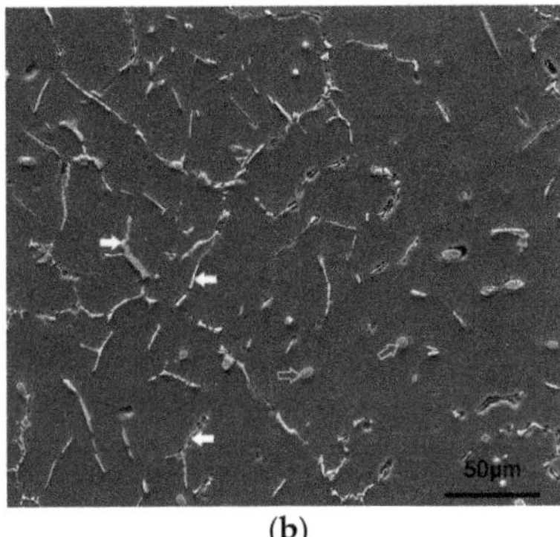

(b)

Figure 6. SEM images showing constituent particles in the as-cast recycled Al alloys containing (**a**) 0.1 wt % Fe and (**b**) 0.5 wt % Fe.

Table 4. Statistical analysis result of the precipitates limited to a size range from 1 µm to 200 µm.

Fe Content (wt %)	0.1	0.5
Numbers of lager intermetallic particles (1–200 µm) in Figure 6 [mm^{-2}]	2068	4826
Area [%]	1.839	3.706
Max diameter [µm]	18.20	18.04
Average diameter [µm]	3.425	3.339

It is of significance to control the microstructure of Fe-containing Al alloys before annealing and cold rolling. Lee et al. [22] discussed how the microstructure and texture after hot rolling affected the microstructure, texture and surface roughness of the cold rolled sheet. As the Fe-containing phases can be dissolved in the Al matrix during hot rolling, breaking the relatively larger Fe-containing phases by hot rolling becomes more important. The recycled Al alloy containing higher Fe content exhibits larger peak stress and activation energy than that containing lower Fe content; therefore, the recycled Al alloy containing a higher Fe content needs larger forming force and/or rolling reduction in the hot rolling process, which is beneficial to breaking the larger Fe-containing dispersed phases and can contribute to the higher strength and greater ductility of the cold rolled recycled Al alloy sheet containing higher Fe content. The effect of the Fe content on the mechanical properties of the cold rolled recycled Al alloy sheet will be reported elsewhere in future.

5. Conclusions

Casting ingots of recycled Al alloys containing 0.1 wt % Fe and 0.5 wt % Fe were prepared for hot compression tests in a temperature range from 673 K to 823 K and a wide strain rate range from 0.01 s^{-1} to 10 s^{-1}. Both Al alloys exhibited dynamic recrystallization during hot compression, which is evident from the wavy true stress vs. true strain curves with a decreasing trend after reaching the peak. A model is established for the peak stress and the parameters in the model have been identified. A higher Fe content leads to a larger peak stress in the recycled Al alloy, which results from the following two aspects: (1) more/larger Fe-containing dispersed phases in the Al alloy containing higher Fe content; and (2) the deformation of Fe-containing dispersed phase is more difficult to deform at higher Fe content. These two aspects also increase the resistance of dislocation movement and dynamic recrystallization, and therefore, the 0.5 wt % Fe containing Al alloy requires greater activation energy (154.5 kJ·mol^{-1}) than the 0.1 wt % Fe containing Al alloy (140.5 kJ·mol^{-1}). As a result, the Al alloy containing higher Fe content needs larger rolling force and/or rolling reduction in hot rolling, which may help to break the Fe-containing dispersed phases and consequently to increase the strength and ductility of the cold rolled Fe-containing recycled Al alloy sheets. The mechanical properties of the cold rolled recycled Al alloy sheets containing different Fe contents will be published elsewhere.

Supplementary Materials: The following are available online at http://www.mdpi.com/2075-4701/7/7/262/s1, Figure S1: (a) ln $\dot{\varepsilon}_N$ vs. σ_P curves at 673 K and 723 K and (b) ln $\dot{\varepsilon}_N$ vs. ln σ_P curves at 773 K and 823 K of the recycled Al alloy containing 0.1 wt % Fe, Figure S2: ln $\dot{\varepsilon}_N$ vs. ln[sin h($\alpha\sigma_P$)] curves of the recycled Al alloy containing 0.1 wt % Fe, Figure S3: ln[sin h($\alpha\sigma_P$)] vs. $1/T$ curves of the recycled Al alloy containing 0.1 wt % Fe, Figure S4: Correlation between ln[sin h($\alpha\sigma_P$)] and ln Z of the recycled Al alloy containing 0.1 wt % Fe, Figure S5: (a) ln $\dot{\varepsilon}_N$ vs. σ_P curves at 673 K and 723 K and (b) ln $\dot{\varepsilon}_N$ vs. ln σ_P curves at 773 K and 823 K of the recycled Al alloy containing 0.5 wt % Fe, Figure S6: ln $\dot{\varepsilon}_N$ vs. ln[sin h($\alpha\sigma_P$)] curves of the recycled Al alloy containing 0.5 wt % Fe, Figure S7: ln[sin h($\alpha\sigma_P$)] vs. $1/T$ curves of the recycled Al alloy containing 0.5 wt % Fe, Figure S8: correlation between ln[sin h($\alpha\sigma_P$)] and ln Z of the recycled Al alloy containing 0.5 wt % Fe.

Acknowledgments: The authors would like to acknowledge the financial contribution from the National Science & Technology Pillar Program during the 12th Five-year Plan Period (grant No. 2011BAG03B00). The authors also thank the reviewers of Metals for the suggestive comments which improve the technical content of this paper.

Author Contributions: Guimin Lu, Hongzhou Lu and Mingtu Ma conceived and designed the experiments; Zeran Hou and Hongzhou Lu performed the experiments; Hongzhou Lu analyzed the data; Hongzhou Lu wrote the paper.

Conflicts of Interest: The authors declare no conflict of interest.

References

1. Choate, W.T.; Green, J.A.S. Modeling the impact of secondary recovery (recycling) on the U.S. Aluminum supply and nominal energy requirements. In *Light Metals*, 2004th ed.; Tabereaux, A.T., Ed.; John Wiley and Sons Ltd.: Warrendale, PA, USA, 2004; pp. 913–918.

2. Das, S.K. Designing aluminium alloys for a recycling friendly world. *Mater. Sci. Forum* **2006**, *519–521*, 1239–1244. [CrossRef]

3. Sarkar, J.; Kutty, T.R.G.; Conlon, K.T.; Wilkinson, D.S.; Embury, J.D.; Lloyd, D.J. Tensile and bending properties of AA5754 aluminum alloys. *Mater. Sci. Eng. A* **2001**, *316*, 52–59. [CrossRef]

4. Das, S.K.; Green, J.A.S.; Kaufman, J.G. The development of recycle-friendly automotive aluminum alloys. *JOM* **2007**, *59*, 47–51. [CrossRef]

5. Gesing, A.; Berry, L.; Dalton, R.; Wolanski, R. Assuring continued recyclability of automotive aluminum alloys: Grouping of wrought alloys by color, X-ray absorption and chemical composition-based sorting. In Proceedings of the TMS 2002 Annual Meeting: Automotive Alloys and Aluminum Sheet and Plate Rolling and Finishing Technology Symposia, Warrendale, PA, USA, 18–21 Feburary 2002; pp. 3–15.

6. Pan, F.; Zhang, D. *Aluminium Alloy and Application*; Chemical Industry Press: Beijing, China, 2006. (In Chinese)

7. Staley, J.T.; Lege, D.J. Advances in aluminium alloy products for structural applications in transportation. *J. Phys. IV* **1993**, *3*, 179–190. [CrossRef]

8. Shabestari, S.G. The effect of iron and manganese on the formation of intermetallic compounds in aluminum-silicon alloys. *Mater. Sci. Eng. A* **2004**, *383*, 289–298. [CrossRef]

9. Matsubara, H.; Izawa, N.; Nakanishi, M. Macroscopic segregation in Al-11 mass % Si slloy containing 2 mass % Fe solidified under centrifugal force. *J. Jpn. Inst. Light Met.* **1998**, *48*, 93–97. [CrossRef]

10. Joon-Pyou, P.; Sassa, K.; Asai, S. Elimination of iron in molten Al-Si alloys by electromagnetic force. *J. Jpn. Inst. Light Met.* **1995**, *59*, 312–318.

11. Haghdadi, N.; Cizek, P.; Beladi, H.; Hodgson, P.D. A novel high-strain-rate ferrite dynamic softening mechanism facilitated by the interphase in the austenite/ferrite microstructure. *Acta Mater.* **2017**, *126*, 44–57. [CrossRef]

12. Min, J.; Hector, L.G., Jr.; Zhang, L.; Sun, L.; Carsley, J.E.; Lin, J. Plastic instability at elevated temperatures in a trip-assisted steel. *Mater. Des.* **2016**, *95*, 370–386. [CrossRef]

13. Ji, H.S.; Ji, H.K.; Wagoner, R.H. A plastic constitutive equation incorporating strain, strain-rate, and temperature. *Int. J. Plast.* **2010**, *26*, 1746–1771.

14. Jonas, J.J.; Sellars, C.M.; Tegart, W.J.M. Strength and structure under hot-working conditions. *Int. Mater. Rev.* **1969**, *14*, 1–24. [CrossRef]

15. Huang, C.-Q.; Diao, J.-P.; Deng, H.; Li, B.-J.; Hu, X.-H. Microstructure evolution of 6016 aluminum alloy during compression at elevated temperatures by hot rolling emulation. *Trans. Nonferr. Met. Soc. China* **2013**, *23*, 1576–1582. [CrossRef]

16. Shi, H.; Mclaren, A.J.; Sellars, C.M.; Shahani, R.; Bolingbroke, R. Constitutive equations for high temperature flow stress of aluminium alloys. *Mater. Sci. Technol.* **1997**, *13*, 210–216. [CrossRef]

17. Zener, C.; Hollomon, J.H. Effect of strain rate upon plastic flow of steel. *J. Appl. Phys.* **1944**, *15*, 22–32. [CrossRef]

18. Jeniski, R.A.; Thanaboonsombut, B.; Sanders, T.H. The effect of iron and manganese on the recrystallization behavior of hotrolled and solution-heat-treated aluminum alloy 6013. *Metall. Mater. Trans. A* **1996**, *27*, 19–27. [CrossRef]

19. Huang, X.; Zhang, H.; Han, Y.; Wu, W.; Chen, J. Hot deformation behavior of 2026 aluminum alloy during compression at elevated temperature. *Mater. Sci. Eng. A* **2010**, *527*, 485–490. [CrossRef]

20. Jin, N.; Zhang, H.; Han, Y.; Wu, W.; Chen, J. Hot deformation behavior of 7150 aluminum alloy during compression at elevated temperature. *Mater. Charact.* **2009**, *60*, 530–536. [CrossRef]

21. Haghdadi, N.; Zarei-Hanzaki, A.; Abedi, H.R. The flow behavior modeling of cast A356 aluminum alloy at elevated temperatures considering the effect of strain. *Mater. Sci. Eng. A* **2012**, *535*, 252–257. [CrossRef]

22. Lee, K.J.; Woo, K.D. Effect of the hot-rolling microstructure on texture and surface roughening of Al-Mg-Si series aluminum alloy sheets. *Met. Mater. Int.* **2011**, *17*, 689–695. [CrossRef]

Article

Influence of Sludge Particles on the Fatigue Behavior of Al-Si-Cu Secondary Aluminium Casting Alloys

Lorella Ceschini [1], Alessandro Morri [2], Stefania Toschi [2,*], Anton Bjurenstedt [3] and Salem Seifeddine [4]

[1] Department of Civil, Chemical, Environmental and Materials Engineering, University of Bologna, 40136 Bologna, Italy; lorella.ceschini@unibo.it

[2] Department of Industrial Engineering, University of Bologna, 40136 Bologna, Italy; alessandro.morri4@unibo.it

[3] Swerea SWECAST, 55322 Jönköping, Sweden; anton.bjurenstedt@swerea.se

[4] Department of Materials and Manufacturing, School of Engineering, Jönköping University, 55111 Jönköping, Sweden; salem.seifeddine@ju.se

* Correspondence: stefania.toschi3@unibo.it

Received: 22 March 2018; Accepted: 11 April 2018; Published: 14 April 2018

Abstract: Al-Si-Cu alloys are the most widely used materials for high-pressure die casting processes. In such alloys, Fe content is generally high to avoid die soldering issues, but it is considered an impurity since it generates acicular intermetallics (β-Fe) which are detrimental to the mechanical behavior of the alloys. Mn and Cr may act as modifiers, leading to the formation of other Fe-bearing particles which are characterized by less harmful morphologies, and which tend to settle on the bottom of furnaces and crucibles (usually referred to as sludge). This work is aimed at evaluating the influence of sludge intermetallics on the fatigue behavior of A380 Al-Si-Cu alloy. Four alloys were produced by adding different Fe, Mn and Cr contents to A380 alloy; samples were remelted by directional solidification equipment to obtain a fixed secondary dendrite arm spacing (SDAS) value (~10 μm), then subjected to hot isostatic pressing (HIP). Rotating bending fatigue tests showed that, at room temperature, sludge particles play a detrimental role on fatigue behavior of T6 alloys, diminishing fatigue strength. At elevated temperatures (200 °C) and after overaging, the influence of sludge is less relevant, probably due to a softening of the α-Al matrix and a reduction of stress concentration related to Fe-bearing intermetallics.

Keywords: high pressure die casting; Al-Si-Cu alloys; iron; sludge; intermetallics; fatigue behavior

---

1. Introduction

Aluminum casting alloys are frequently used in several industrial fields, especially in the automotive industry, due to their specific properties in minimizing vehicles' weight, and therefore, the level of emissions it produces. In this regard, high-pressure die casting (HPDC) is nowadays a common process used to produce complex parts for cars and motorbikes. For this particular process, Al-Si-Cu alloys are the most commonly used materials. Since they are usually secondary alloys (i.e., produced from scrap materials), their chemical composition can vary considerably, especially the Fe, Mn ad Cr content.

Fe is generally considered an impurity for casting alloys, since it induces the formation of brittle and acicular Fe-based particles (β-Al$_5$FeSi) which exert a detrimental role on the mechanical behavior of castings, acting as stress concentrators [1–5]. On the other hand, in HPDC alloys, high levels of Fe are generally added (~1 wt %); this helps in reducing die-soldering issues [6–8]. Other alloying elements can act as modifiers of undesirable β phases, fostering the formation of intermetallic compounds characterized by their less harmful morphologies [3,4,6,9–11]. Mn, for example, is the most common alloying element to counteract the formation of acicular β phases, while promoting the formation of α particles. A Fe:Mn ratio of less than 2 is usually recommended to avoid the formation of β phases [6,12]. Also, Cr possesses similar qualities if other alloying elements are present [6,13]. Together with Fe and Si, Mn and Cr may induce the generation of primary polygonal intermetallics, called sludge, prior to the formation of dendrites [6]. The generic formula α-$Al_x(Fe,Mn,Cr)_ySi_z$ is commonly accepted and used to describe the stoichiometry of sludge particles. Due to their high specific density, sludge particles generally tend to segregate to the bottom of the melt, settling on the furnace floor [6,13]. Alloy chemical composition, holding temperature and solidification rate may significantly influence amount, size and shape of such intermetallics [6,14–17].

Some models have been developed to predict the formation of sludge. The following empirical equation, for instance, formulated by Gobrecht [14] and Jorstad [15], relates Fe, Mn and Cr contents to the so-called Sludge Factor (SF):

$$SF = (1 \times wt \% \, Fe) + (2 \times wt \% \, Mn) + (3 \times wt \% \, Cr).$$

The SF may thus vary strongly depending on the chemical composition of the alloy; the higher the SF, the higher the amount of sludge present in the microstructure. The effect of sludge particles on the mechanical behavior of Al-Si-Cu alloys is still not well understood. Despite some studies which focused upon the influence of sludge particles on the tensile properties of Al-Si alloys [18–20], little work has been carried out about the relationship between sludge and the fatigue behavior of alloys at room and high temperatures.

The aim of this study is to evaluate the effect of Fe, Mn and Cr on the microstructure and mechanical behavior of A380 (Al-Si-Cu) casting alloy, typically used in HPDC process. In particular, the study is focused on the relationship between Fe, Mn and Cr levels, Fe-rich sludge particles and their effect on fatigue strength of alloys both at room and high temperature, on artificially aged and overaged specimens.

2. Materials and Methods

The reference alloy ("A") is a commercial EN AC-46000, commonly used in HPDC processes. In order to evaluate the effect of SF and sludge particles on the mechanical behavior of such an alloy, four alloys with different Fe, Mn and Cr contents, thus with different SF, were produced (A, B, C and D). Alloys were melted in an electric furnace. The base alloy and master alloys for Fe, Mn and Cr addition were molten at 800 °C; eutectic Si chemical modification by Sr addition (250 ppm) was carried out at 775 °C; the temperature was maintained at this value for 10 min then lowered to 720 °C; the molten alloy was then poured into a permanent die and pre-heated to 250 °C, in order to obtain cylindrical bars. Chemical compositions of the alloys were evaluated by an optical emission spectrometer (OES) SpectroMax. Results of chemical analysis and calculated SF are presented in Table 1.

Table 1. Alloys' chemical compositions measured by OES (wt %, except for Sr* in ppm) and corresponding SF.

Alloy	Si	Cu	Mg	Fe	Mn	Cr	Zn	Ni	Sr*	Al	SF
A	9.30	2.79	0.05	0.74	0.25	0.03	0.96	0.04	220	Bal.	1.3
B	9.40	2.77	0.05	1.17	0.25	0.03	0.91	0.04	270	Bal.	1.8
C	9.23	2.65	0.04	1.29	0.53	0.15	0.86	0.04	250	Bal.	2.8
D	9.30	2.64	0.04	1.59	0.80	0.18	0.80	0.04	260	Bal.	3.7

Die cast bars were then remelted using directional solidification equipment (a detailed description of the equipment can be found in [21]), aiming to produce samples characterized by a fixed secondary dendrite arm spacing (SDAS) value of 10 μm. To this end, the furnace temperature was set at 710 °C, maintained for 30 min, then cooled down by water. The furnace itself was moved upwards at a speed of 3 mm/s, in order to obtain the target SDAS value, thus inducing the required cooling rate. During the process, an Ar gas shield was blown into the machine to protect samples from oxidation. This process yields a lower defect content, since the solidification front pushes oxides towards the top of samples during solidification. Samples were then subjected to hot isostatic pressing (HIP), in order to reduce the internal porosities and their associated detrimental effect on fatigue strength. T6 heat treatment was carried out on HIPped samples as follows: (i) solution treatment at 495 °C for 3 h; (ii) quenching in water at 50 °C; (iii) artificial aging at 210 °C for 2 h.

Microstructural analyses were carried out by Axio Imager optical microscope (OM; Zeiss, Oberkochen, Germany) and Zeiss Evo-50 scanning electron microscope (SEM; Zeiss, Oberkochen, Germany) equipped with an energy dispersive spectroscopy probe (EDS; Oxford INCA 350, Oxfordshire, UK). Optical micrographs were processed by Image-Pro Plus software (Media Cybernetics, Rockville, MD, USA) to collect data on intermetallic particles (IM). Samples for microstructural characterization were prepared according to ASTM E3-01 [22], and etched with a 0.5% HF solution (for SDAS calculation, general microstructure analysis) and 10% H_2SO_4 (for analyses on sludge particles).

Rotating bending fatigue tests were carried out both at room and high temperatures (200 °C). Room temperature tests were carried out on artificially aged samples (T6 state), while high temperature fatigue tests were carried out on overaged samples (hereafter referred to as OA). Overaging treatment—consisting of soaking samples at 200 °C for 48 h—was carried out after T6 treatment and prior to fatigue testing. Fatigue tests were performed according to ISO 1143 (samples size and geometry are reported in Figure 1); data were analyzed using the staircase method, according to UNI 3964 [23], to calculate fatigue strength ($\sigma f_{50\%}$, the stress yielding to the 50% probability of failure); rotating frequency and run out condition were set to 50 Hz and 2×10^6 cycles respectively.

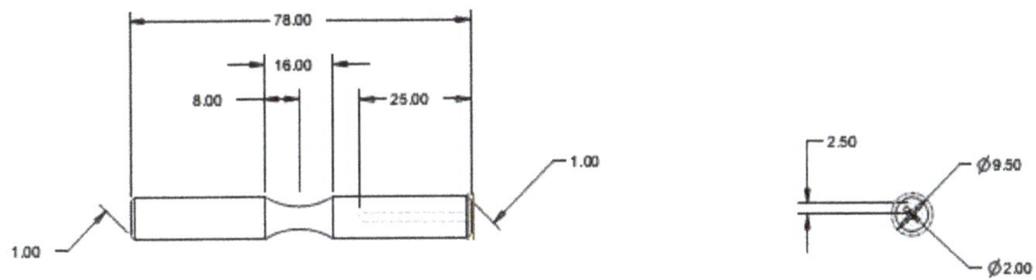

Figure 1. Geometry of fatigue samples; dimensions are expressed in mm.

A stress step of 10 MPa was used. High temperature fatigue tests were carried out by means of a resistance furnace containing the rotating specimens. Samples run for 90 min before applying the load, aiming to reach the steady test temperature (200 °C), as a result of a preliminary thermal control calibration study. The hardness of samples was measured through Brinell tests with a Galileo tester (Officine Galileo, Firenze, Italy) according to ASTME10-08 [24], using a 2.5 mm diameter steel ball and 62.5 kg as a load (HB10 scale). Fracture surfaces of specimens were analyzed by SEM to investigate crack initiation cause and failure mechanisms.

3. Results and Discussion

3.1. Microstructural Analyses

The microstructure of the investigated materials—alloys A, B, C and D—is typical of Al-Si casting alloys consisting of primary α-Al dendrites surrounded by the Al-Si eutectic structure (Figure 2). SDAS evaluation confirmed that the production process allowed us to obtain samples with the desired SDAS value of 10 μm. Average SDAS values calculated for each investigated alloy are reported in Table 2.

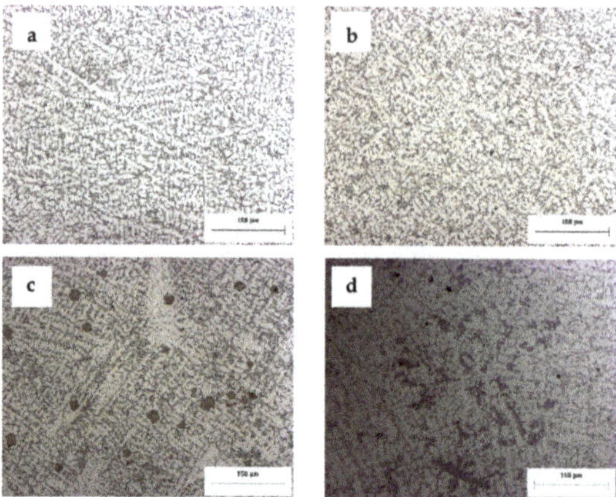

Figure 2. Low-magnification optical images showing the microstructure of alloys (**a**) A; (**b**) B; (**c**) C and (**d**) D, consisting in α-Al dendrites, Al-Si eutectic and intermetallic particles.

Table 2. Average SDAS value calculated on T6 alloys.

Alloy	Average SDAS (μm)
A	9.7 ± 2.0
B	9.8 ± 1.7
C	9.7 ± 1.3
D	9.6 ± 1.6

On metallographic samples, no internal voids related to gas pores or micro-shrinkage were observed, thanks to the densification action of HIP process.

Different intermetallic particles which remained undissolved by solution treatment were observed both in the interdendritic areas, and within α-Al, such as the acicular β-Al$_5$FeSi phases, the α-Al$_{15}$(Fe, Mn)$_3$Si$_2$ particles, and the Fe-based coarse sludge, to contain Fe, Mn and Cr (Figure 3).

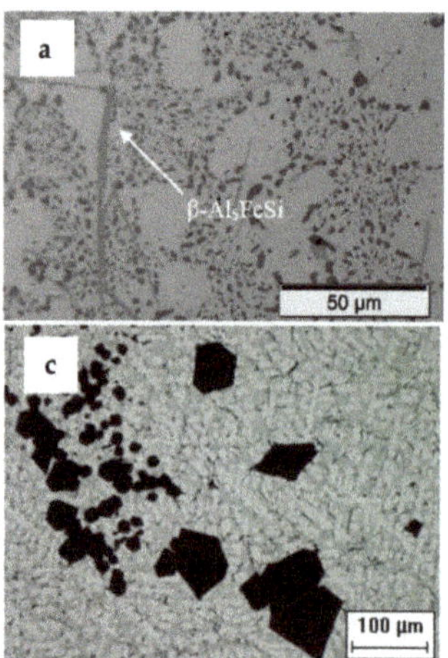
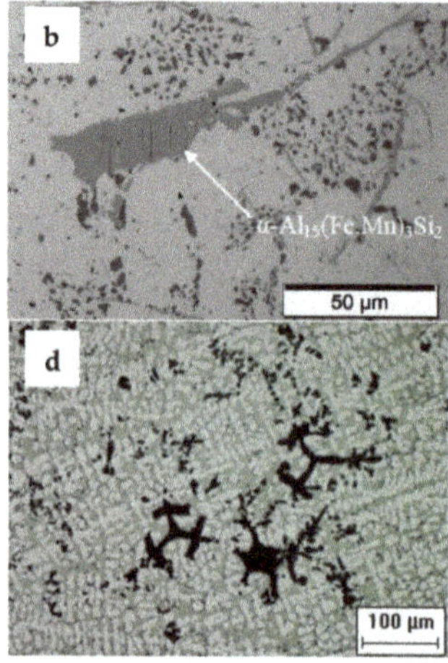

Figure 3. Optical images of intermetallic particles: (**a**) β-Fe phases; (**b**) α-Fe phase; (**c**) blocky-like and (**d**) star-like sludges; (a, b from [25]).

An example of EDS spectrum analysis performed on the sludge particles found in alloy C is reported in Figure 4.

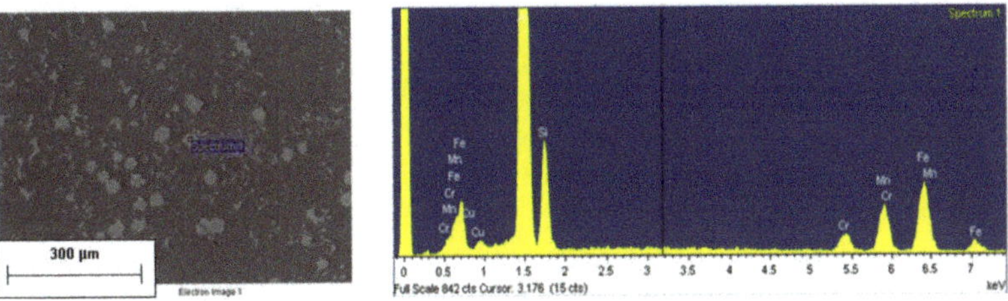

Figure 4. EDS spectrum of a sludge particle observed in C-T6 alloy [25].

Alloy chemical composition and SF played a relevant role on the type and amount of intermetallic particles formed in the four alloys. Results of the optical images elaboration are reported in Figure 5.

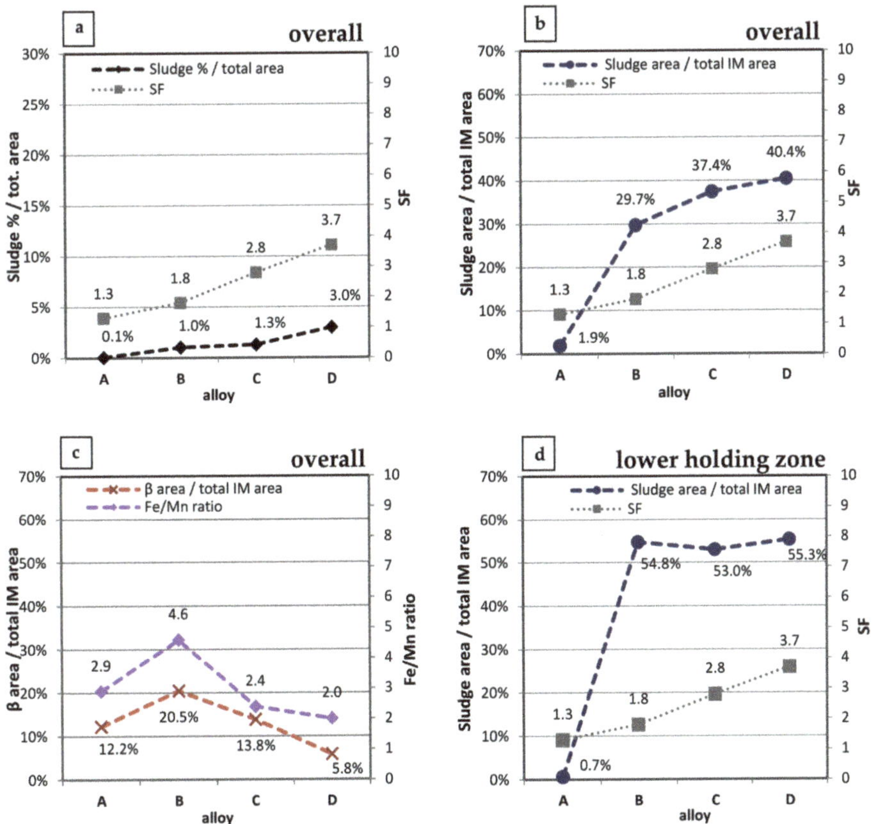

Figure 5. Results of optical images analyses, showing (**a**) the relation between SF and percentage of area covered by sludge particles; (**b**) correspondence between SF and sludge IM fraction calculated on the overall length of the sample; (**c**) variation of β-Fe IM fraction with Fe/Mn ratio and (**d**) the relation between sludge IM fraction and SF in the lower holding zone of fatigue specimens.

Increasing the SF leads to an overall increase in the area fraction of sludge particles, as represented in Figure 5a. In the base alloy A, corresponding to the lowest SF (1.3), sludge particles cover only 0.1% of the total area, while in alloy D, characterized by the maximum SF (3.7), sludge area is 3.0%. Correspondingly, in alloy A only 1.9% of all the intermetallic particles are represented by sludge; the percentage drastically increases for alloys B, C and D, reaching the maximum fraction of 40.4% in alloy D (Figure 5b).

It is generally known that a Fe:Mn ratio lower than 2 is necessary to prevent the formation of acicular β-Fe [6,12]. As a result, since the nominal Fe/Mn ratio was always ≥2, β-Al5FeSi phases were observed in all the investigated alloys. It is also interesting to note that the IM fraction of β phases in the four alloys is related to the Fe/Mn ratio (Figure 5c): the higher Fe/Mn ratio, the higher is the β-Fe IM fraction. The maximum Fe/Mn ratio (4.6) was alloy B, containing the highest IM fraction of β phases, corresponding to ~20% of all intermetallic particles.

Uneven distribution of sludge particles was observed along the length of samples. Since samples were remelted using the directional solidification technique in a vertical position, a higher amount of sludge phases were found in the lower holding zone of fatigue samples as a result of heavy intermetallic segregation in the bottom of solidified bars. Image analyses showed that in this area, differently from the gauge length, more than 50% of intermetallics in alloys B, C and D were sludge particles (Figure 5d). This confirms the tendency of intermetallic sludge to settle to the bottom of furnaces or crucibles, as reported in literature [6,8,13]. It should be pointed out that such uneven and elevated concentration of sludge particles was circumscribed to the low holding zone, not affecting the gauge length, which is subjected to fatigue loading.

3.2. Mechanical Characterization

Results of hardness tests on T6 and overaged (OA) alloys are summarized in Figure 6. No significant difference was observed in the T6 condition, since alloys A, B, C and D presented comparable hardness (ranging from 93 to 97 HB10). A certain decrease of hardness was registered in all the alloys after overaging: long-term high temperature exposure induced coarsening of strengthening precipitates and consequent α-Al softening, thus resulting in a loss of strength.

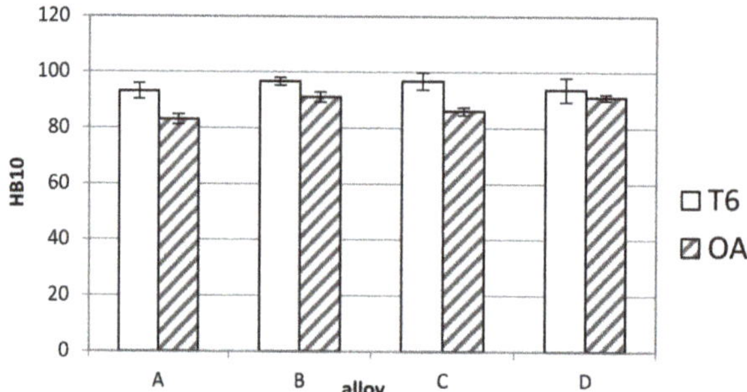

Figure 6. Brinell hardness (HB10 scale) measured on T6 and overaged (OA) alloys.

Results of rotating bending fatigue tests are summarized in Figure 7. It can be seen that SF plays a detrimental role on fatigue behavior. At room temperature, $\sigma f_{50\%}$ of T6 alloys gradually decreased with an increase in the SF; a loss of about 22% in fatigue strength was registered from alloy A to alloy D. It is thought that this behavior is related to the presence of sludge intermetallic particles. As known, coarse Fe-bearing intermetallic particles could hinder dislocation movement, leading to a piling-up at the α-Al matrix-intermetallics interface, and consequent stress concentration.

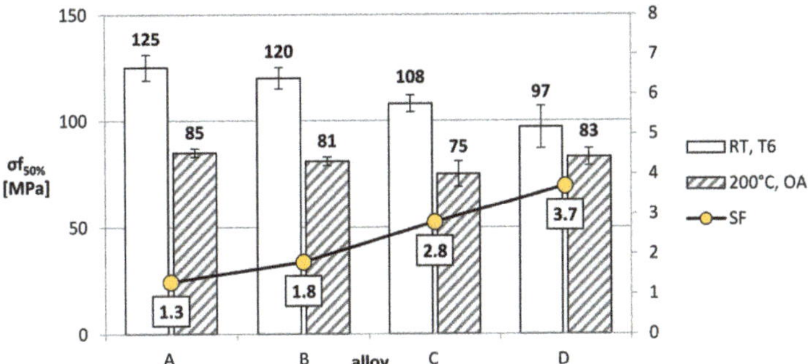

Figure 7. Fatigue strength (σf$_{50\%}$) and SF of the investigated alloys, tested at room temperature in the T6 condition, and at 200 °C after T6 and overaging (48 h at 200 °C).

Despite the fact that the applied cyclic load is lower than the yielding level, local yielding would occur in any case due to microscale effects related to inhomogeneous deformation of microstructural constituents [26,27]. As a result of such interaction, local internal stresses and damage rate increase, leading to decohesion between intermetallics and matrix, particle cracking, and related crack nucleation. Since under low applied stresses the fatigue behaviour is mainly governed by crack nucleation stage, the presence of coarse intermetallic compounds such as the α-Al$_x$(Fe,Mn,Cr)$_y$Si$_z$ sludge particles should significantly reduce the fatigue strength of the alloys, as reported in [26–28]. A marked decrease in fatigue strength was registered at high temperature after overaging in comparison to room temperature and in the T6 state. This is ascribed both to the coarsening of strengthening precipitates induced by long-term high temperature exposure leading to an overall decrease of mechanical strength, and to the softening of α-Al matrix, known to cause reduction of UTS and, correspondingly, fatigue strength [29–31]. Nevertheless, it should be pointed out that at high temperature and after overaging, the effect of SF on fatigue behavior was almost negligible, since the σf$_{50\%}$ of alloys A, B, C and D are basically comparable. It is thought that, at high testing temperature, a softening of the matrix may induce a reduction of stress concentrations associated with intermetallic compounds, thus reducing the gap between the fatigue strength of alloys with different SF. Similar findings were reported by the authors in [29].

3.3. Fractographic Analyses

An analysis of fracture surfaces revealed that cracks nucleated always near the sample surface, where the highest stresses induced by rotating fatigue loading were registered. In alloys A and B, both at room and high temperature, cracks were nucleated mainly in the correspondence of casting defects. Superficial and sub-superficial voids are, in fact, not easily closed by the HIP process.

As a result, in alloys A and B, fatigue cracks were initiated by superficial or sub-superficial pores, oxides and microshrinkages. An example of a sub-superficial pore nucleating fatigue crack is reported in Figure 8a. In alloys C and D, characterized by higher amount of sludge particles in comparison to A and B, a relationship between sludge and crack nucleation was observed. At room temperature in the T6 condition, cracks were in fact generated by superficial sludge particles (Figure 8b), confirming the link between Fe-bearing intermetallics and crack nucleation.

At high temperatures, on the other hand, for the alloys C and D, different crack nucleation causes were identified, such as sludge particles, casting defects, as well as multiple initiators (e.g., sludge and pore in different locations of the sample surface, as shown in Figure 8c). This occurrence reflects the higher sensitivity of T6 alloys to stress concentration induced by sludge particles in comparison to the overaged alloys tested at high temperature, as inferred in Section 3.2. From the analysis of nucleation causes, it seems that a defect critical size exists, as widely discussed by Wang et al. [26,27]: in all the investigated alloys, superficial/sub-superficial pores generating fatigue cracks were characterized by a minimum size of ~30 μm, while sludge particles causing crack nucleation on the sample fracture were generally characterized by a minimum size of 40–50 μm. Although more accurate analyses should be carried out on this respect, this fact could indicate that, at SDAS and loading conditions of the present investigation, pores are more detrimental than Fe-based sludges to fatigue life.

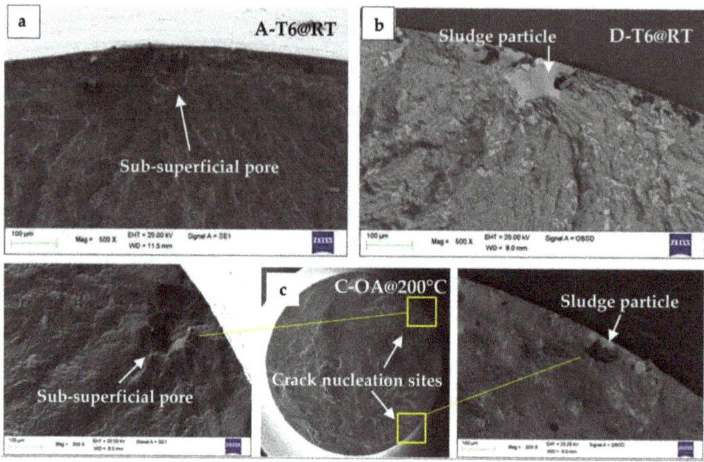

Figure 8. SEM images of fracture surfaces: (**a**) sub-superficial pore inducing crack nucleation in A-T6 alloy; (**b**) crack nucleated on sample surface by a sludge particle in D-T6 alloy; (**c**) multiple crack nucleation in C-OA alloy tested at 200 °C, caused by sub-superficial pore and superficial particle.

In accordance with the results of microstructural analyses, different kind of intermetallics were found on propagation areas of the different alloys. Alloys A and B were characterized by β-Fe needles (Figure 9a), while alloys C and D presented a high amount of coarse sludge particles (Figure 9b).

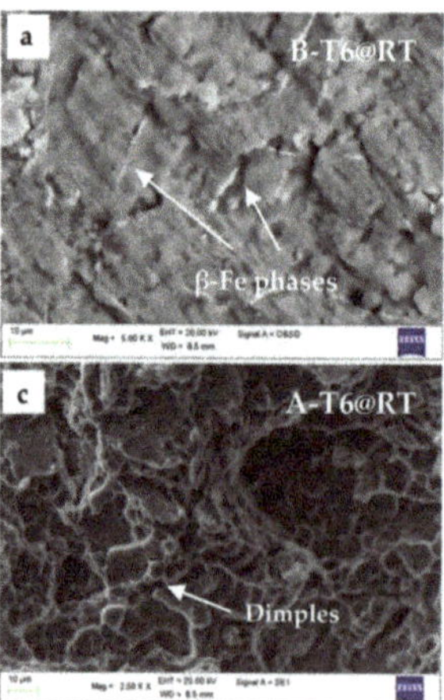

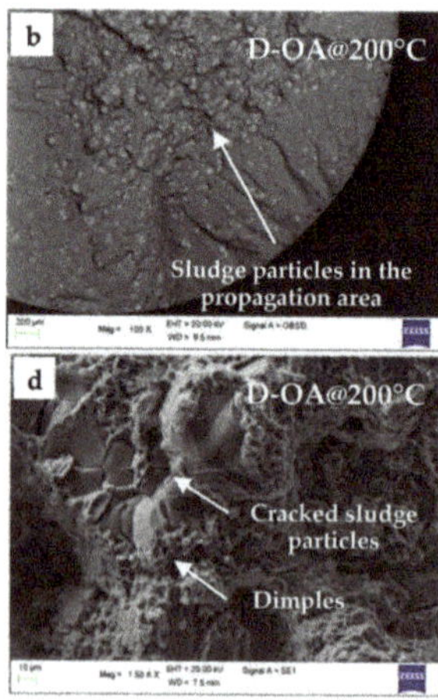

Figure 9. SEM images of fracture surfaces, showing (**a**) β-Fe phases in the propagation area of a B-T6 sample; (**b**) a dense population of sludge particles in the propagation area of D-OA alloy; (**c**) ductile morphology of final failure region in A-T6 alloy and (**d**) final failure area of D-OA alloy tested at 200 °C characterized by cracked sludge particles and dimples.

The final failure region of T6 and overaged alloys was characterized, both at room and high temperature, by a typical ductile morphology. In alloys A and B, only dimples and small intermetallics were observed (Figure 9c), whereas in alloys C and D, a large amount of cracked and decohesed sludge particles were found (Figure 9d).

4. Conclusions

- In the investigated Al-Si-Cu alloys, sludge factor (SF) is a good indicator of the tendency to form coarse, Fe-based sludge particles containing Fe, Mn and Cr; this was reflected in the different sludge area fractions in the tested alloys.

- A segregated microstructure was observed in samples of alloys B, C and D, characterized by SF of 1.8, 2.8 and 3.7 respectively. The heavy sludge particles tend to settle towards the bottom of the samples. The highest IM fraction of β-Al$_5$FeSi phases was found in alloy B, characterized by the highest Fe:Mn ratio.

- Fatigue behavior is negatively affected by SF; at room temperature and in the T6 condition, a decrease of 22% in fatigue strength was registered from alloy A (base alloy) to alloy D, characterized by the highest SF and containing the maximum area fraction of sludge particles.

- At high temperature, overaged alloys presented an overall decrease in fatigue strength in comparison to the T6 condition at room temperature, because of a coarsening of strengthening precipitates and α-Al softening. Nevertheless, a less marked effect of SF was registered on fatigue strength.

- While at room temperature the cracks in alloys C and D nucleated mainly in correspondence with sludge particles, at 200 °C different nucleation causes were observed, namely: sludge, casting defects and multiple crack initiators. It is thought that, at high temperature, softening of the matrix causes a reduction of stress concentration which is induced by intermetallic sludge particles, thereby reducing their detrimental effect on fatigue life.

Acknowledgments: The authors gratefully acknowledge Marcello Gobbi for his experimental Master Thesis work.

Author Contributions: University of Jönköping produced samples for mechanical characterization and carried out part of the work on microstructural analyses; fatigue testing and fractographic analyses were carried out at University of Bologna; all the authors contributed to data evaluation, discussion and paper writing.

Conflicts of Interest: The authors declare no conflict of interest.

References

1. Crepeau, P. Effect of iron in Al-Si casting alloys: A critical review. *AFS Trans.* **1995**, *103*, 361–366.
2. Couture, A. Iron in aluminum casting alloys—A literature survey. *Int. J. Cast Met. Res.* **1981**, *6*, 9–17.
3. Dinnis, C.; Taylor, J.; Dahle, A. As-cast morphology of iron-intermetallics in Al–Si foundry alloys. *Scr. Mater.* **2005**, *53*, 955–958. [CrossRef]
4. Ma, Z.; Samuel, A.M.; Samuel, F.H.; Doty, H.W.; Valtierra, S. A study of tensile properties in Al–Si–Cu and Al-Si-Mg alloys: Effect of β-iron intermetallics and porosity. *Mater. Sci. Eng. A* **2008**, *490*, 36–51. [CrossRef]
5. Seifeddine, S.; Svensson, I.L. Prediction of mechanical properties of cast aluminium components at various iron contents. *Mater. Des.* **2010**, *31*, 6–12. [CrossRef]
6. Shabestari, S.G. The effect of iron and manganese on the formation of intermetallic compounds in aluminum-silicon alloys. *Mater. Sci. Eng. A* **2004**, *383*, 289–298. [CrossRef]
7. Shankar, S.; Apelian, D. Die soldering: Mechanism of the interface reaction between molten aluminum alloy and tool steel. *Metall. Mater. Trans. B* **2002**, *33*, 465–476. [CrossRef]
8. Makhlouf, M.; Apelian, D. *Casting Characteristics of Aluminum Die Casting Alloys*; Report Performed under DOE Contract Number DEFC07-99ID13716; Worcester Polytechnic Institute (US): Worcester, MA, USA, 2002.
9. Seifeddine, S.; Johansson, S.; Svensson, I.L. The influence of cooling rate and manganese content on the β-Al5FeSi phase formation and mechanical properties of Al-Si-based alloys. *Mater. Sci. Eng. A* **2008**, *490*, 385–390. [CrossRef]

10. Ceschini, L.; Boromei, I.; Morri, A.; Seifeddine, S.; Svensson, I.L. Microstructure, tensile and fatigue properties of the Al-10%Si-2%Cu alloy with different Fe and Mn content cast under controlled conditions. *J. Mater. Process. Technol.* **2009**, *209*, 5669–5679. [CrossRef]

11. Hwang, J.Y.; Doty, H.; Kaufman, M.J. The effects of Mn additions on the microstructure and mechanical properties of Al-Si-Cu casting alloys. *Mater. Sci. Eng. A* **2008**, *488*, 496–504. [CrossRef]

12. Taylor, J.A. The Effect of Iron in Al-Si Casting Alloys. In Proceedings of the 35th Australian Foundry Institute National Conference on Casting Concepts, Adelaide, South Australia, 31 October–3 November 2004.

13. Shabestari, S.G.; Gruzleski, J.E. Gravity segregation of complex intermetallic compounds in liquid aluminum-silicon alloys. *Metall. Mater. Trans. A* **1995**, *26*, 999–1006. [CrossRef]

14. Gobrecht, J. Settling-out of Fe, Mn and Cr in Al-Si casting alloys. *Giesserei* **1975**, *62*, 263–266.

15. Jorstad, J. Understanding sludge. *Die Cast. Eng.* **1986**, *30*, 30–36.

16. Wang, L.; Makhlouf, M.; Apelian, D. Aluminium die casting alloys: Alloy composition, microstructure, and properties-performance relationships. *Int. Mater. Rev.* **1995**, *40*, 221–238. [CrossRef]

17. Ferrero, S.; Timelli, G.; Fabrizi, A. Evolution of sludge particles in secondary die-cast aluminium alloys as function of Fe, Mn and Cr contents. *Mater. Chem. Phys.* **2015**, *153*, 168–179. [CrossRef]

18. Bjurenstedt, A.; Seifeddine, S.; Jarfors, A.E.W. The effects of Fe-particles on the tensile properties of Al-Si-Cu alloys. *Metals* **2016**, *6*, 314. [CrossRef]

19. Ji, S.; Yang, W.; Gao, F.; Watson, D.; Fan, Z. Effect of iron on the microstructure and mechanical property of Al-Mg-Si-Mn and Al-Mg-Si die cast alloys. *Mater. Sci. Eng. A* **2012**, *564*, 130–139. [CrossRef]

20. Timelli, G.; Bonollo, F. The influence of Cr content on the microstructure and mechanical properties of AlSi$_9$Cu$_3$ (Fe) die-casting alloys. *Mater. Sci. Eng. A* **2010**, *528*, 273–282. [CrossRef]

21. Seifeddine, S. Characteristics of Cast Aluminium-Silicon Alloys: Microstructures and Mechanical Properties. Linköping studies in Science and Technology: Dissertations; Ph.D. Thesis, Linköping University, Linköping, Sweden, 2006.

22. ASTM E3-01. *Standard Practice for Preparation of Metallographic Specimens*; ASTM International: West Conshohocken, PA, USA, 2007.

23. UNI Standard. *Mechanical Testing of Metallic Materials Fatigue Testing at Room Temperature*; UNI 3964-85; Ente Nazionale Italiano di Unificazione: Milan, Italy, 1985.

24. ASTM E10-08. *Standard Test Method for Brinell Hardness of Metallic Materials*; ASTM International: West Conshohocken, PA, USA, 2007.

25. Gobbi, M. *Influence of Sludge Particles on the Mechanical Properties of Al-Si-Cu-Mg Secondary Casting Aluminium Alloy*; University of Bologna: Bologna, Italy, 2014.

26. Wang, Q.G.; Apelian, D.; Lados, D.A. Fatigue behavior of A356-T6 aluminum cast alloys. Part I. Effect of casting defects. *J. Light Met.* **2001**, *1*, 73–84. [CrossRef]

27. Wang, Q.G.; Apelian, D.; Lados, D.A. Fatigue behavior of A356/357 aluminum cast alloys. Part II—Effect of microstructural constituents. *J. Light Met.* **2001**, *1*, 85–97. [CrossRef]

28. Ceschini, L.; Boromei, I.; Morri, A.; Seifeddine, S.; Svensson, I.L. Effect of Fe content and microstructural features on the tensile and fatigue properties of the Al-Si10-Cu2 alloy. *Mater. Des.* **2012**, *36*, 522–528. [CrossRef]

29. Ceschini, L.; Morri, A.; Toschi, S.; Seifeddine, S. Room and high temperature fatigue behaviour of the A354 and C355 (Al-Si-Cu-Mg) alloys: Role of microstructure and heat treatment. *Mater. Sci. Eng. A* **2016**, *653*, 129–138. [CrossRef]

30. Ceschini, L.; Morri, A.; Morri, A.; Toschi, S.; Johansson, S.; Seifeddine, S. Effect of microstructure and overaging on the tensile behavior at room and elevated temperature of C355-T6 cast aluminum alloy. *Mater. Des.* **2015**, *83*, 626–634. [CrossRef]

31. Ceschini, L.; Morri, A.; Toschi, S.; Johansson, S.; Seifeddine, S. Microstructural and mechanical properties characterization of heat treated and overaged cast A354 alloy with various SDAS at room and elevated temperature. *Mater. Sci. Eng. A* **2015**, *648*, 340–349. [CrossRef]

Article

Effect of Anode Pulse-Width on the Microstructure and Wear Resistance of Microarc Oxidation Coatings

Zhen-Wei Li * and Shi-Chun Di

School of Mechatronics Engineering, Harbin Institute of Technology, Harbin 150001, China;
13904605946@126.com
* Correspondence: 14b908070@hit.edu.cn; Tel.: +86-0451-8641-7672

Received: 3 May 2017; Accepted: 27 June 2017; Published: 30 June 2017

Abstract: Microarc oxidation (MAO) coatings were prepared on 2024-T4 aluminum alloys using a pulsed bipolar power supply at different anode pulse-widths. After the MAO coatings were formed, the micropores and microcracks on the surface of the MAO coatings were filled with Fluorinated ethylene propylene (FEP) dispersion for preparing MAO self-lubricating composite coatings containing FEP. The effect of the anode pulse-width on the microstructure and wear resistance of the microarc oxidation coatings was investigated. The wear resistance of the microarc oxidation self-lubricating composite coatings was analyzed. The results revealed that the MAO self-lubricating composite coatings integrated the advantages of wear resistance of the MAO ceramic coatings and a low friction coefficient of FEP. Compared to the MAO coatings, the microarc oxidation self-lubricating composite coatings exhibited a lower friction coefficient and lower wear rates.

Keywords: 2024-T4 aluminum alloys; microarc oxidation; anode pulse-width; FEP; adhesion strength; wear resistance

1. Introduction

Aluminum alloys, due to their advantages of high strength to weight ratio, lightweight [1], proper corrosion resistance, and great workability, have received much attention and are extensively applied in various industries [2]. Unfortunately, the application of aluminum alloys is seriously restricted by their low surface hardness, high and unstable friction coefficient, and poor wear-resistance [3–5]. In the past decades, many surface modification technologies have been developed and employed to improve the hardness and wear resistance of aluminum alloys. However, most of these technologies require high temperatures and are complicated to employ. These technologies are also not environmentally friendly [6,7]. Microarc oxidation (MAO), namely plasma electrolytic oxidation (PEO) [8], is an effective technique for the surface treatment of aluminum alloys [9,10]. MAO could be used to prepare ceramic-like oxide coatings on the surface of valve metals [11,12]. The ceramic coatings prepared by MAO have high hardness, good adhesion to the substrate, excellent wear resistance, etc. [13–15].

The microstructure and properties of MAO coatings are influenced by many factors such as the electrolyte composition [16], power supply modes [17], electrical parameters [18], oxidation time [19], and additives [20]. Hussein et al. [21] investigated the effect of the current mode (unipolar, bipolar) on the wear properties of MAO coatings on AM60B magnesium alloys. Wu et al. [22] found that the composition, structure, and physical and chemical properties of MAO coatings on the surface of Ti alloys can be extensively modified by changing the anodic and cathodic voltages. Li et al. [23] discovered that the wear resistance of MAO coatings can be significantly improved by increasing the cathodic voltage. Moreover, the MAO ceramic coatings produced also behave as high friction coefficient layers, which limit the extensive engineering applications for MAO technology. The wear resistance of MAO coatings could be improved through decreasing the friction coefficient, which has received increased attention. Wang et al. [7] prepared MAO self-lubricating Al_2O_3/PTFE composite coatings on the surface of aluminium alloys, and these MAO composite coatings exhibited excellent self-lubricating behaviour. Yin et al. [24] prepared self-lubricating alumina-graphite composite coatings using a one-step PEO process in the appropriate graphite-dispersed electrolyte solution. This MAO composite coating showed excellent performance. Mu et al. [20] prepared MAO self-lubricating TiO_2/graphite composite coatings on TI6Al4V alloys. Qin et al. [25] prepared MAO self-lubricating TiO_2/MoS_2 composite coatings on TI6Al4V alloys.

The microarc oxidation self-lubricating composite coatings could possess relatively improved hardness and higher anti-wear attributes as well as self-lubricating performance coupled with a lower friction coefficient. MAO ceramic coatings typically contain many crater-like micropores of various sizes and a small number of microcracks. This creates the possibility to deposit small sized lubricants into these micropores and microcracks to form microarc oxidation self-lubricating composite coatings. Fluorinated ethylene propylene (FEP) has low shear strength and is often used as a solid lubricant. Moreover, FEP dispersion has good liquidity.

In this study, 2024-T4 aluminum alloys were treated through microarc oxidation using a pulsed bipolar current mode at different anode pulse-widths. After the MAO coatings were formed, the micropores and microcracks on the surface were filled with FEP dispersion for preparing MAO self-lubricating composite coatings containing FEP. The microstructure and properties of the MAO coatings, and the wear resistance of the microarc oxidation self-lubricating composite coatings were investigated. The adhesion strength between the MAO coatings and the substrates was also analyzed.

2. Experimental Details

2.1. Sample Preparation

The MAO coatings were prepared on 2024-T4 aluminum alloys. The main chemical composition (mass fraction, wt %) of the 2024-T4 aluminum alloys is 3.8–4.9% Cu, 0.5% Si, 0.5% Fe, 0.3–0.9% Mn, 1.2–1.8% Mg, 0.25% Zn, 0.10% Cr, 0.15% Ti, balanced with Al. The main mechanical properties of the 2024-T4 aluminum alloys are a yield stress (0.2% offset) of $\sigma_{0.2}$ = 325 MPa, a tensile strength of σ_b = 470 MPa, a Vickers hardness of 130 Hv, an elongation of δ = 20%, and a Young's modulus of 69 GPa. The sample size was 30 mm $\times$ 15 mm $\times$ 2 mm. Samples were polished using abrasive papers from 600# to 2000#, degreased by ultrasound in acetone, rinsed in deionized water, and then dried in warm air. The alternating current (AC) pulse microarc oxidation unit (Harbin Institute of Technology, Harbin, China) was used in this experiment. The sample was used as the anode and the electrolytic tank was used as the cathode. The silicate electrolyte was prepared in deionized water. The composition of the electrolyte was (6 g/L) Na_2SiO_3, (1.5 g/L) KOH, and was of high purity. A pulsed bipolar constant current power supply was used in this experiment, and the schematic of the pulse output is shown in Figure 1. The MAO process parameters are listed in Table 1. All samples were processed for 30 min. I^+ is the positive current and I^- is the negative current. The mixing pump was opened and the electrolyte was stirred continuously during the MAO treatment. The cycle cooling system was opened to ensure that the electrolyte temperature was less than 40 °C.

The variation curves of the voltages are shown in Figure 2. As shown in Figure 2I, the anode and cathode voltages increased as the MAO processing time increased during microarc oxidation. The resistance of MAO coatings increased as the MAO coatings grew, resulting in an increase of the anode and cathode voltages under the constant current mode during MAO. Moreover, Figure 2II,III show that the growth rates of the anode and cathode voltages increased in association with the anode pulse-width, which may be related to the growth rate of the MAO coatings during the process.

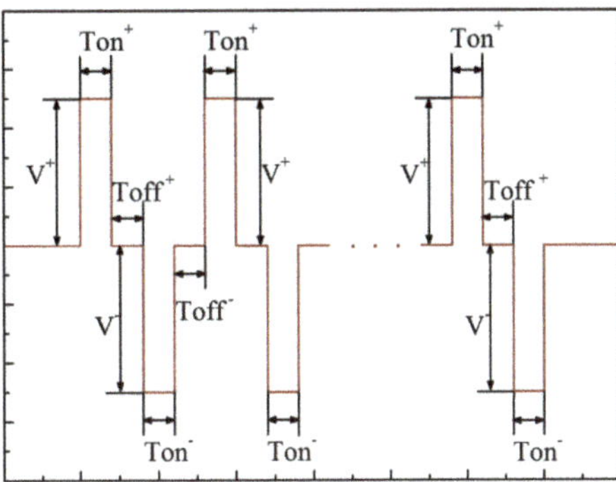

Figure 1. Schematic of the pulse output with a bipolar constant current power supply.

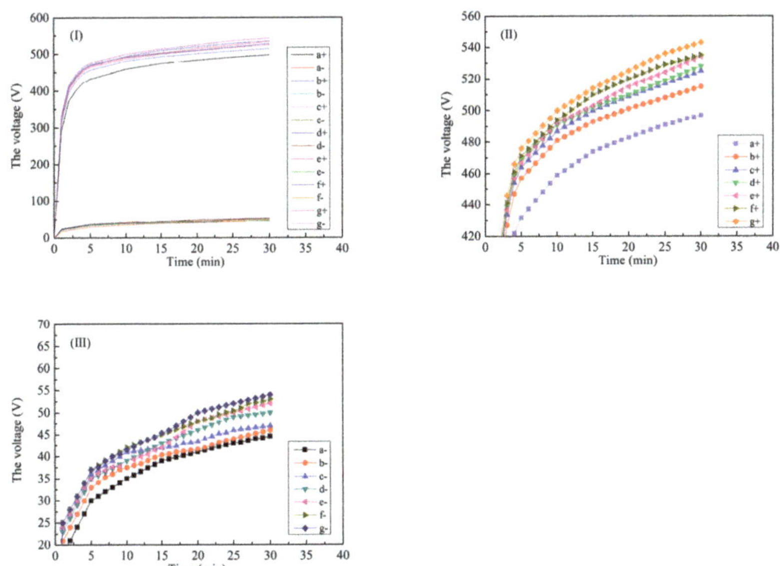

Figure 2. Variation curves of the voltage at different anode pulse-widths during microarc oxidation. (**I**) variation curves of anode and cathode voltages; (**II**) variation curves of anode voltages; (**III**) variation curves of cathode voltages; (a): 1000 µs; (b): 1500 µs; (c): 2000 µs; (d): 2500 µs; (e): 3000 µs; (f): 3500 µs; (g): 4000 µs.

Table 1. Microarc oxidation process parameters for coatings formed on aluminum alloys at different anode pulse-widths.

Labels	Ton⁺ (µs)	Toff⁺ (µs)	Ton⁻ (µs)	Toff⁻ (µs)	V⁺ (V)	V⁻ (V)	I⁺ (A/dm²)	I⁻ (A/dm²)	Time (min)
a (a1)	1000	300	2000	300	Figure 2	Figure 2	37	9	30
b (b1)	1500	300	2000	300	Figure 2	Figure 2	37	9	30
c (c1)	2000	300	2000	300	Figure 2	Figure 2	37	9	30
d (d1)	2500	300	2000	300	Figure 2	Figure 2	37	9	30
e (e1)	3000	300	2000	300	Figure 2	Figure 2	37	9	30
f (f1)	3500	300	2000	300	Figure 2	Figure 2	37	9	30
g (g1)	4000	300	2000	300	Figure 2	Figure 2	37	9	30

After the microarc oxidation coatings were formed, the coated samples were rinsed with deionized water and dried in warm air. The coated samples were immersed in a water-based FEP dispersion. The immersion time was 15 min. The micropores and microcracks on the surface of the MAO coatings were filled with the FEP dispersion. After immersing for 15 min, the samples were heated for 40 min at 220 °C and the FEP dispersion solidified. The FEP was embedded into the micropores and microcracks of the MAO ceramic coatings, which formed the MAO self-lubricating composite coatings containing FEP. The MAO coatings are those from "a" to "g", while the MAO-FEP coatings are labeled with a1 and, therefore, are those from "a1" to "g1".

2.2. Testing and Characterization

The morphologies of the microarc oxidation ceramic coatings formed using the pulsed bipolar power supply at different anode pulse-widths and the morphologies of the microarc oxidation self-lubricating composite coatings were observed by a FEI Sirion scanning electron microscope (SEM, FEI Quanta 200F, FEI, Eindhoven, The Netherlands). The coating thickness was measured via SEM. The surface roughness of the microarc oxidation ceramic coatings was measured by an OLS3000 laser confocal microscope (OLS, Suzhou, China). The phase structure of the microarc oxidation ceramic coatings was analyzed by an X' Pert PRO X-ray diffractometer (XRD, Philips, Amsterdam, The Netherlands).

The adhesion strength of the microarc oxidation coatings was evaluated using a CSM microscratch tester (CSM, Neuchatel, Switzerland). In this test, the load on a diamond Rockwell indenter with a tip radius of 100 μm was linearly increased from 0 to 30 N at a normal loading speed of 30 N·min^{-1} as the diamond is drawn across the surface of the MAO coatings. The scratch length was 5 mm. The critical load values (L_c) were determined using supplementary data graphics including acoustic emission and friction coefficient.

The hardness was measured by a Vickers hardness tester (Huayin, Laizhou, China): the load was 10 N and the dwelling time was 10 s. For each sample, the tests were carried out six times. The friction coefficient of the MAO coatings and the microarc oxidation self-lubricating composite coatings was measured by a CETR-UMT-2 ball-on-disk wear tester (CETR, Campbell, CA, USA). WC balls with a diameter of 5 mm and a surface roughness greater than 0.05 μm were used as counterface materials. All wear tests were conducted with WC balls in sliding contact with a load of 10 N at a fixed sliding speed. The rotating diameter and rotating speed of the wear tester were 6 mm and 200 rpm, respectively. The test time of each sample measured on the wear tester was 20 min. The width of the wear tracks were measured via SEM and an OLS3000 laser confocal microscope. The depth and volume of the wear tracks were detected by an OLS3000 laser confocal microscope. The average value of the three profiles on the wear tracks was used to calculate the wear rate. The computation formula of the wear rate is reported in Equation (1) [20], where Q is the wear rate, V_W is the wear volume (mm^3), P is the applied load (N), and S is the sliding distance (m).

$$Q = \frac{V_W}{PS} \tag{1}$$

3. Results and Discussion

3.1. Microstructure of MAO Coatings and MAO Self-Lubricating Composite Coatings

The surface morphologies of the microarc oxidation ceramic coatings prepared on the surface of 2024-T4 aluminum alloys are shown in Figure 3. MAO ceramic coatings contained many crater-like micropores of various sizes and a small number of microcracks. The porosity of the MAO ceramic coatings increased significantly and the MAO coatings became more porous as the anode pulse-width increased from 1000 to 4000 μs. These micropores were formed by molten "oxide magma" and the gas bubbles ejected from discharge channels. The rapid solidification of the molten oxide magma ejected from discharge channels in the electrolyte, however, also produced thermal stress, which caused microcracks [26].

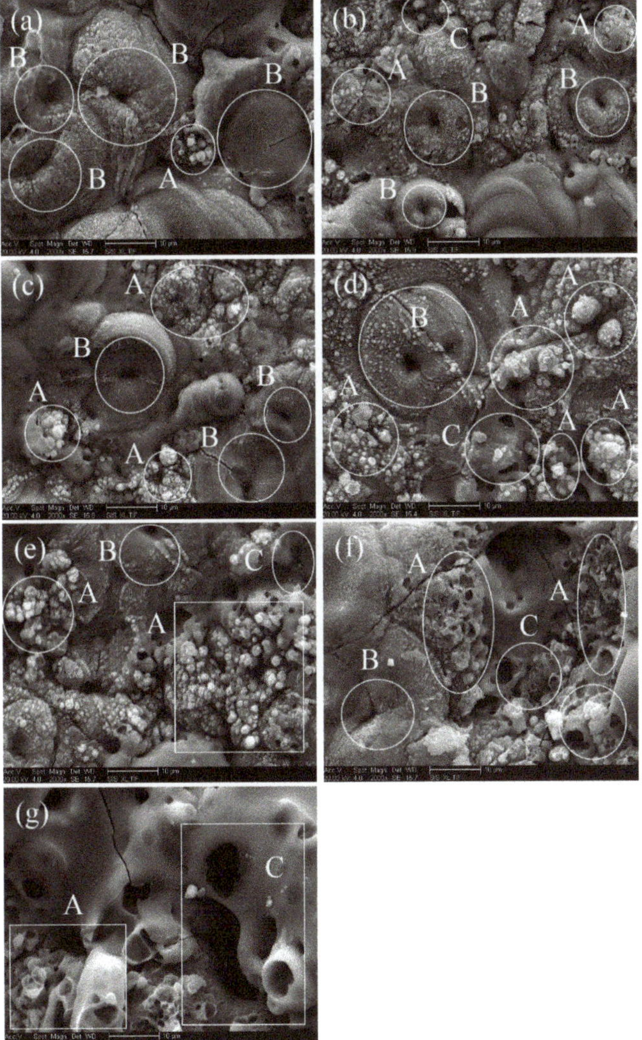

Figure 3. Surface morphologies of the microarc oxidation coatings formed on aluminum alloys at different anode pulse-widths. (**a**) 1000 μs; (**b**) 1500 μs; (**c**) 2000 μs; (**d**) 2500 μs; (**e**) 3000 μs; (**f**) 3500 μs; (**g**) 4000 μs.

As shown in Figure 3, A-type holes, B-type holes, and C-type holes were clearly exhibited on the surface of the MAO coatings. A-type holes and C-type holes showed a porous foam-like structure. B-type holes looked like flat pancakes and showed a more compact structure. According to previous studies, three typical discharge events occurred in the MAO process. A-type discharge events occurred in holes on the surface of the MAO coatings. B-type discharge events penetrated through the MAO coatings. C-type discharge events occurred in the relatively deep holes on the surface of the MAO coatings. A-type holes, B-type holes and C-type holes were formed by A-type discharge events, B-type discharge events, and C-type discharge events, respectively [27]. Therefore, it can be inferred that with the increase in the anode pulse-width, A-type discharge events and C-type discharge events increase. At the same time, B-type discharge events decrease and MAO coatings became more porous.

The surface roughness of microarc oxidation ceramic coatings is shown in Figure 4. The surface roughness of the microarc oxidation coatings increased significantly with the increase in the anode pulse-width. As shown in Figure 3, the foam-like structures increased in size and quantity with the increase in the anode pulse-width, which increased the surface roughness of the microarc oxidation coatings.

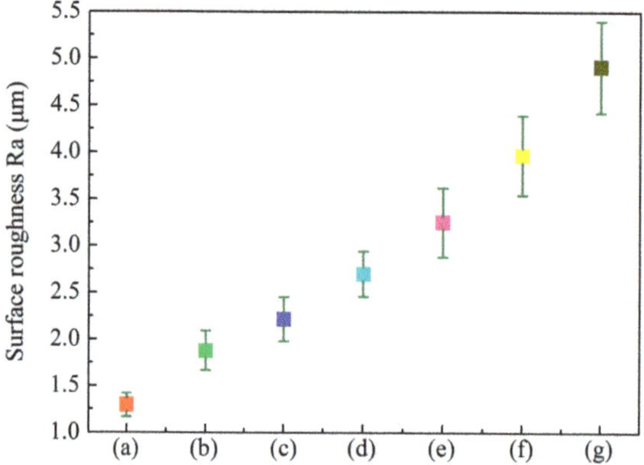

Figure 4. The surface roughness of microarc oxidation coatings formed on aluminum alloys at different anode pulse-widths. (a): 1000 µs; (b): 1500 µs; (c): 2000 µs; (d): 2500 µs; (e): 3000 µs; (f): 3500 µs; (g): 4000 µs.

The cross-sectional morphologies of the microarc oxidation ceramic coatings are shown in Figure 5. Figure 5 shows the defects and micropores (region D). In the MAO coatings, the defects and micropores increased significantly with an increase in the anode pulse-width, and the MAO coatings became more porous. A-type discharge events and C-type discharge events did not penetrate through the entire MAO coating. This did not result in the formation of a new oxide. A-type discharge events and C-type discharge events increased with the increase in the anode pulse-width. The defects and micropores in the MAO coatings were not completely filled with the molten oxide magma, which rapidly solidified and formed MAO coatings with higher porosity.

The thickness of the microarc oxidation ceramic coatings is shown in Figure 6. The thickness of the MAO ceramic coatings increased with an increase in anode pulse-width. As shown in Figure 3, the foam-like structures increased in size and quantity with the increase in the anode pulse-width, which increased the growth rates of the MAO coatings. Moreover, the non-uniformity of the A-type discharge events and C-type discharge events on the surface of the MAO coatings, the pressure in the discharge channels, and the ejected height of the molten oxide magma increased along with the anode pulse-width. This increased the growth rates and the non-uniformity of the thickness of the MAO coatings.

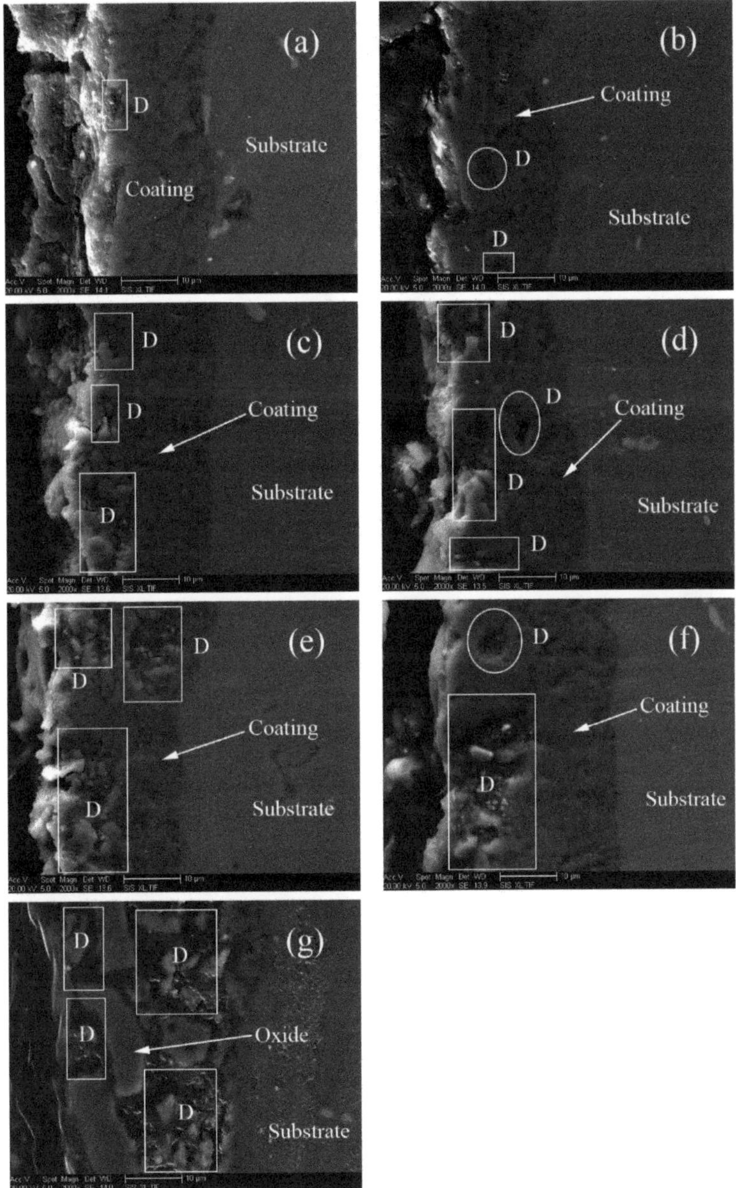

Figure 5. The cross-section morphologies of microarc oxidation coatings formed on aluminum alloys at different anode pulse-widths. (**a**) 1000 μs; (**b**) 1500 μs; (**c**) 2000 μs; (**d**) 2500 μs; (**e**) 3000 μs; (**f**) 3500 μs; (**g**) 4000 μs.

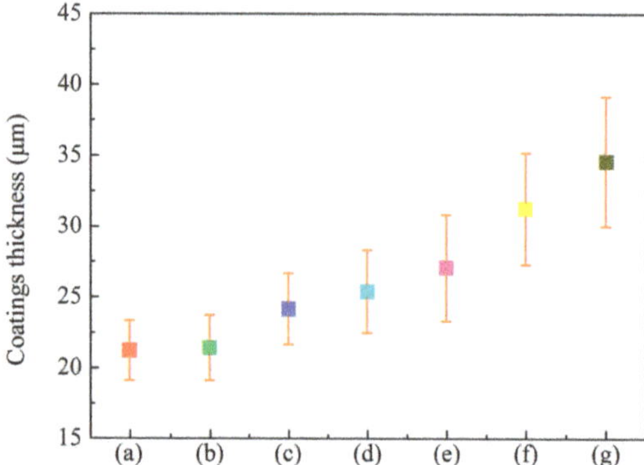

Figure 6. The thickness of microarc oxidation coatings formed on aluminum alloys at different anode pulse-widths. (a): 1000 µs; (b): 1500 µs; (c): 2000 µs; (d): 2500 µs; (e): 3000 µs; (f): 3500 µs; (g): 4000 µs.

The surface morphologies of the microarc oxidation self-lubricating composite coatings are shown in Figure 7. The micropores and microcracks on the surface of the MAO ceramic coatings were filled with the FEP dispersion.

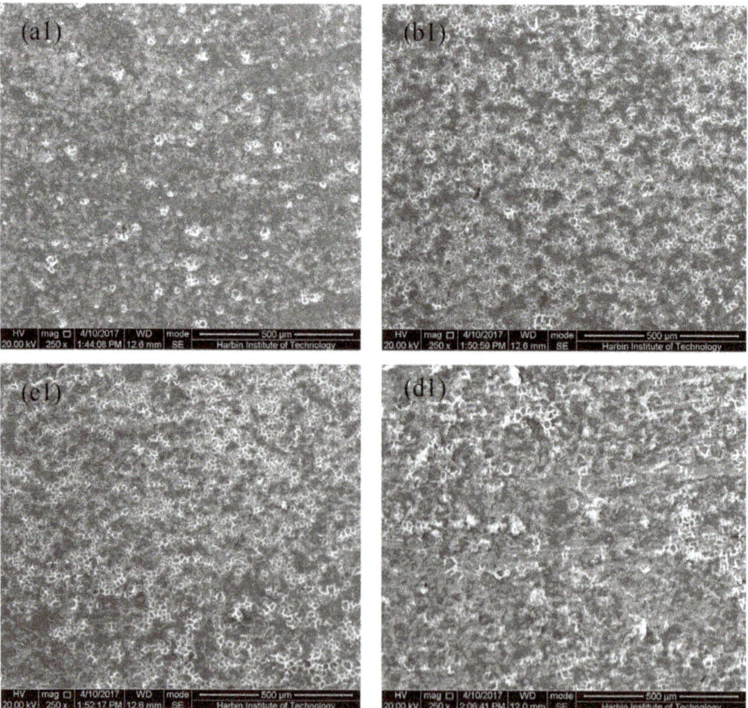

Figure 7. *Cont.*

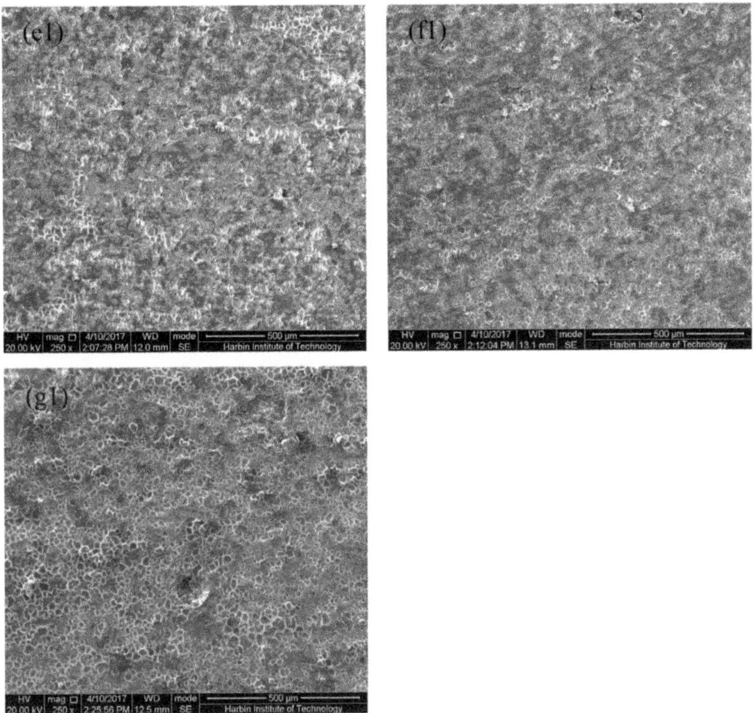

Figure 7. The surface morphologies of the microarc oxidation self-lubricating composite coatings formed on aluminum alloys at different anode pulse-widths. (**a1**) 1000 μs; (**b1**) 1500 μs; (**c1**) 2000 μs; (**d1**) 2500 μs; (**e1**) 3000 μs; (**f1**) 3500 μs; (**g1**) 4000 μs.

3.2. Effect of Anode Pulse-Width on the Phase Structure of MAO Ceramic Coatings

The phase structure of the microarc oxidation ceramic coatings is shown in Figure 8. The XRD analysis results revealed that the MAO ceramic coatings were made up of γ-Al_2O_3, and a small amount of α-Al_2O_3 and mullite ($3Al_2O_3 \cdot 2SiO_2$ and $2Al_2O_3 \cdot SiO_2$). Figure 9 shows the structure of mullite. Three typical γ-Al_2O_3 peaks are marked with 1, 2, and 3 in Figure 8. The intensity of the three typical γ-Al_2O_3 peaks is also listed in Table 2. The intensity of the γ-Al_2O_3 peaks first increased and then decreased with the increase in the anode pulse-width. As shown in Figure 6, the thickness of the MAO coatings increased with the increase in the anode pulse-width. The intensity of the γ-Al_2O_3 peaks increased with the increase in the anode pulse-width.

B-type discharge events penetrated through the entire MAO coating and can promote the formation of the new molten oxide magma on the interface of the coatings/substrates [27]. However, B-type discharge events and the quantity of the formed molten oxide magma on the coatings/substrates interface decreased with the increase in the anode pulse-width. When the anode pulse-width exceeded 1500 μs, the intensity of the γ-Al_2O_3 peaks decreased with the increase in the anode pulse-width.

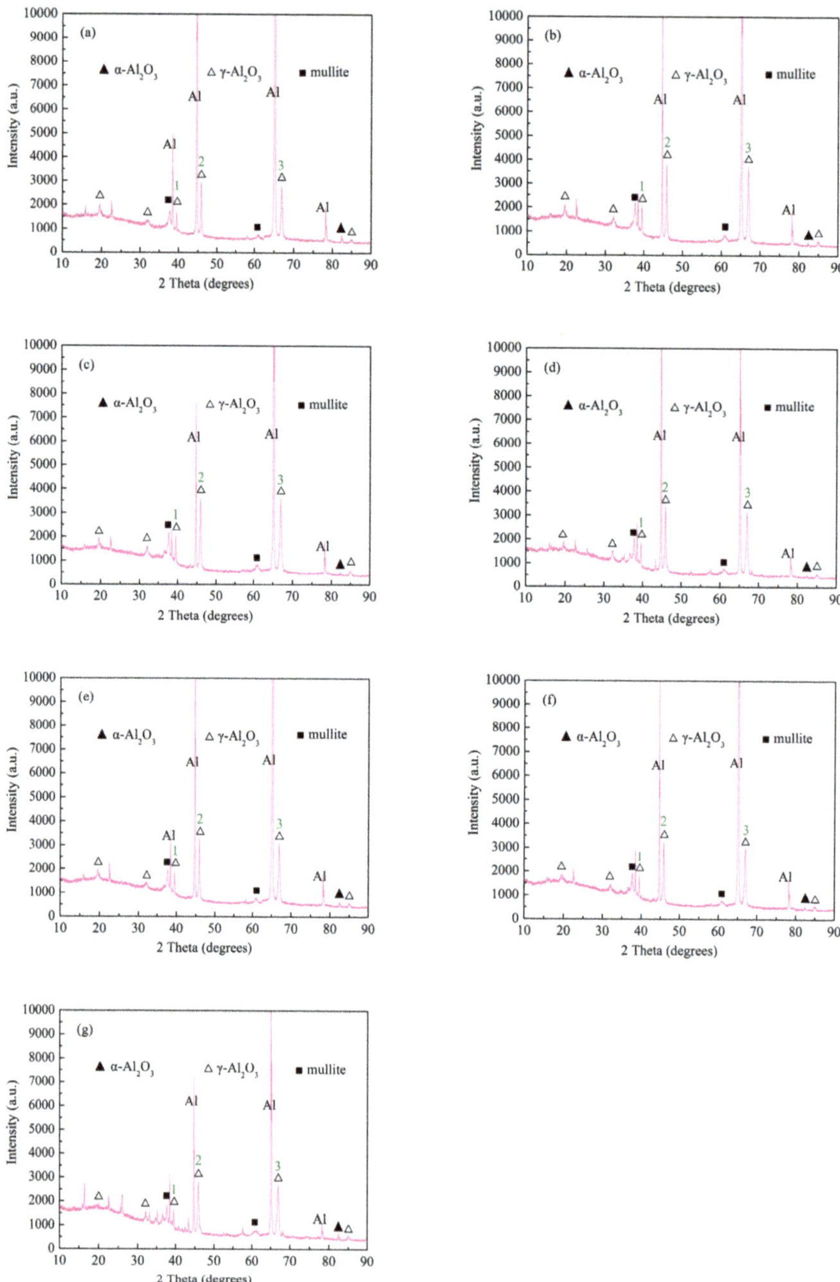

Figure 8. X-ray diffractometer (XRD) patterns of the microarc oxidation coatings formed on aluminum alloys at different anode pulse-widths. (**a**) 1000 μs; (**b**) 1500 μs; (**c**) 2000 μs; (**d**) 2500 μs; (**e**) 3000 μs; (**f**) 3500 μs; (**g**) 4000 μs.

Figure 9. The structure of mullite.

Table 2. The intensity of γ-Al$_2$O$_3$ peaks in the microarc oxidation (MAO) coatings formed on aluminum alloys at different anode pulse-widths.

Labels	Intensity (a.u., arbitrary unit)						
	a	b	c	d	e	f	g
1	1720	2110	2032	1895	1879	1818	1671
2	2988	3789	3607	3404	3335	3253	2853
3	2789	3731	3606	3163	3099	2991	2701

3.3. Effect of Anode Pulse-Width on the Adhesion Strength of MAO Ceramic Coatings

Figures 10 and 11 show the SEM morphologies and supplementary data graphics of scratches on the microarc oxidation coatings, respectively. The critical loads at which the acoustic emission and friction coefficient increased sharply are indicated on the graphs. The corresponding critical load values are summarized in Table 3, and they decreased with the increase in the anode pulse-width. Tekin et al. [28] suggests that the bonding strength between coatings and substrates, and the cohesion of inter-particles can be improved by decreasing the micropores' size and porosity on the surface of the microarc oxidation coatings. The compactness of the microarc oxidation coatings, the bonding strength between the coatings and substrates, as well as the cohesion of inter-particles decreased with the increase in the anode pulse-width.

With the increase of the MAO coating porosity, the area of the mechanical meshing region on the interface between the coatings and substrates as well as the inter-particles in the MAO coatings decreased, which decreased the critical load values.

Moreover, with the increase of the MAO coating porosity, the distance between the coatings and substrates as well as the inter-particles in the MAO coatings increased, which decreased the Van der Waals forces between the coatings and substrates as well as the inter-particles. This decreased the critical load values.

Besides, as shown in Figure 5, the micropores and other defects (region D) on the MAO coatings increased with the increase in the anode pulse-width, which contributed to the formation of the crack sources. These crack sources expand in the direction of the interface between the MAO coatings and substrates and create exfoliation under the applied load during scratch testing. This decreased the critical load values.

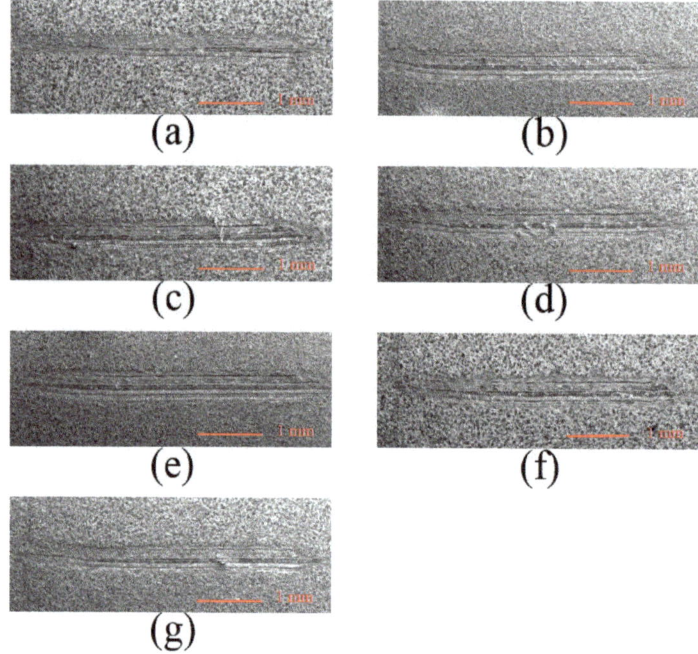

Figure 10. The scanning electron microscope (SEM) morphologies of scratches on the microarc oxidation coatings formed on aluminum alloys at different anode pulse-widths. (**a**) 1000 μs; (**b**) 1500 μs; (**c**) 2000 μs; (**d**) 2500 μs; (**e**) 3000 μs; (**f**) 3500 μs; (**g**) 4000 μs.

Table 3. Critical load values of MAO coatings formed on aluminum alloys at different anode pulse-widths.

Coatings	L_c (N)
A (Figure 10a)	11.58
B (Figure 10b)	10.50
C (Figure 10c)	8.83
D (Figure 10d)	8.17
E (Figure 10e)	7.17
F (Figure 10f)	6.83
G (Figure 10g)	6.67

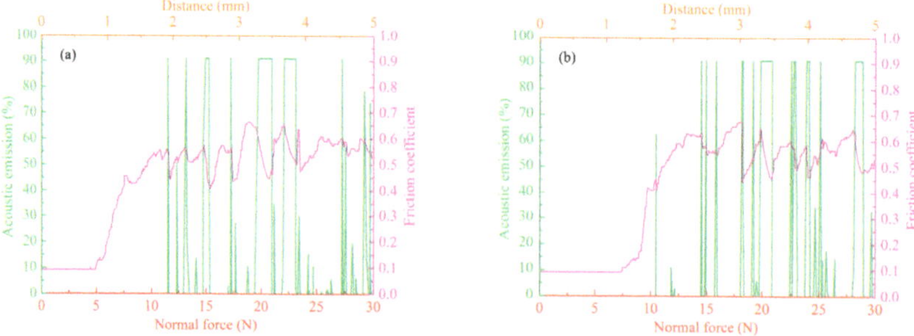

Figure 11. *Cont.*

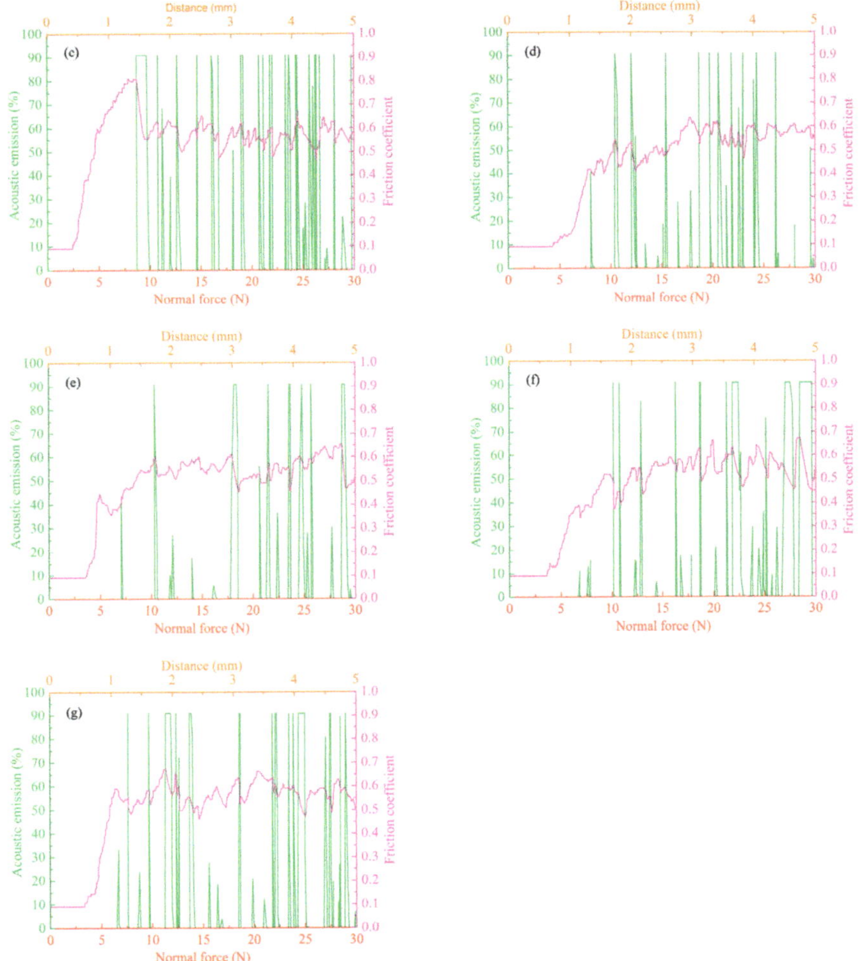

Figure 11. The supplementary data graphics of scratches on the microarc oxidation coatings formed on aluminum alloys at different anode pulse-widths. (**a**) 1000 μs; (**b**) 1500 μs; (**c**) 2000 μs; (**d**) 2500 μs; (**e**) 3000 μs; (**f**) 3500 μs; (**g**) 4000 μs.

3.4. Effect of Anode Pulse-Width on the Wear Resistance of MAO Ceramic Coatings

The hardness of the microarc oxidation ceramic coatings is shown in Figure 12. With the increase in the anode pulse-width, the compactness and hardness of the MAO ceramic coatings decreased gradually.

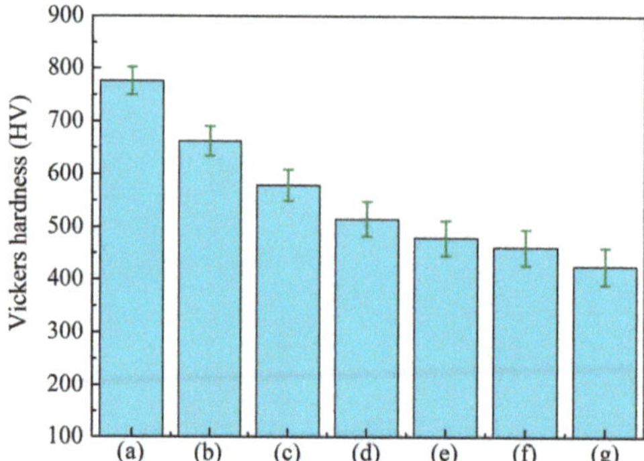

Figure 12. The hardness of the microarc oxidation coatings formed on aluminum alloys at different anode pulse-widths. (a): 1000 μs; (b): 1500 μs; (c): 2000 μs; (d): 2500 μs; (e): 3000 μs; (f): 3500 μs; (g): 4000 μs.

Figure 13 shows the friction coefficient of the microarc oxidation coatings against the WC balls at room temperature during the wear testing. The friction coefficient of the MAO ceramic coatings ranged from 0.60–0.85, determined by the performance of the ceramic itself. Moreover, the friction coefficient was significantly affected by the cracks and pores on the MAO coatings. The wave peaks on the surface of the WC balls easily penetrated into the surface of the MAO coatings due to the high porosity of the MAO coatings. This increased the contact area between the WC balls and MAO coatings as well as the friction coefficient. The uneven distribution of the cracks and pores on the surface of the MAO coatings increased the fluctuation range of the friction coefficient. The ease of penetration by the wave peaks on the surface of the WC balls created serious abrasive wear and produced a significant quantity of wear debris during the wear testing. The microcracks in the MAO coatings easily expanded and formed exfoliation under the periodically applied load during wear testing. The wear debris and exfoliation were involved in the friction process, which significantly affected the friction coefficient.

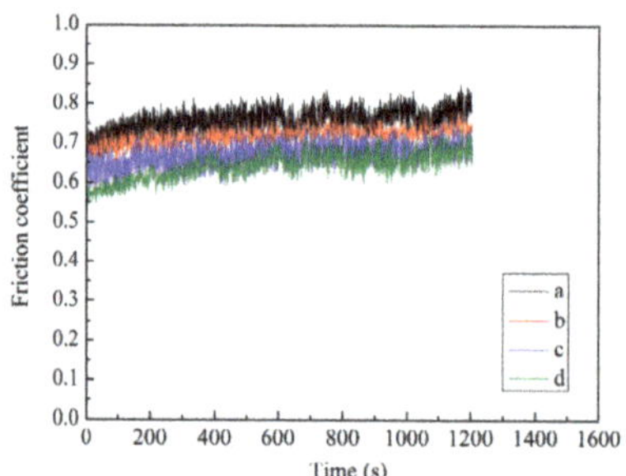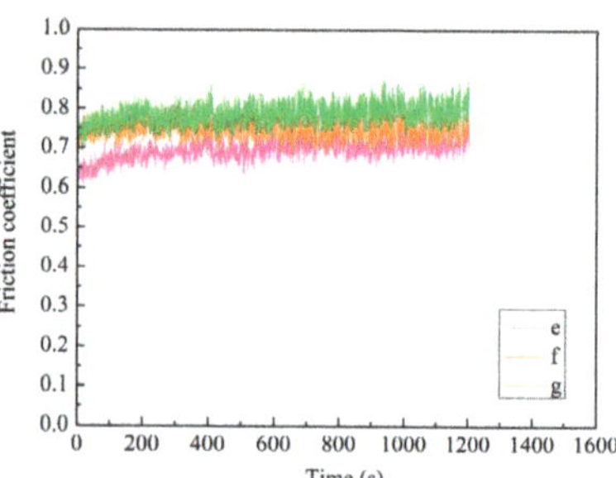

Figure 13. Friction coefficient of the microarc oxidation coatings formed on aluminum alloys at different anode pulse-widths. (a): 1000 μs; (b): 1500 μs; (c): 2000 μs; (d): 2500 μs; (e): 3000 μs; (f): 3500 μs; (g): 4000 μs.

Figure 14 shows the morphologies of the wear tracks for the microarc oxidation coatings against WC balls at room temperature after wear testing. The morphologies of the wear tracks for the MAO coatings exhibited furrows and a wide range of exfoliation. This revealed that the dominant wear mechanisms of the microarc oxidation ceramic coatings were abrasive and fatigue wear. Moreover, the furrows and the region of exfoliation increased in size with the increase in the anode pulse-width.

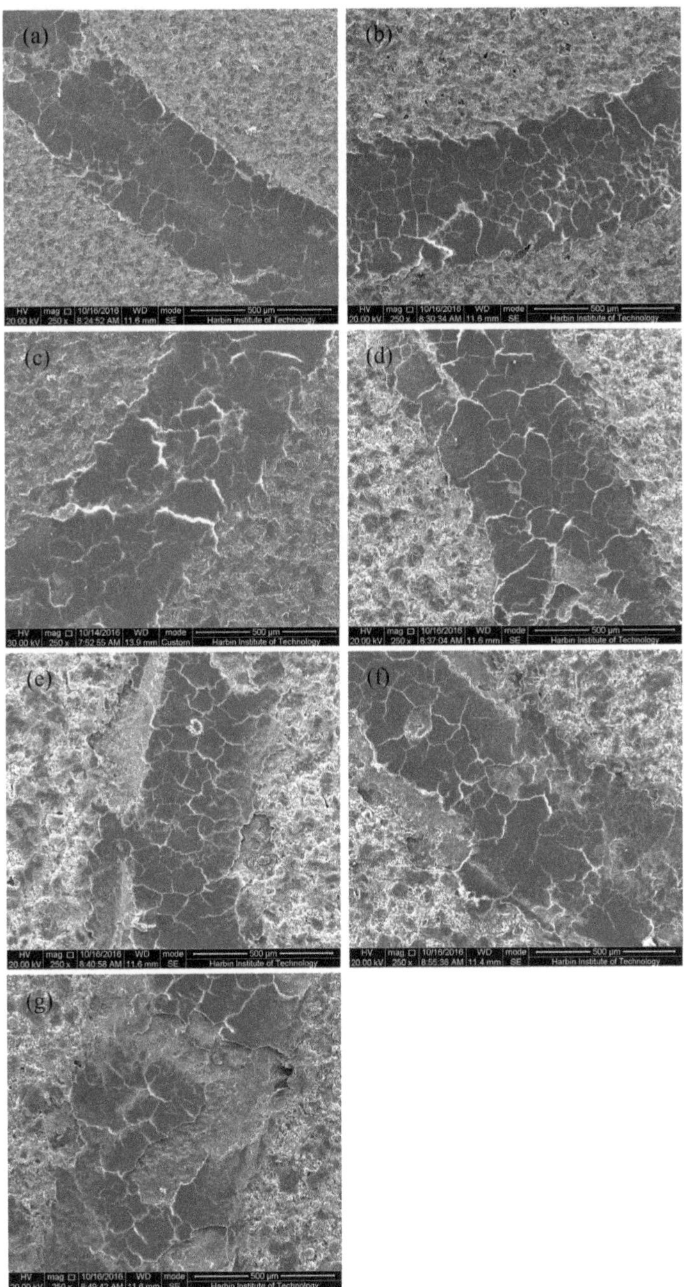

Figure 14. The morphologies of wear tracks for the microarc oxidation coatings formed on aluminum alloys at different anode pulse-widths. (**a**) 1000 μs; (**b**) 1500 μs; (**c**) 2000 μs; (**d**) 2500 μs; (**e**) 3000 μs; (**f**) 3500 μs; (**g**) 4000 μs.

The compactness and hardness of the MAO coatings decreased with the increase in the anode pulse-width. This could make the microbulges on the surface of the WC balls penetrate into the surface of MAO coatings more easily, and increased the abrasive wear during wear testing. Moreover, as shown in Figure 3, some microcracks existed on the surface of the MAO coatings and these microcracks easily expanded and formed exfoliation under the periodically applied load during wear testing.

As shown in Figure 5, many micropores and defects (region D) in the MAO coatings contributed to the formation of crack sources. These crack sources expand and create exfoliation under the periodically applied load during wear testing. The porosity and defects in the MAO coatings increased and the compactness of the MAO coatings decreased with the increase in the anode pulse-width. This increased the formation of crack sources and exfoliation during wear testing.

Table 4 shows the wear rates of the MAO coatings. The wear rates of the MAO coatings increased significantly with the increase in the anode pulse-width. As shown in Figures 3 and 5, the porosity and defects of the MAO coatings increased with the increase in the anode pulse-width, which significantly increased the furrows, exfoliation, and the wear rates of the MAO coatings during wear testing.

Table 4. Wear rates of the MAO coatings formed on aluminum alloys at different anode pulse-widths.

Coatings	Wear Time (s)	Wear Track Width (μm)	Wear Track Depth (μm)	Wear Rate (mm^3·N^{-1}·m^{-1})
a	1200	420.56	3.19	0.13×10^{-5}
b	1200	490.65	3.73	0.15×10^{-5}
c	1200	560.75	4.26	0.18×10^{-5}
d	1200	654.21	4.97	0.20×10^{-5}
e	1200	700.94	5.32	0.21×10^{-5}
f	1200	724.30	5.50	0.22×10^{-5}
g	1200	794.39	6.03	0.24×10^{-5}

3.5. Tribological Properties of MAO Self-Lubricating Composite Coatings

Figure 15 shows the friction coefficient of the MAO self-lubricating composite coatings against WC balls at room temperature during wear testing. As shown in Figure 15, compared to the MAO coatings, the MAO self-lubricating composite coatings exhibited a lower friction coefficient. The friction coefficient was in the range of 0.09–0.25. Moreover, the friction coefficient of the MAO self-lubricating composite coatings decreased with increased anode pulse-width. It can be inferred that the porosity of the MAO coatings increased with increased anode pulse-width, the FEP in micropores and microcracks on the surface of the MAO ceramic coatings increased, and the lubrication was sufficient during wear testing, leading to the decreased friction coefficients for the MAO self-lubricating composite coatings with increased anode pulse-width.

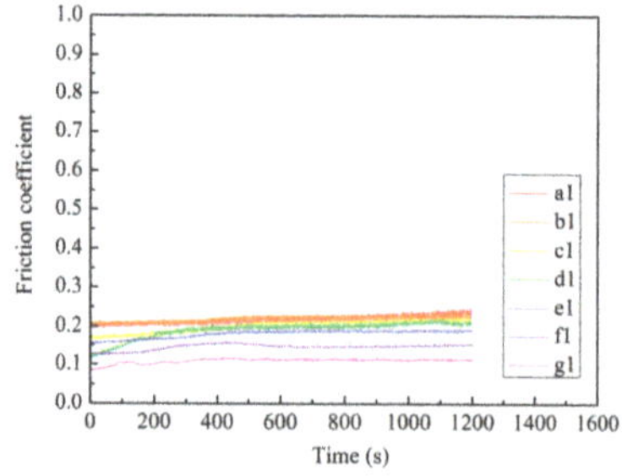

Figure 15. Friction coefficient of the microarc oxidation self-lubricating composite coatings formed on aluminum alloys at different anode pulse-widths. (a1): 1000 μs; (b1): 1500 μs; (c1): 2000 μs; (d1): 2500 μs; (e1): 3000 μs; (f1): 3500 μs; (g1): 4000 μs.

Figure 16 shows the morphologies of the wear tracks for the MAO self-lubricating composite coatings against WC balls at room temperature after wear testing. As shown in Figure 16, the morphologies of the wear tracks for the MAO coatings did not exhibit a significant quantity of furrows or exfoliation. Thus, the dominant wear mechanism of the MAO self-lubricating composite coatings was mild abrasive wear.

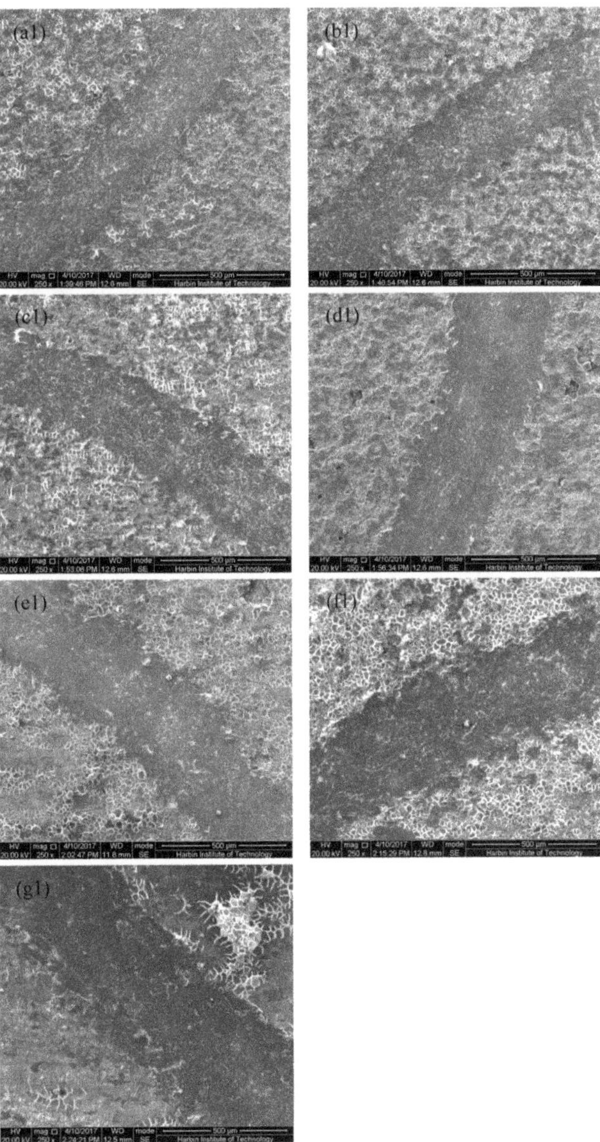

Figure 16. The morphologies of wear tracks for the MAO self-lubricating composite coatings formed on aluminum alloys at different anode pulse-widths. (**a1**) 1000 μs; (**b1**) 1500 μs; (**c1**) 2000 μs; (**d1**) 2500 μs; (**e1**) 3000 μs; (**f1**) 3500 μs; (**g1**) 4000 μs.

Table 5 shows the wear rates of the MAO self-lubricating composite coatings. Compared to the MAO coatings, the MAO self-lubricating composite coatings exhibited lower wear rates, and the wear rates of the MAO self-lubricating composite coatings first decreased and then increased with the increased anode pulse-width. When the anode pulse-width was 2000 µs, the wear rate of the MAO self-lubricating composite coatings was the lowest. Because the porosity of the MAO coatings increased as the anode pulse-width increased from 1000 to 2000 µs, the FEP in the micropores and microcracks on the surface of the MAO ceramic coatings increased and the lubrication was sufficient during wear testing.

Table 5. Wear rates of the MAO self-lubricating composite coatings formed on aluminum alloys at different anode pulse-widths.

Coatings	Wear Time (s)	Wear Track Width (µm)	Wear Track Depth (µm)	Wear Rate ($mm^3 \cdot N^{-1} \cdot m^{-1}$)
a1	1200	380.47	2.92	0.12×10^{-5}
b1	1200	357.10	2.81	0.11×10^{-5}
c1	1200	333.74	2.73	0.10×10^{-5}
d1	1200	403.83	3.16	0.13×10^{-5}
e1	1200	427.20	3.24	0.13×10^{-5}
f1	1200	473.92	3.69	0.15×10^{-5}
g1	1200	520.65	4.15	0.17×10^{-5}

However, when the anode pulse-width exceeded 2000 µs, the wear rates of the MAO self-lubricating composite coatings increased with the increased anode pulse-width, because the porosity of the MAO coatings increased, and the compactness as well as the hardness of the MAO ceramic coatings decreased with increased anode pulse-width. This led to increased wear rates for the MAO self-lubricating composite coatings with increased anode pulse-width.

4. Conclusions

The 2024-T4 aluminum alloys were treated with microarc oxidation in the silicate system electrolyte using a pulsed bipolar power supply at different anode pulse-widths. After MAO, the micropores and microcracks on the surface of the MAO ceramic coatings were filled with FEP dispersion to prepare the micro-arc oxidation self-lubricating composite coatings containing FEP. The hardness and wear resistance of the MAO coatings decreased as the anode pulse-width increased. Compared to the MAO coatings, the friction coefficient and wear rates of the MAO self-lubricating composite coatings sharply decreased. Moreover, the wear rates of the MAO self-lubricating composite coatings first decreased and then increased with increased anode pulse-width. When the anode pulse-width was 2000 µs, the wear rate of the MAO self-lubricating composite coating was the lowest. The MAO self-lubricating composite coatings integrated the advantages of wear resistance of the MAO coatings and a low friction coefficient of FEP, possessing superior tribological properties.

Acknowledgments: The authors gratefully acknowledge the Micro/Nano Technology Research Center, Harbin Institute of Technology for device support.

Author Contributions: Zhen-Wei Li and Shi-Chun Di conceived and designed the experiments; Zhen-Wei Li performed the experiments; Zhen-Wei Li and Shi-Chun Di analyzed the data; Shi-Chun Di contributed reagents/materials/analysis tools; Zhen-Wei Li wrote the paper.

Conflicts of Interest: The authors declare no conflicts of interest.

References

1. Wang, X.S.; Guo, X.W.; Li, X.D.; Ge, D.Y. Improvement on the fatigue performance of 2024-T4 alloy by synergistic coating technology. *Materials* **2014**, *7*, 3533–3546. [CrossRef]

2. Wang, P.; Li, J.P.; Guo, Y.C.; Yang, Z.; Wang, J.L. Ceramic coating formation on high Si containing Al alloy by PEO process. *Surf. Eng.* **2016**, *32*, 428–434. [CrossRef]

3. Rao, R.N.; Das, S.; Mondal, D.P.; Dixit, G. Effect of heat treatment on the sliding wear behavior of aluminium alloy (Al–Zn–Mg) hard particle composite. *Tribol. Int.* **2010**, *43*, 330–339. [CrossRef]

4. Tseng, C.C.; Lee, J.L.; Kuo, T.h.; Kuo, S.N.; Tseng, K.H. The influence of sodium tungstate concentration and anodizing conditions on microarc oxidation (MAO) coatings for aluminum alloy. *Surf. Coat. Technol.* **2012**, *206*, 3437–3443. [CrossRef]

5. Polat, A.; Makaraci, M.; Usta, M. Influence of sodium silicate concentration on structural and tribological properties of microarc oxidation coatings on 2017A aluminum alloy substrate. *J. Alloys Compd.* **2010**, *504*, 519–526. [CrossRef]

6. Nimura, K.; Sugawara, T.; Jibiki, T.; Ito, S.; Shima, M. Surface modification of aluminum alloy to improve fretting wear properties. *Tribol. Int.* **2016**, *93*, 702–708. [CrossRef]

7. Wang, Z.J.; Wu, L.; Qi, Y.L.; Cai, W.; Jiang, Z.H. Self-lubricating Al_2O_3/PTFE composite coating formation on surface of aluminium alloy. *Surf. Coat. Technol.* **2010**, *204*, 3315–3318. [CrossRef]

8. Tsutumi, Y.; Niinomi, M.; Nakai, M.; Shimabukuro, M.; Ashida, M.; Chen, P.; Doi, H.; Hanawa, T. Electrochemical surface treatment of a β-titanium alloy to realize an antibacterial property and bioactivity. *Metals* **2016**, *6*, 76. [CrossRef]

9. Yavari, S.A.; Necula, B.S.; Fratila-Apachitei, L.E.; Duszczyk, J.; Apachitei, I. Biofunctional surfaces by plasma electrolytic oxidation on titanium biomedical alloys. *Surf. Eng.* **2016**, *32*, 411–417. [CrossRef]

10. Han, O.S.; Hwang, M.J.; Song, Y.H.; Song, H.J.; Park, Y.J. Effects of surface structure and chemical composition of binary Ti alloys on cell differentiation. *Metals* **2016**, *6*, 150. [CrossRef]

11. Mioč, U.B.; Stojadinović, S.; Nedić, Z. Characterization of bronze surface layer formed by microarc oxidation process in 12-tungstophosphoric acid. *Materials* **2010**, *3*, 110–126. [CrossRef]

12. Cheng, Y.L.; Xue, Z.G.; Wang, Q.; Wu, X.Q.; Matykina, E.; Skeldon, P.; Thompson, G.E. New findings on properties of plasma electrolytic oxidation coatings from study of an Al-Cu-Li alloy. *Electrochim. Acta* **2013**, *107*, 358–378. [CrossRef]

13. Chen, M.A.; Ou, Y.C.; Yu, C.Y.; Xiao, C.; Liu, S.Y. Corrosion performance of epoxy/BTESPT/MAO coating on AZ31 alloy. *Surf. Eng.* **2016**, *32*, 38–46. [CrossRef]

14. Guan, Y.J.; Xia, Y.; Xu, F.T. Interface fracture property of PEO ceramic coatings on aluminum alloy. *Surf. Coat. Technol.* **2008**, *202*, 4204–4209. [CrossRef]

15. Martin, J.; Leone, P.; Nomine, A.; Renaux, D.V.; Henrion, G.; Belmonte, T. Influence of electrolyte ageing on the plasma electrolytic oxidation of aluminium. *Surf. Coat. Technol.* **2015**, *269*, 36–46. [CrossRef]

16. Shokouhfar, M.; Dehghanian, C.; Baradaran, A. Preparation of ceramic coating on Ti substrate by Plasma electrolytic oxidation in different electrolytes and evaluation of its corrosion resistance. *Appl. Surf. Sci.* **2011**, *257*, 2617–2624. [CrossRef]

17. Jin, F.Y.; Chu, P.K.; Xu, G.D.; Zhao, J.; Tang, D.; Tong, H.H. Structure and mechanical properties of magnesium alloy treated by micro-arc discharge oxidation using direct current and high-frequency bipolar pulsing modes. *Mater. Sci. Eng. A* **2006**, *435–436*, 123–126. [CrossRef]

18. Bayati, M.R.; Moshfegh, A.Z.; Golestani-Fard, F. Effect of electrical parameters on morphology, chemical composition, and photoactivity of the nano-porous titania layers synthesized by pulse-microarc oxidation. *Electrochim. Acta* **2010**, *55*, 2760–2766. [CrossRef]

19. Montazeri, M.; Dehghanian, C.; Shokouhfar, M.; Baradaran, A. Investigation of the voltage and time effects on the formation of hydroxyapatite-containing titania prepared by plasma electrolytic oxidation on Ti–6Al–4V alloy and its corrosion behavior. *Appl. Surf. Sci.* **2011**, *257*, 7268–7275. [CrossRef]

20. Mu, M.; Zhou, X.J.; Xiao, Q.; Liang, J.; Huo, X.D. Preparation and tribological properties of self-lubricating TiO_2/graphite composite coating on TI6Al4V alloy. *Appl. Surf. Sci.* **2012**, *258*, 8570–8576. [CrossRef]

21. Hussein, R.O.; Northwood, D.O.; Su, J.F.; Xie, X. A study of the interactive effects of hybrid current modes on the tribological properties of a PEO (plasma electrolytic oxidation) coated AM60B Mg-alloy. *Surf. Coat. Technol.* **2013**, *215*, 421–430. [CrossRef]

22. Wu, H.H.; Lu, X.Y.; Long, B.H.; Wang, X.Q.; Wang, J.B.; Jin, Z.S. The effects of cathodic and anodic voltages on the characteristics of porous nanocrystalline titania coatings fabricated by microarc oxidation. *Mater. Lett.* **2005**, *59*, 370–375. [CrossRef]

23. Li, Q.B.; Liang, J.; Liu, B.X.; Peng, Z.J.; Wang, Q. Effects of cathodic voltages on structure and wear resistance of plasma electrolytic oxidation coatings formed on aluminium alloy. *Appl. Surf. Sci.* **2014**, *297*, 176–181. [CrossRef]

24. Yin, B.; Peng, Z.J.; Liang, J.; Jin, K.J.; Zhu, S.Y.; Yang, J.; Qiao, Z.H. Tribological behavior and mechanism of self-lubricating wear-resistant composite coatings fabricated by one-step plasma electrolytic oxidation. *Tribol. Int.* **2016**, *97*, 97–107. [CrossRef]

25. Qin, Y.K.; Xiong, D.S.; Li, J.L. Characterization and friction behavior of LST/PEO duplex-treated Ti6Al4V alloy with burnished MoS$_2$ film. *Appl. Surf. Sci.* **2015**, *347*, 475–484. [CrossRef]

26. Wang, S.Y.; Si, N.C.; Xia, Y.P.; Liu, L. Influence of nano-SiC on microstructure and property of MAO coating formed on AZ91D magnesium alloy. *Trans. Nonferr. Met. Soc. China* **2015**, *25*, 1926–1934. [CrossRef]

27. Hussein, R.O.; Xie, X.; Northwood, D.O.; Yerokhin, A.; Matthews, A. Spectroscopic study of electrolytic plasma and discharging behavior during the plasma electrolytic oxidation (PEO) process. *J. Phys. D Appl. Phys.* **2010**, *43*, 105203. [CrossRef]

28. Tekin, K.C.; Malayoglu, U.; Shrestha, S. Tribological behaviour of plasma electrolytic oxide coatings on Ti6Al4V and cp-Ti alloys. *Surf. Eng.* **2016**, *32*, 435–442. [CrossRef]

 *metals*

Article

Study in Wire Feedability-Related Properties of Al-5Mg Solid Wire Electrodes Bearing Zr for High-Speed Train

Bo Wang [1], Songbai Xue [1,*], Chaoli Ma [1], Jianxin Wang [2] and Zhongqiang Lin [3]

[1] College of Materials Science and Technology, Nanjing University of Aeronautics and Astronautics, Nanjing 210016, China; wangbo4175@126.com (B.W.); machaoli006@163.com (C.M.)

[2] Jiangsu Provincial Key Laboratory of Advanced Welding Technology, Jiangsu University of Science and Technology, Zhenjiang 212003, China; wangjx_just@126.com

[3] Zhejiang Yuguang Aluminum Material Co., Ltd., Jinhua 321200, China; linzhongqiang666@163.com

* Correspondence: xuesb@nuaa.edu.cn; Tel.: +86-025-8489-6070

Received: 24 September 2017; Accepted: 20 November 2017; Published: 23 November 2017

Abstract: This work offers an analysis of the wire feedability-related properties of Al-5Mg solid wire electrodes bearing Zr. Effects of Zr content on microstructures and mechanical properties of the Al-5Mg alloys were studied. Experimental results have demonstrated that α-Al dendrites of the as-cast Al-5Mg alloy are refined, and the tensile strength, microhardness and roughness of the 1.2 mm wire electrode are improved with an appropriate addition of Zr. In addition, the tensile strength and elongation of the welded joints welded using Al-5Mg wire electrodes bearing Zr reach the maximum value when 0.12% Zr is added into the wire alloy. However, when excess Zr is added, α-Al phases of the wire alloy and welded joint are coarsened, and the mechanical properties are deteriorated. Moreover, the structure and principle of a novel apparatus, which can enhance the feedability of the wire electrode, are introduced and the apparatus can achieve the rough and fine adjustments of cast and helix of the wire electrode.

Keywords: Al-5Mg wire electrode; Zr; wire feedability; microstructure; mechanical properties

---

1. Introduction

Aluminum and its alloys are widely used in aerospace, railway, shipbuilding, pressure vessels and other fields due to their high specific strength, good corrosion resistance, and so on [1,2]. Among these applications, 5XXX, 6XXX and 7XXX aluminum alloys are often applied to the manufacture of high-speed train car bodies [3–5], and lots of research on advanced welding techniques for welding aluminum car bodies have been carried out at home and abroad [6–8]. With the development of robot welding automation technologies, the gas metal arc welding (GMAW) method has drawn much attention. During the GMAW process, the internal and external quality of welding consumables affect the shaping and quality of a weld. However, compared with imported aluminum wire electrodes, it is considered that domestic ones used for automatic and semi-automatic GMAW still have many urgent problems to be solved. These problems are serious arc wander and frequent downtime due to wire feeding problems [9].

The issue of aluminum wire electrode feedability has been studied by several authors [10–12]. Padilla et al. [10] reported that a key factor in the performance of GMAW was the feedability of the welding wire through the wire liner, and for aluminum wire electrodes, fluctuations in the wire feed speed (WFS) of even 1% could result in irregular arc lengths, oscillatory voltage and current levels, and the degradation of the overall weld quality. In part, variations in the target WFS were caused by adverse conditions during welding. These included damaging effects such as the "stick-slip" motion of the welding wire, the premature wear of the contact tube and, in general, the movement of the hose package during welding. Previous research has pointed out that, during an automatic GMAW process, considering only the characteristics of the aluminum wire electrode, the stability of arc is mainly determined by the diameter, stiffness (column strength and elongation), surface quality and the winding characteristics of the wire electrode [9,11–14]. The Al-5Mg wire electrode, which is based on Mg as the main alloying element, is a kind of general-purpose welding consumable for welding 5XXX, 6XXX and 7XXX alloys. Tang et al. [9] studied the effect of the column strength of the ER5356 wire electrode with a diameter of 1.2 mm on the wire feedability during welding. They found that the wire electrode with a high column strength (448 MPa) could offer very good rigidity and was easier to push through a GMAW torch. In addition, ER5356 wire electrode with higher column strength has also been proved to feed better and yield welds that are stronger and more ductile compared with ER4043 wire electrode [13,14]. However, there still exists a huge gap between imported Al-Mg wire electrodes and domestic ones in terms of chemical composition control, melt refining, wire drawing and surface finishing technologies.

The method of microalloying plays a vital role in improving the mechanical properties of the aluminum alloy. Researches [15–19] have revealed the influences of some modification elements such as Zr, Sc and Er on the microstructure and mechanical properties of the aluminum wires and/or the welded joints. Al-Zr alloys have already been used according to international standard IEC 62004 by a number of widely known companies, such as Lamifil (Hemiksem, Belgium), 3M (Minnesota Mining and Manufacturing; St. Paul, MN, USA), J-Power Systems (Hitachi; Ibaraki, Japan) and others during the development of a heat-resistant aluminum wire for overhead power transmission lines [15]. Chao et al. [16] studied the mechanical properties and electrical conductivities of aluminum conductive wires of Al-0.16Zr, Al-0.16Sc, Al-0.12Sc-0.04Zr (mass %) and pure Al (99.996%) with a diameter of 9.5 mm. They found that the separate addition of 0.16% Sc and 0.16% Zr to pure Al improved the ultimate tensile strength but reduced the electrical conductivity, and a similar trend was found in the Al-0.12Sc-0.04Zr alloy. As for Al-Mg alloy, Taendl et al. [20] also reported that a balanced addition of small amounts of Sc and Zr was considered as one of the most promising approaches to increasing the specific strength of Al-Mg alloys, while maintaining the beneficial materials properties. With regard to the microalloying of the Al-5Mg welding wire, Norman et al. [21] researched the influence of the Sc concentration in the fusion zone of GMAW welds of typical 7XXX aluminum aerospace alloys; they noted that the Sc-bearing welding wire greatly outperformed all the conditional commercial welding wires (ER5087, 5180 and 5039), both in terms of weld strength and ductility. Zhang et al. [19] indicated that the addition of Sc into the aluminum alloy could ostentatiously boost the strength of alloy, while the low cost element Er shared similar functions with Sc in the aluminum alloy, so they studied the microstructure characteristics of welded joints of 7A52 welded by fiber laser, and using Al-Mg-Mn welding wire and Al-Mg-Mn-Zr-Er welding wire, respectively. They found that a huge amount of fine equiaxed grains was formed in the weld zone of Zr and Er micro-alloying Al-Mg-Mn welding wire. Xu et al. [18] also investigated the influence of the combinative addition of Er and Zr on the microstructure and mechanical properties of the GTAW welded joints of 7A52. By comparing the reference [18,19], although the thicknesses of the aluminum alloy plates were different (4 mm and 20 mm, respectively) and the welding methods were different (GTAW and fiber laser welding, respectively), a significant grain refinement was found in the weld zone of the two welded joints due to the formation of fine Al_3Er, Al_3Zr and $Al_3(Zr, Er)$.

The abovementioned investigations focused mainly on the influence of the separate addition of Zr or the combinative addition of Zr and Er (or Sc) on the microstructures and the mechanical properties of aluminum conductive wires (pure aluminum) or the welded joints of 7XXX alloy. However, very few studies have been focused on the influence of the separate addition of Zr used as microalloying element on the wire feedability related properties of the Al-5Mg wire electrode. Therefore, the microstructure and mechanical properties of the Al-5Mg wire alloys bearing different contents of Zr were investigated in this study. The produced wires with diameter of 1.2 mm were used to welding the 5083-H112 plate. The 5083 alloy was selected as the welding coupons according to the European Standard EN 14532-3:2004 [22] and according to the specifications of approval testing for Al-5Mg welding wire specified by the Deutsche Bahn (DB) and the American Bureau of Shipping (ABS). The optimum Zr content in Al-5Mg wire alloy was obtained by the analysis of effects of Zr on the wire feedability-related properties of Al-5Mg-xZr wire electrodes and mechanical properties of resulting welded joints. Furthermore, a novel adjustable apparatus to enhance the wire feedability of the wire electrode was also introduced and the principle, structure and application results of the apparatus were simply analyzed.

2. Materials and Methods

Wire alloy for experiments was prepared using pure aluminum (99.7% Al), magnesium ingot, metal additives (75% metal) of manganese, titanium and chromium. The trace element Zr was added in the form of Al-10Zr master alloy at 720–750 °C, and the melt was purged with hexachloroethane (C_2Cl_6) and pure argon. The chemical compositions of wire electrodes are listed in Table 1, and the composition was measured using a SPECTRO® MAXx04D spectrometer (SPECTRO Analytical Instruments GmbH, Kleve, Germany).

Table 1. Chemical compositions of wire electrodes (wt %).

No.	Material	Mg	Mn	Cr	Ti	Zr	Al
1#	Al-5Mg	4.93	0.124	0.06	0.08	0.0003	balance
2#	Al-5Mg-0.06Zr	5.03	0.122	0.06	0.08	0.058	balance
3#	Al-5Mg-0.12Zr	5.05	0.123	0.06	0.08	0.123	balance
4#	Al-5Mg-0.24Zr	4.98	0.121	0.06	0.08	0.236	balance
5#	Al-5Mg-0.5Zr	5.07	0.122	0.06	0.08	0.52	balance

The casting process was conducted using a continuous casting and direct water-cooling method. The produced cast rods with a diameter of approximately 8.0 mm subsequently underwent these procedures, which included annealing-rolling-annealing-drawing-shaving-cleaning, to prepare the 1.2 mm diameter wire electrodes. The base metal was 5083-H112 alloy with dimensions of 10 mm × 175 mm × 350 mm intended for GMAW. Prior to welding, the base material coupons were wire brushed and degreased with acetone. The welding process was carried out according to the standard developed by ABS: "ABS-2017: Rules for Materials and Welding-Part 2: Aluminum and Fiber Reinforced Plastics (FRP)". An AOTAI P-MIG 500 welder (Aotai Electric, Jinan, China) was employed to weld the AA5083 plate.

Tensile specimens were prepared from the 1.2 mm wire electrodes and from the welded coupons. Each wire electrode used for tensile test was 200 mm long and the gauge length was 100 mm. Full-size transverse tensile specimens were machined from the welded plates according to ASTM E8 so that the weld was centered in the gauge section (the gauge length was 100 mm), and the loading axis was normal to the welding direction. To avoid the effect of weld profile on the stress concentration during the tensile test, the weld reinforcement was grinded with a polisher (Fujian Hitachi Koki Co., Ltd., Fuzhou, China) prior to being tested. Tensile tests were performed respectively on a WDW-20E electronical universal testing machine and a WAW-300B hydraulic one (Zhejiang Jingyuan Mechanical Equipment Co., Ltd., Jinhua, China). The elongation of the wire electrode and the resulting welded joint was the ratio of the variation in the gauge length to the original gauge length value. The Vicker's hardness measurements were conducted with 1.0 mm spacing on the wire electrodes and the cross section of these welded joints using an HXS-1000A microhardness tester (Teshi Testing Technology (Shanghai) Co., Ltd., Shanghai, China), under a load of 2.94 N for 15 s. Microstructures of Al-5Mg-xZr alloys and the resulting welds were characterized using a TV-400D optical microscopy (Shanghai Tuanjie Instrument Manufacturing Co., Ltd., Shanghai, China) and a ZEISS ΣIGMA field-emission scanning electron microscope coupled to an energy dispersion X-ray (FE-SEM/EDX; Carl Zeiss, Oberkochen, Germany).

3. Results and Discussion

3.1. Microstructures of As-Cast Al-5Mg-xZr Wire Alloys

Figures 1 and 2 show the optical microstructures and the scanning electron microscopy (SEM) microstructures of as-cast Al-5Mg-xZr wire alloys, respectively. As shown in Figures 1a and 2a, the alloy (no alloying Zr) was mainly composed of α-Al matrix, granule Mg_5Al_8 precipitated phase and a small amount of Mn- or Cr-bearing intermetallic compounds (IMCs). Most of Mg element was dissolved in the α-Al matrix. However, the α-Al grains, which occupied a large volume fraction in the microstructure, were coarse and unevenly distributed. When adding 0.06 wt % Zr into the as-cast alloy, α-Al dendrites were refined (Figure 1b). With the increase of Zr content, the refinement effect on α-Al dendrites was further enhanced. When the addition of Zr in the wire alloy was 0.12%, refined equiaxed α-Al dendrites were obtained, the size of which was the smallest (Figures 1c and 2b). However, with a further increase in Zr content (up to 0.24% Zr), the size of α-Al grains became large.

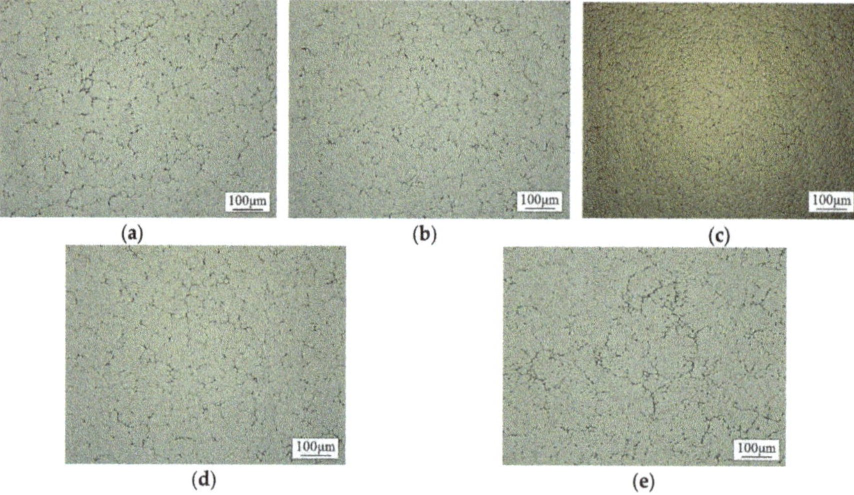

Figure 1. Optical microstructures of as-cast Al-5Mg-*x*Zr alloys: (**a**) *x* = 0; (**b**) *x* = 0.06; (**c**) *x* = 0.12; (**d**) *x* = 0.24; (**e**) *x* = 0.5.

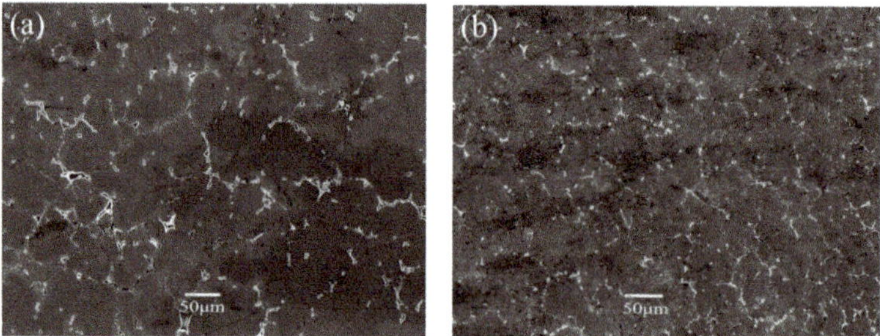

Figure 2. Scanning electron microscopy (SEM) microstructures of as-cast Al-5Mg-*x*Zr alloys: (**a**) *x* = 0; (**b**) *x* = 0.12.

As shown in Figure 1e, when the content of Zr in the alloy reached 0.5%, α-Al dendrites in the matrix were obviously coarsened, and the microstructure uniformity of the as-cast alloy was seriously deteriorated. Figure 3 shows the SEM microstructure at high magnification of the as-cast Al-5Mg-0.5Zr alloy. Coarse lath-like IMCs were found on the surface of the α-Al phase. A further EDS (Energy Dispersive Spectroscopy) analysis of the lath-like IMC (Point A; IMC: Intermetallic Compound) showed that the lath-like IMC consisted of element Al and Zr and the atomic ratio of Al to Zr was close to 3:1. Thus, the lath-like phase could be determined as the Al_3Zr phase.

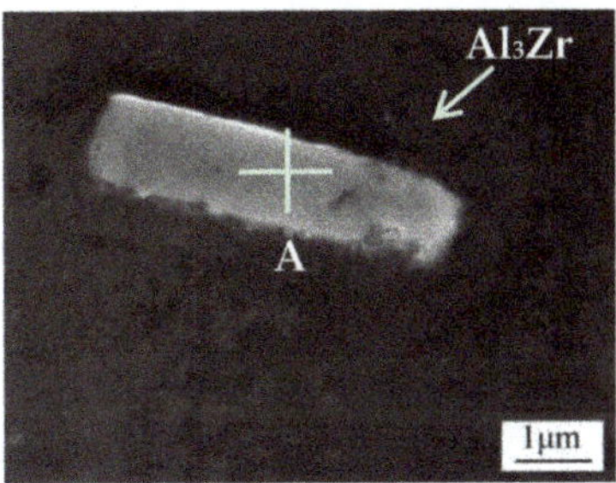

Figure 3. SEM image of the lath-like IMC (Intermetallic Compound) bearing Zr.

The refining effect of element Zr on α-Al dendrites in the wire alloy can be explained through the following analysis: when no Zr was added, Ti in the alloy chemically reacted with Al, producing fine Al_3Ti phases with uniform distribution. Al_3Ti phases were effective heterogeneous substrates for nucleation of α-Al dendrites and, hence, led to the refinement of α-Al phases. However, Ma et al. [23] had discovered that the refining effect of Ti addition alone on Al-Mg alloy was limited, and excessive Ti would result in the serious segregation of Ti in the alloy, which deteriorated the refining effect of Ti on α-Al dendrites. When adding a proper amount of Zr into the wire alloy (Figure 1b,c and Figure 2b), finely dispersed Al_3Zr particles, which kept coherent with Al matrix, were produced from the peritectic reaction of Zr and Al [24]. According to the theory of heterogeneous nucleation, the combined effect of Al_3Zr particles and Al_3Ti phases greatly increased the amounts of heterogeneous substrates and then significantly enhanced the refining effect on the alloy. In addition, according to the Al-Zr binary phase diagram, the maximum solid solubility of Zr in Al was only 0.07%. Thus, during the solidification of liquid Al-5Mg-0.12Zr alloy, element Zr was enriched in front of the solid-liquid interface and led to constitutional supercooling, which promoted the necking of α-Al dendrites, increased the amounts of secondary dendrites and then refined α-Al grains of the wire alloy. However, when excess Zr was added, a large number of Al_3Zr IMCs would form in the alloy, which reduced the enrichment of Zr at the front of solid-liquid interface, resulting in the coarsening of the microstructure.

3.2. Mechanical Properties of 1.2 mm Al-5Mg-xZr Wire Electrodes

Figure 4a shows the tensile strength of Al-5Mg-*x*Zr wire electrodes with a diameter of 1.2 mm. As can be seen, the tensile strength and elongation of Al-5Mg wire electrodes were both improved by adding trace Zr. These results show that when the content of Zr was 0.12%, the mechanical properties of the wire achieved the best, and the tensile strength and elongation were increased by 4.8% and 30.4%, respectively, compared with those of the Zr-free wire. Figure 5 shows the tensile curves of the Zr-free and Al-5Mg-0.12Zr wire electrode with the diameter of 1.2 mm. As can be seen from Figure 5, the tensile curves of the two wire electrodes showed typical yield behavior and the fracture mechanism of the two kinds of alloys was a ductile fracture due to significant plastic deformation. However, with a further addition of Zr, the tensile strength and elongation of the wire electrode declined. When the content of Zr was 0.5%, the mechanical properties of the wire electrode were even lower than those of the Zr-free wire.

Figure 6 shows the longitudinal-sectional metallographic structure of the 1.2 mm Al-5Mg-0.12Zr wire electrode. As can be seen from Figure 6, an obvious orientation was observed in the microstructure of the finished wire electrode, and the grains were refined evidently by wire drawing. Under the same process conditions (drawing and annealing), according to the hereditary microstructure, due to the refinement of the original as-cast structure of wire alloy by adding an appropriate amount of Zr, the microstructure of the finished Al-5Mg-0.12Zr wire electrode was still better than that of others in the grain size. In addition, when annealed at 410 ± 10 °C, recrystallization accompanied with softening would occur in the semi-finished wire alloys with a deformed microstructure. Researchers have found that the addition of Zr and/or Sc elements could cause the resistance to recrystallization due to Zener pinning effect from secondary precipitates [15,16,25]. Therefore, it could be deduced that Al_3Zr phases, which were distributed dispersively in the matrix, inhibited the recrystallization of the annealed semi-finished wire alloys. According to the Hall–Petch formula, the refinement of the microstructure could improve the mechanical strength of the alloy. Crack initiation and propagation did not easily occur in the alloy with fine grain structure, thus the alloy showed high elongation. On the other hand, when excess Zr was added, a large number of lath-like IMCs appeared and the effect of grain refinement strengthening and the grain pinning effect on recrystallization was weakened, which resulted in the decline in the tensile strength and elongation of the finished wire electrode.

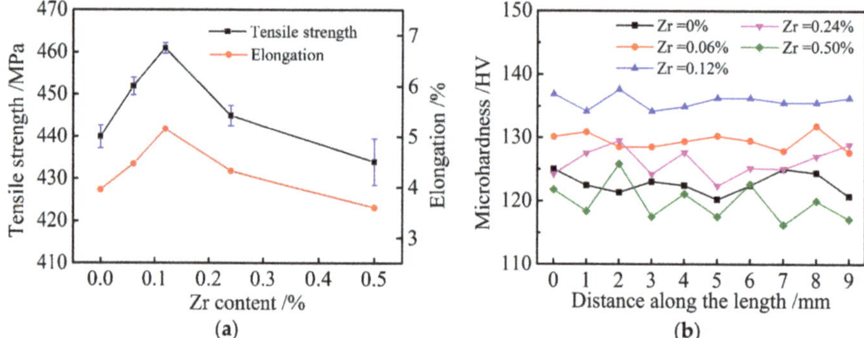

Figure 4. Mechanical properties of 1.2 mm Al-5Mg-*x*Zr wire electrodes: (**a**) tensile strength; (**b**) microhardness.

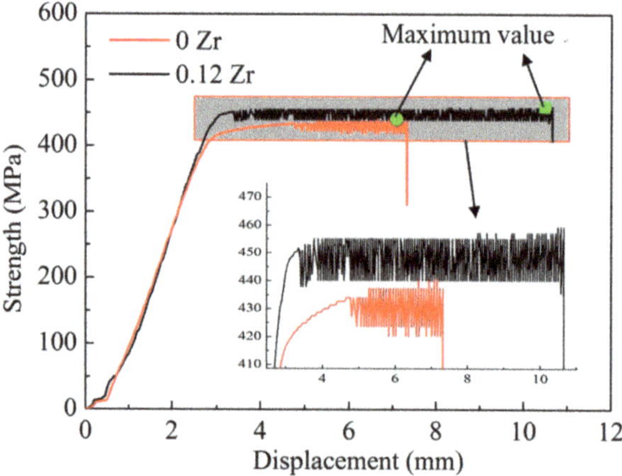

Figure 5. Tensile curves of 1.2 mm Al-5Mg and Al-5Mg-0.12Zr wire electrodes.

Figure 6. Longitudinal-sectional microstructure of the 1.2 mm Al-5Mg-0.12Zr wire electrode.

The microhardness results of 1.2 mm Al-5Mg-*x*Zr wire electrodes are shown in Figure 4b. After adding an appropriate amount of Zr, the average hardness of the wire electrode increased, and the fluctuation value of hardness along its length was small. However, for the Zr-free wire electrode or the wire electrode with excess Zr addition, the average value of hardness was small and the fluctuation of hardness was large. The hardness evolution of Al-5Mg-*x*Zr wire electrodes is mainly determined by the constituent of the alloy, morphology, size and distribution of α-Al phases, Mg_5Al_8 particles and so on. For Al-5Mg-*x*Zr alloy, the finer the α-Al grain, the higher the hardness of the wire. According to the structure heredity of aluminum alloy, the as-cast Al-5Mg-0.12Zr alloy has fine equiaxed α-Al grains in its microstructure, thus its average hardness was the highest. In addition, the hardness of the alloy has a certain relationship with its surface roughness and the friction coefficient of the alloy is represented by:

$$\mu = \frac{\tau}{H} + \tan\theta + \mu_p \tag{1}$$

where μ is the friction coefficient of the alloy, τ is the ultimate shear strength, H is the hardness, θ is the angle of friction surface, and μ_p is the component when the plow effect is considered. According to Equation (1), keeping other variables constant, the higher the H, the lower the μ. Therefore, the addition of Zr can not only refine the microstructure of the wire alloy, increase the tensile strength and hardness of the wire electrode, but also reduce the friction coefficient of wire surface, enhance the wire feedability of the wire electrode during the GMAW process.

3.3. Roughness of 1.2 mm Al-5Mg-xZr Wire Electrodes

The shaving process is extremely important in removing the oxide that forms on the surface of the wire alloys. Figure 7 shows the surface topographies of Al-5Mg-*x*Zr (*x* = 0, 0.12, 0.5) wires, which were mechanical shaved with shaving dies made of synthetic diamond. As can be seen, the surface quality of the Zr-free wire was not high, and there was a small number of longitudinal scratches on the surface of the wire. According to the standard BS EN 14532-3 [22], longitudinal scratches can be acceptable, but could adversely affect the welding characteristics of the wire. The surface quality of Al-5Mg-0.12Zr wire was the best and the surface finish was smooth, uniform and free of scratches (Figure 7b). However, when the content of Zr was 0.5%, the dull matt surface of the wire was full of scratches and depressions, which was unacceptable according to BS EN 14532-3 [22].

Abrasive wearing is the main wearing mechanism of the shaving die. During the shaving process, with the deformation of the Al matrix, most of IMCs embedded in the Al matrix of the wire were transferred into the Al chips or the surface of the shaved wire. With regard to the Al-5Mg-0.12Zr wire, because that fine α-Al dendrites of the wire alloy were evenly distributed and the fluctuation of wire hardness was very small, a large and uniform plastic deformation appeared in the Al matrix of the wire during the shaving process, which led to a relatively good surface quality of the wire. Whereas when 0.5% Zr was added, a large number of coarse IMCs in the microstructure of the wire were easy to slip toward the surrounding coarse Al matrix, some even were stripped from the Al matrix, leaving lots of irregularly distributed pits in various sizes on the surface of the wire. A small and uneven plastic deformation appeared in the Al matrix due to the coarse α-Al phases and IMCs, which easily caused stress concentration and resulted in crack initiation and propagation. In addition, the cutting edge of the die was under uneven stress and showed significant wear due to the strong friction between the shaving die and coarse IMCs. Cutting chips were easy to stick on the worn cutting edge of the shaving die, and thus produced many imperfections on the surface of the wire, deteriorating the surface quality of the wire.

Figure 7. Roughness of 1.2 mm Al-5Mg-xZr wire electrodes: (**a**) $x = 0$; (**b**) $x = 0.12$; (**c**) $x = 0.5$.

3.4. Mechanical Properties of Welded Joints Welded Using 1.2 mm Al-5Mg-xZr Wires

Figures 8 and 9 shows the optical microstructures of different zones of welded joints of 5083 alloy welded using 1.2 mm Al-5Mg-xZr wire electrodes. As shown in Figure 8a, a considerable number of black precipitated phases was homogeneously distributed along a certain direction, which indicated that the base metal was heavily rolled. As for the heat-affected zone (HAZ), its wide range was mainly composed of fibrous tissues, which were elongated along the rolling direction and were almost the same as those of base metal (BM) (Figure 8b). As can be seen from Figure 9a, the microstructure of the weld center mainly consisted of coarse α-Al phases and Mg_5Al_8 particles. Most of these particles were distributed in the grains, and others were precipitated in grain boundaries. When adding an appropriate amount of Zr, the microstructure of the weld center was refined (Figure 9b,c). The size of α-Al grains of the weld center reached the minimum value when 0.12% Zr was added. The refinement of the microstructure of the weld center was due to the large number of fine Al_3Zr, Al_3Ti and $Al_3(Zr, Ti)$ particles [26–28], which formed during the solidification of welding pool and acted as substrates for heterogeneous nucleation of α-Al phases. Figure 10a also showed that the tensile strength of the welded joint welded using Al-5Mg-xZr wire reached the maximum value of 312.6 MPa with the adding of 0.12% Zr. Figure 10b shows the microhardness distributions of the welded joints with Al-5Mg (Zr-free) and Al-5Mg-0.12Zr wire electrode, respectively. As shown in Figure 10b, the minimum hardness value was located in the weld zone (WZ), which indicated that the frail position of the joints was located at the HAZ. After adding 0.12% Zr, the minimum hardness of the WZ increased, and the average hardness of the WZ increased. The microhardness evolution of the WZ can be explained by the abovementioned observation and analysis of the microstructure in Figure 9.

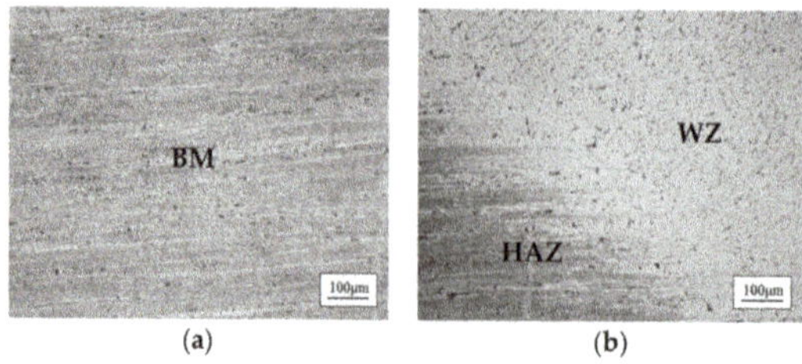

Figure 8. Optical microstructures of different zones of welded joints at low magnification that were welded using 1.2 mm Al-5Mg-0.12Zr wire electrodes: (**a**) BM (base metal); (**b**) HAZ (heat-affected zone) and WZ (weld zone).

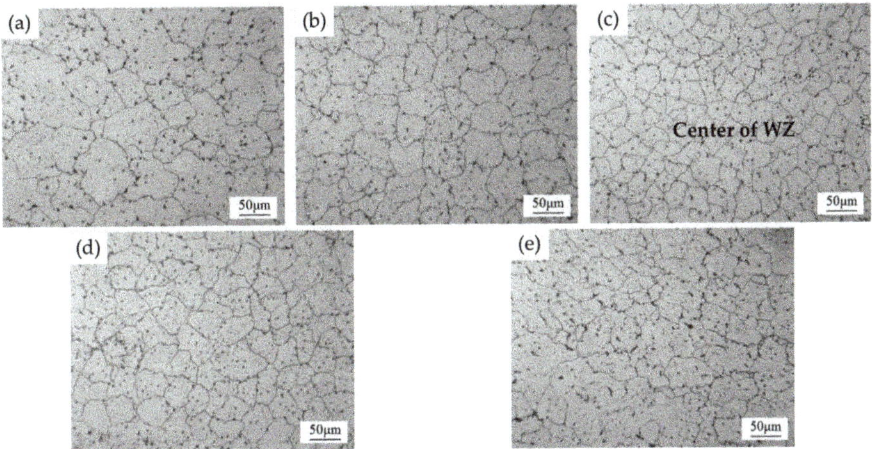

Figure 9. Optical microstructures of the weld center of welded joints at high magnification welded using 1.2 mm Al-5Mg-*x*Zr wire electrodes: (**a**) *x* = 0; (**b**) *x* = 0.06; (**c**) *x* = 0.12; (**d**) *x* = 0.24; (**e**) *x* = 0.5.

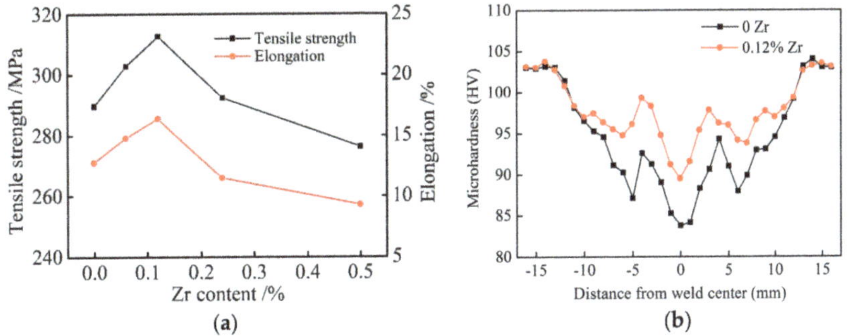

Figure 10. Mechanical properties of the welded joints with 1.2 mm Al-5Mg-*x*Zr wire electrodes: (**a**) tensile strength (fracture location: weld zone); (**b**) microhardness distribution across the welded joints.

3.5. Novel Apparatus for Adjusting Wire Feedability-Related Properties during Drawing Process

After further increasing the Zr content to 0.24% (Figure 9d), the α-Al grains of the weld center became coarse and were varied in size. Furthermore, when the Zr content was 0.5%, large lath-like IMCs were found in the interior of α-Al grains of the weld. An EDS analysis of the black lath-like phase in Figure 9e was performed and the data showed that the lath-like IMC was also an Al₃Zr particle. Combined with the microstructure evolution (Figure 9) and the results of the mechanical properties including the tensile strength and microhardness (Figure 10) of the welded joints, it can be concluded that there is an optimum addition for element Zr. Adding excessive Zr element caused the segregation of the Zr atom in the weld, leading to the formation of large Al₃Zr IMCs and deteriorating the refining effect. In addition, Abdel-Hamid and Zaid [29] had revealed that when Zr and Ti elements existed simultaneously in the aluminum alloy and the content of Zr was more than the peritectic point, the excess Zr element could react with the original Al₃Ti particles and form large Al₃(Ti$_x$Zr$_{1-x}$) IMCs, which resulted in the poisoning of Ti modification, and seriously deteriorated the microstructure and the mechanical properties of the welded joint. The above mechanism analysis can be used for explaining the fact that the tensile strength of the welded joint welded using Al-5Mg-0.5Zr wire was even lower than that of the welded joint welded using Zr-free wire. Due to the grain refinement strengthening mechanism owing to the addition of Zr element, the change trend of the elongation of the welded joint was in good agreement with that of the tensile strength of the welded joint.

3.5.1. Cast and Helix of Wire Electrode on a Spool

With regard to the Al-5Mg-0.12Zr wire electrode designed for high-speed automatic GMAW, the wire feedability of the wire electrode is dependent not only on the strength, hardness, roughness of the wire, but also on the winding state of the wire, which can be characterized by cast and helix of wire electrode on a spool. The most commonly used packaging of GMAW wire electrodes is the spool with an outside diameter of 300 mm. As shown in Figure 11, a sample of wire electrode should be taken from the spool large enough to form a loop when it is cut from the package, and then left unrestrained on the floor. Cast is essentially the diameter of the loop while helix is the rise of the wire electrode off of a flat surface. According to the winding requirements for GMAW solid wire electrodes specified by the American Welding Society (AWS), the cast should be not less than 381 mm and the helix should be less than 25 mm at any location [30].

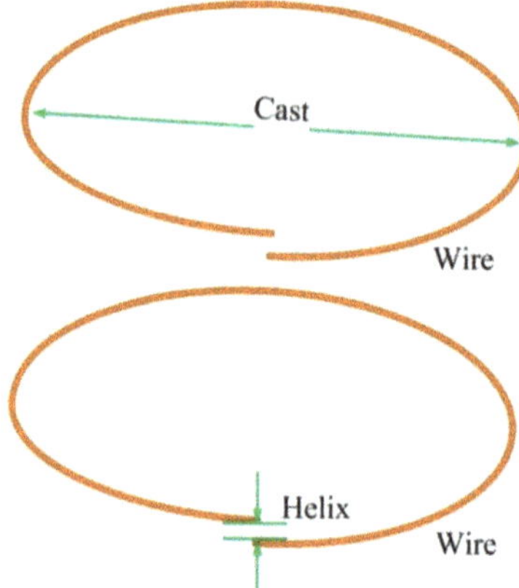

Figure 11. Cast and helix in a gas metal arc welding (GMAW) wire electrode.

3.5.2. Principle of the Novel Adjustable Apparatus

Figure 12 shows the plastic deformation of the wire electrode during drawing. As can be seen, drawing operations involve pulling wire through a die by means of a tensile force applied to the exit side of the die and the plastic flow is caused by compression force, arising from the reaction of the wire with the die. The last drawing die which causes the wire to produce plastic deformation can be defined as the exit die. As shown in Figure 13, there are two characteristic angles between an exit die and a pulling capstan. According to the literature [11], angle α in vertical plane affects the cast of the wire, angle β in horizontal plane affects the helix of the wire. When the angle α becomes small or large, the corresponding curvature ρ of the wire will become large or small, leading to a corresponding decrease or increase in the cast of the wire. When the angle α is too small, a shape of "∞" would be found in the wire electrode unrestrained on the floor. The cause of the helix is the residual stress of the wire, which is induced by the uneven load on the wire during drawing. As shown in Figure 13b, angle β is equal to 90° when the wire is under balanced load conditions. While when the angle β is smaller than or greater than 90°, uneven load produces uneven plastic deformation in the wire, which leads the drawn wire to present a left-hand or right-hand spring-like shape.

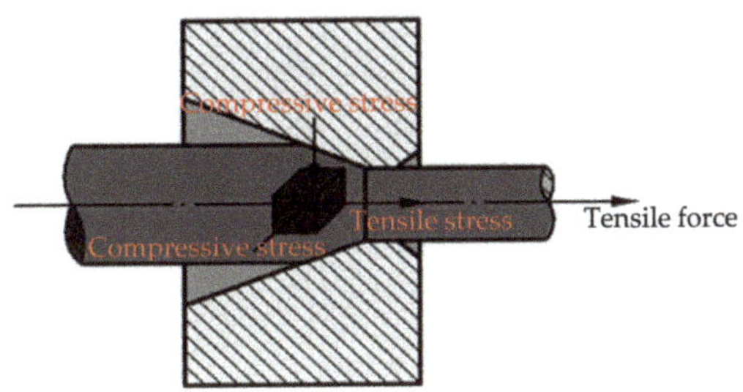

Figure 12. Schematic representation of drawing deformation process of wire electrode.

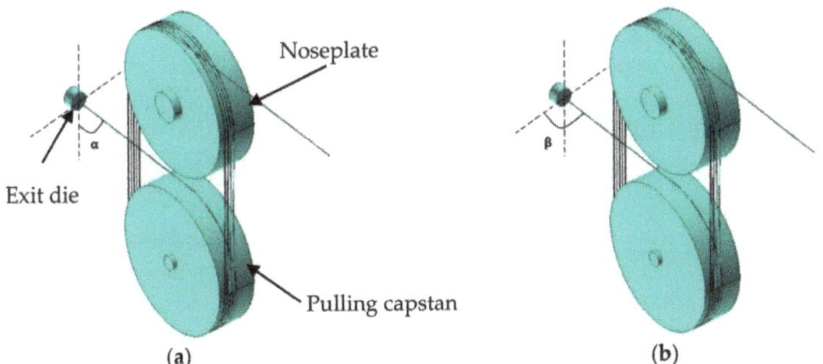

Figure 13. Schematic illustration of characteristic angles between an exit die and a pulling capstan: (**a**) angle α in the vertical plane; (**b**) angle β in the horizontal plane.

3.5.3. Structure and Application of the Novel Adjustable Apparatus

In practice, without considering the quality factor of the exit die itself, the helix is caused by the poor alignment of the exit die with capstan. Figures 14 and 15 show the physical picture and the schematic diagram of the novel apparatus for adjusting the cast and helix of the wire electrode, respectively. As shown in Figure 15a, part No. 1 is well located on the external wall of a finish drawing machine. Install part No. 2, 3, 4 in sequence after the installation of part No. 1 is completed. Then the core part No. 5 (Figure 15b) through the center holes of parts No. 1–4 is fixed with four screws. The exit die is placed into the part No. 5.

Figure 14. Physical picture of the novel apparatus for adjusting the cast and helix of wire electrodes.

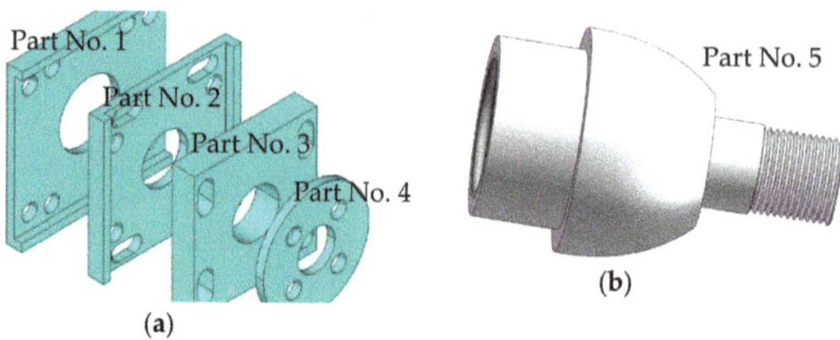

Figure 15. Schematic diagram of the novel apparatus: (**a**) basic parts of the apparatus; (**b**) core part of the apparatus.

According to the structure and principle of the apparatus, the rough adjustment of angle β can be achieved by the horizontal sliding of part No. 2 in the slideway of part No. 1, and the rough adjustment of angle α can be achieved by the vertical sliding of part No. 3 in the slideway of part No. 2. The fine adjustments of angle α and β are mainly achieved by the slack level of threaded connection of the four screws, which are numbered as ①–④ in a clockwise order (Figure 14).

The results of the adjustment of the Al-5Mg-0.12Zr wire electrode cast are shown in Figure 16, and the cast increased gradually during the adjustment of angle α, which was performed by keeping the adjustment of ① synchronized with that of ②, and the adjustment of ③ synchronized with that of ④. Figure 17 shows the adjustment results of helix. Keep the adjustment of ① synchronized with that of ④, and keep the adjustment of ② synchronized with that of ③. When the screw-in depth of ① and ④ was much smaller than that of ② and ③, the angle β was far less than 90°, the wire showed a left-hand spring-like structure and the helix was large in value (Figure 17a). By further increasing the screw-in depth of ① and ④ and decreasing that of ② and ③ at the same time, the left hand helix was reduced to 65 mm, but the value was still unqualified (Figure 17b), which still meant the angle β < 90°. When the angle β was adjusted to 90° by changing the screw-in depths of ①–④ to some extent, the helix of the wire was 0 mm (Figure 17c), which met the AWS requirements for wire feedability. However, further increasing the screw-in depth of ① and ④ on the basis of 0 mm helix, the wire began to show a right-hand structure and the helix was increased again (Figure 17d), which meant the angle β > 90°. It can be concluded from the results of adjustment of the wire (Figures 16 and 17) that the apparatus can effectively adjust the cast and helix of the wire and lead the wire to satisfy the AWS requirements for wire feedability.

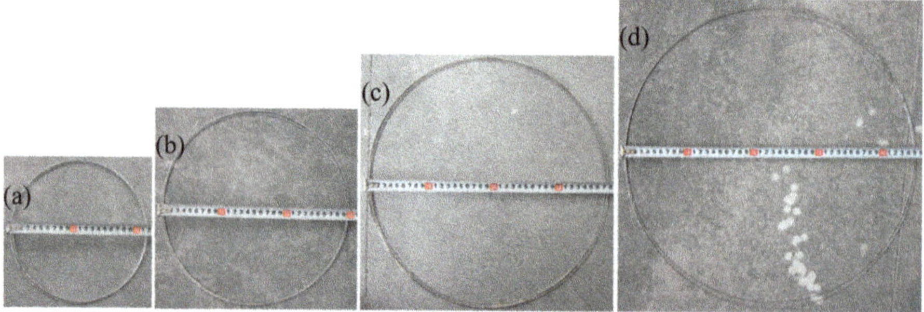

Figure 16. Measured values of cast during adjusting process: (**a**) cast = 215 mm; (**b**) cast = 295 mm; (**c**) cast = 375 mm; (**d**) cast = 450 mm.

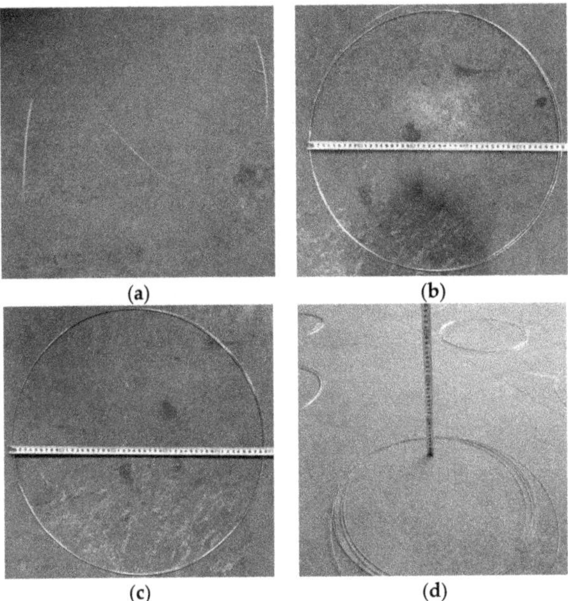

Figure 17. Measured values of helix during adjusting process: (**a**) left hand, helix = ∞; (**b**) left hand, helix = 65 mm; (**c**) helix = 0 mm; (**d**) right hand, helix = 215 mm.

4. Conclusions

The effect of Zr on the microstructure and mechanical properties of Al-5Mg wire alloys and the resulting welded joints was investigated. In addition, factors that affect the wire feedability of the wire electrodes were also analyzed, and a novel adjustable apparatus was introduced and discussed. The conclusions are as follows:

(1) The addition of 0.12 wt % Zr could refine α-Al dendrites of the as-cast Al-5Mg wire alloy due to the increased amount of heterogeneous substrates for nucleation of α-Al grains, while an excess addition of Zr could result in the coarsening of the α-Al dendrites.

(2) The tensile strength, microhardness, and surface roughness of 1.2 mm Al-5Mg wire electrodes, as well as the tensile strength and elongation of the resulting welded joints, were improved by adding an appropriate amount of Zr, which indicated that Zr addition could improve not only the mechanical properties of welded joints, but also the wire feedability of the Al-5Mg wire electrode bearing Zr.

(3) The cast and helix of wire electrode are two key indices for wire feedability of the Al-5Mg wire electrode, which are affected by the two characteristic angles α and β between the exit die and the pulling capstan. The novel apparatus, which consisted of five parts and four screws, could achieve the rough and fine adjustments of the angle α and β, producing a perfect alignment of exit die with capstan to ensure consistency in wire characteristics.

Acknowledgments: This work is financially supported by the National Natural Science Foundation of China (Grant No. 51375233, 51305173), China Postdoctoral Science Foundation (General Financial Grant No. 2014M550289, Special Financial Grant No. 2015T80548), and the Priority Academic Program Development of Jiangsu Higher Education Institutions (PAPD).

Author Contributions: Bo Wang, Songbai Xue and Jianxin Wang conceived and designed the experiments; Bo Wang and Chaoli Ma performed the experiments; Bo Wang and Songbai Xue analyzed the data; Zhongqiang Lin contributed reagents/materials/analysis tools; Bo Wang wrote the paper.

Conflicts of Interest: The authors declare no conflict of interest.

References

1. De Filippis, L.A.C.; Serio, L.M.; Facchini, F.; Mummolo, G.; Ludovico, A.A. Prediction of the vickers microhardness and ultimate tensile strength of AA5754 H111 friction stir welding butt joints using artificial neural network. *Materials* **2016**, *9*, 915. [CrossRef] [PubMed]
2. Haddag, B.; Atlati, S.; Nouari, M.; Moufki, A. Dry machining aeronautical aluminum alloy AA2024-T351: Analysis of cutting forces, chip segmentation and built-up edge formation. *Metals* **2016**, *6*, 197. [CrossRef]
3. Shen, L.; Chen, H.; Che, X.L.; Xu, L.D. Corrosion–fatigue crack propagation of aluminum alloys for high-speed trains. *Int. J. Mod. Phys. B* **2017**, *31*, 1744009. [CrossRef]
4. Kee Paik, J.; Thayamballi, A.K.; Sung Kim, G. The strength characteristics of aluminum honeycomb sandwich panels. *Thin-Walled Struct.* **1999**, *35*, 205–231. [CrossRef]
5. Balasubramanian, V.; Ravisankar, V.; Madhusudhan Reddy, G. Effect of pulsed current welding on mechanical properties of high strength aluminum alloy. *Int. J. Adv. Manuf. Technol.* **2008**, *36*, 254–262. [CrossRef]
6. Kalemba-Rec, I.; Hamilton, C.; Kopyściański, M.; Miara, D.; Krasnowski, K. Microstructure and mechanical properties of friction stir welded 5083 and 7075 aluminum alloys. *J. Mater. Eng. Perform.* **2017**, *26*, 1032–1043. [CrossRef]
7. Wang, B.; Xue, S.B.; Ma, C.L.; Wang, J.X.; Lin, Z.Q. Effects of porosity, heat input and post-weld heat treatment on the microstructure and mechanical properties of TIG welded joints of AA6082-T6. *Metals* **2017**, *7*, 463. [CrossRef]
8. Enz, J.; Riekehr, S.; Ventzke, V.; Huber, N.; Kashaev, N. Fibre laser welding of high-alloyed Al-Zn-Mg-Cu alloys. *J. Mater. Process. Technol.* **2016**, *237*, 155–162. [CrossRef]
9. Tang, L.X.; Huang, X.X.; Li, Y.F.; Wang, X.J. Shallow of aluminum and aluminum alloy MIG welding wire feed stability influence factors. *Weld. Appl.* **2015**, *7*, 41–43.
10. Padilla, T.M.; Quinn, T.P.; Munoz, D.R.; Rorrer, R.A.L. A mathematical model of wire feeding mechanisms in GMAW. *Weld. J.* **2003**, *82*, 100–109.
11. Xu, R.H. Control of pitch and relaxation diameter in welding wire production. *Met. Prod.* **2013**, *39*, 6–7.
12. Anderson, B.E. Lubricated Aluminum Weld Wire and Process for Spooling It. U.S. Patent 4,913,927, 3 April 1990. Available online: http://www.freepatentsonline.com/4913927.pdf (accessed on 21 September 2017).
13. Aluminum GMAW-Gas Metal Arc Welding for Aluminum Guide. Available online: http://www.lincolnelectric.com/assets/global/Products/Consumable_AluminumMIGGMAWWires-SuperGlaze-SuperGlaze5356TM/c8100.pdf (accessed on 22 September 2017).
14. AlcoTec Aluminum Technical Guide. Available online: http://www.esabna.com/shared/documents/litdownloads/alc-10029b_alcotec_technical_guide.pdf (accessed on 22 September 2017).
15. Belov, N.A.; Alabin, A.N.; Matveeva, I.A.; Eskin, D.G. Effect of Zr additions and annealing temperature on electrical conductivity and hardness of hot rolled Al sheets. *Trans. Nonferr. Met. Soc.* **2015**, *25*, 2817–2826. [CrossRef]
16. Chao, R.Z.; Guan, X.H.; Guan, R.G.; Tie, D.; Lian, C.; Wang, X.; Zhang, J. Effect of Zr and Sc on mechanical properties and electrical conductivities of Al wires. *Trans. Nonferr. Met. Soc.* **2014**, *24*, 3164–3169. [CrossRef]
17. Shen, Y.F.; Guan, R.G.; Zhao, Z.Y.; Misra, R.D.K. Ultrafine-grained Al-0.2Sc-0.1Zr alloy: The mechanistic contribution of nano-sized precipitates on grain refinement during the novel process of accumulative continuous extrusion. *Acta Mater.* **2015**, *100*, 247–255. [CrossRef]
18. Xu, Z.; Zhao, Z.H.; Wang, G.S.; Zhang, C.; Cui, J.Z. Microstructure and mechanical properties of the welding joint filled with microalloying 5183 aluminum welding wires. *Int. J. Miner. Metall. Mater.* **2014**, *21*, 577–582. [CrossRef]
19. Zhang, Z.H.; Dong, S.Y.; Wang, Y.J.; Xu, B.S.; Fang, J.X.; He, P. Microstructure characteristics of thick aluminum alloy plate joints welded by fiber laser. *Mater. Des.* **2015**, *84*, 173–177. [CrossRef]
20. Taendl, J.; Orthacker, A.; Amenitsch, H.; Kothleitner, G.; Poletti, C. Influence of the degree of scandium supersaturation on the precipitation kinetics of rapidly solidified Al-Mg-Sc-Zr alloys. *Acta Mater.* **2016**, *117*, 43–50. [CrossRef]
21. Norman, A.F.; Birley, S.S.; Prangnell, P.B. Development of new high strength Al-Sc filler wires for fusion welding 7000 series aluminium aerospace alloys. *Sci. Technol. Weld. Join.* **2003**, *8*, 235–245. [CrossRef]

22. BS EN 14532–3:2004(E), Welding Consumables-Test Methods and Quality Requirements-Part 3: Conformity Assessment of Wire Electrodes, Wires and Rods for Welding of Aluminum Alloys. Available online: http://www.anystandards.com/plus/download.php?open=0&aid=4926&cid=3 (accessed on 11 September 2017).
23. Ma, C.G.; Qi, S.Y.; Li, S.; Xu, H.Y.; He, X.L. Melting purification process and refining effect of 5083 Al-Mg alloy. *Trans. Nonferr. Metal. Soc.* **2014**, *24*, 1346–1351. [CrossRef]
24. Grimes, R.; Dashwood, R.J.; Harrison, A.W.; Flower, H.M. Development of a high strain rate superplastic Al-Mg-Zr alloy. *Mater. Sci. Technol.* **2000**, *16*, 1334–1339. [CrossRef]
25. Ocenasek, V.; Slamova, M. Resistance to recrystallization due to Sc and Zr addition to Al-Mg alloys. *Mater. Charact.* **2001**, *47*, 157–162. [CrossRef]
26. Schempp, P.; Cross, C.E.; Pittner, A.; Oder, G.; Neumann, R.S.; Rooch, H.; Dörfel, I.; Österle, W.; Rethmeier, M. Solidification of GTA aluminum weld metal: Part I—Grain morphology dependent upon alloy composition and grain refiner content. *Weld. J.* **2014**, *93*, 53–59.
27. Yang, D.X.; Li, X.Y.; He, D.Y.; Huang, H. Effect of minor Er and Zr on microstructure and mechanical properties of Al-Mg-Mn alloy (5083) welded joints. *Mater. Sci. Eng. A* **2013**, *561*, 226–231.
28. Knipling, K.E.; Dunand, D.C.; Seidman, D.N. Precipitation evolution in Al-Zr and Al-Zr-Ti alloys during aging at 450–600 °C. *Acta Mater.* **2008**, *56*, 1182–1195. [CrossRef]
29. Abdel-Hamid, A.A.; Zaid, A.I.O. Poisoning of grain refinement of some aluminium alloys. In Proceedings of the Current Advances in Mechanical Design and Production, Seventh Cairo University International MDP Conference, Cairo, Egypt, 15–17 February 2000; pp. 331–338.
30. Gas Metal Arc Welding. Available online: https://www.lincolnelectric.com/assets/global/products/consumable_miggmawwires-superarc-superarcl-56/c4200.pdf (accessed on 15 September 2017).

Article

Optimization of Thermo-Mechanical Processing for Forging of Newly Developed Creep-Resistant Magnesium Alloy ABaX633

Kamineni Pitcheswara Rao [1,*], Chalasani Dharmendra [1],
Yellapregada Venkata Rama Krishna Prasad [2], Norbert Hort [3] and Hajo Dieringa [3]

[1] Department of Mechanical and Biomedical Engineering, City University of Hong Kong, Tat Chee Avenue, Kowloon, Hong Kong; chalasanidharmendra@gmail.com
[2] Independent Researcher, No. 2/B, Vinayaka Nagar, Bengaluru 560024, India; prasad_yvrk@hotmail.com
[3] Magnesium Innovation Centre, Helmholtz Zentrum Geesthacht, Max-Planck-Strasse 1, 21502 Geesthacht, Germany; norbert.hort@hzg.de (N.H.); hajo.dieringa@hzg.de (H.D.)
* Correspondence: mekprao@cityu.edu.hk; Tel.: +852-3442-8409; Fax: +852-3442-0172

Received: 28 October 2017; Accepted: 14 November 2017; Published: 21 November 2017

Abstract: The compressive strength and creep resistance of cast Mg-6Al-3Ba-3Ca (ABaX633) alloy has been measured in the temperature range of 25 to 250 °C, and compared with that of its predecessor ABaX422. The alloy is stronger and more creep-resistant than ABaX422, and exhibits only a small decrease of yield stress with temperature. The higher strength of ABaX633 is attributed to a larger volume fraction of intermetallic particles (Al, Mg)$_2$Ca and Mg$_{21}$Al$_3$Ba$_2$ in its microstructure. Hot deformation mechanisms in ABaX633 have been characterized by developing a processing map in the temperature and strain rate ranges of 300 to 500 °C and 0.0003 to 10 s^{-1}. The processing map exhibits two workability domains in the temperature and strain rate ranges of: (1) 380 to 475 °C and 0.0003 to 0.003 s^{-1}, and (2) 480–500 °C and 0.003 to 0.5 s^{-1}. The apparent activation energy values estimated in the above two domains (204 and 216 kJ/mol) are higher than that for lattice self-diffusion of Mg, which is attributed to the large back-stress that is caused by the intermetallic particles. Optimum condition for bulk working is 500 °C and 0.01 s^{-1} at which hot workability will be maximum. Flow instability is exhibited at lower temperatures and higher strain rates, as well as at higher temperatures and higher strain rates. The predictions of the processing map on the workability domains, as well as the instability regimes are fully validated by the forging of a rib-web (cup) shaped component under optimized conditions.

Keywords: magnesium alloy; compressive strength; hot workability; processing map; hot forging

---

1. Introduction

Magnesium alloys are being developed for automotive applications where strength and creep resistance is of paramount importance [1]. Attempts to improve creep strength of Mg-Al alloys by the addition of rare-earth elements or Ca + Sr or Ca + Ba have led to the development of commercial alloys, like AE42 and AE44, MRI230D and DieMag422 [2]. In these alloys, the formation of intermetallic precipitates, such as Al$_{11}$RE$_3$, Al$_4$RE, Al$_2$RE (AE series), Mg$_2$Ca, Al$_4$Sr, (Mg, Al)$_2$Ca, CaMgSn (Ca + Sr containing alloys), and Mg$_{17}$Ba$_2$, Mg$_{23}$Ba$_6$, Mg$_2$Ba (Ca + Ba containing alloys), are responsible to enhancing the creep resistance. In recent years, Mg-Al-Ba-Ca (ABaX) alloys with higher concentration of alloying elements are being developed [3–5] to obtain enhanced creep resistance, and these include alloys like ABaX422, ABaX633, and ABaX844. In these systems, barium, in combination with Mg and Al, forms a ternary phase Mg$_{21}$Al$_3$Ba$_2$ while Ca forms (Al, Mg)$_2$Ca, both occurring independently at grain boundaries in the cast alloys. While all of these alloys exhibited creep resistance better than AE42, only ABaX633 has shown significantly higher creep resistance than both MRI230D and ABaX422 [2,4]. The effect of Ba and Ca addition to Mg-Al alloys on strength and elongation showed that the tensile strength increases in the order ABaX211, ABaX422, and ABaX633 [4]. However, high concentrations of alloying elements give rise to several problems of chemical segregation, non-uniform precipitate distribution, micro-porosity, and casting defects. To mitigate these problems, it is essential to subject the cast alloys to thermo-mechanical processing at higher temperatures. The aim of the present investigation is to develop technology for hot working of ABaX633 alloy using processing map technique and validate the results by conducting hot forging. ABaX633 has been chosen since this gives optimum creep resistance as well as microstructural control during casting.

2. Experimental Details

The magnesium alloy Mg-6Al-3Ba-3Ca billet was prepared by melting pure elemental metals under SF_6 and argon gas cover. After melt has reached 720 °C, it was held at that temperature for 10 min before casting it into a preheated steel mold of 100 mm diameter and 300 mm height with about 50 mm riser. The cast billet was sliced into disks, from which cylindrical specimens of 10 mm diameter and 15 mm of height were machined for performing compression tests. A hole of 1 mm diameter and 5 mm depth was machined at mid-height of the specimen to reach the center. A thermocouple was inserted into this hole for controlling the heating of the specimen and measuring its instantaneous temperature during the entire compression test. The testing procedure was described in detail earlier [6]. The specimens were deformed under uniaxial compression in the temperature from 300 to 500 °C at 40 °C intervals and at six strain rates in the range of 0.0003 to 10 s^{-1}. In order to achieve constant true strain rates during compression, an exponential decay in the actuator speed of the servo hydraulic machine (DARTEC, M1000/RK, Bournemouth, UK) was applied. Graphite powder mixed with grease as carrier was used as the lubricant at the specimen-die interfaces. The specimens were deformed up to a true strain of about 1 and were quenched in water. The deformed specimens were sectioned in the center parallel to the compression axis for microstructural observation. After cold mounting in plastic molds, the cut surfaces of the specimens were ground, polished, and etched with dilute nitric acid for obtaining the microstructures. Compression tests were also conducted at a strain rate of 0.001 s^{-1} to obtain the mechanical properties of the alloy in the temperature range of 25 to 250 °C at 25 °C intervals. Experimental details of the forging experiments designed to validate the results of the processing map were described in an earlier publication [7], and a brief description was included later in the relevant section.

3. Results and Discussion

3.1. Characterization of Initial Alloy

The chemical composition of the ABaX633 alloy is shown in Table 1. The optical micrograph of as-cast microstructure of ABaX633 alloy is shown in Figure 1a. The average grain size is about 22 μm. The microstructure exhibits dark and light grey colored phases at the grain boundaries. Figure 1b shows the scanning electron micrograph, and the energy dispersive spectroscopy (JEOL 5600 SEM, JEOL Ltd., Akishima, Tokyo) analysis revealed that the black lamellar phase is (Al, Mg)$_2$Ca and the lighter grey phase is $Mg_{21}Al_3Ba_2$ [4].

Table 1. The chemical composition of the ABaX633 alloy.

Al	Ba	Ca	Other Elements	Mg
6.39	2.37	2.74	0.021 Si, 0.012 Sr, 0.0013 Cu, 0.018 Fe, 0.0012 Ni	Balance

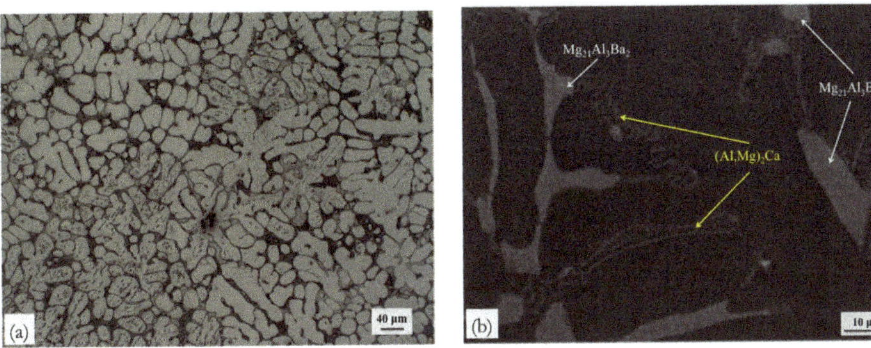

Figure 1. Initial microstructure of the ABaX633 alloy in the as-cast condition: (**a**) Optical micrograph; (**b**) SEM image.

The macro- and the micro-hardness values recorded on ABaX633 alloy are compared with that of another alloy (ABaX422) belonging to the same family of alloys, but with lower concentration of alloying elements. The results are shown in Table 2, which reveals that the ABaX633 is harder and the micro-hardness of each of the individual phases is higher. This may be attributed to the higher concentration of the alloying elements.

Table 2. Macro and micro-hardness (HV) values for ABaX422 and ABaX633 alloys.

Hardness	ABaX422	ABaX633
Macro-hardness	51	65
Micro-hardness (Matrix)	59	67
Micro-hardness ((Al, Mg)$_2$Ca)	95	112
Micro-hardness (Mg$_{21}$Al$_3$Ba$_2$)	166	187

3.2. Creep and Compressive Strength

Compression creep tests at a high temperature of 200 °C were performed on cylindrical specimens (6 mm diameter and 15 mm length) under various constant stresses of 60, 70, 80, and 100 MPa for ABaX633 alloy and the creep rates that were obtained are shown in Figure 2. For comparison purpose, data on its predecessor alloy ABaX422 is also included in the figure. It is evident that ABaX633 alloy exhibited better creep resistance, i.e., lower creep rate, than ABaX422 at all the applied stress levels.

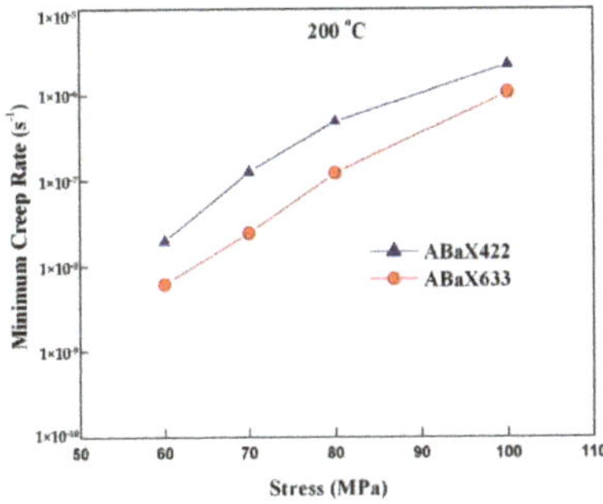

Figure 2. Creep rates at 200 °C under different compressive stresses for ABaX633 and ABaX422 alloys.

To evaluate the compressive strength of the alloy, compression tests were conducted in the temperature range of 25 to 250 °C at a strain rate of 0.0001 s^{-1}. The ultimate compressive strength (UCS) and yield strength (YS) of the ABaX633 alloy are shown in Figure 3, and are compared with its predecessor, ABaX422 [8]. While the YS decreases only slightly with temperature, UCS drops steeply for both of the alloys. The relative increase in the strength of ABaX633 may be attributed to a higher solid solution strengthening by Al and the larger volume fraction of Ba and Ca containing intermetallic phases at the grain boundaries.

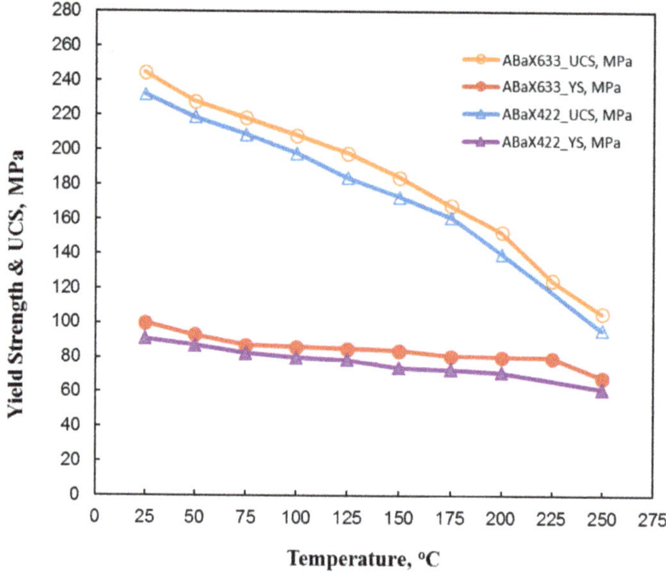

Figure 3. Ultimate compressive strength (UCS) and yield strength (YS) of ABaX633 and ABaX422 alloys.

3.3. Stress-Strain Curves

The true stress-true strain curves obtained on ABaX633 alloy at three different temperatures and all of the strain rates are shown in Figure 4a–c. At the lowest test temperature of 300 °C, the flow curves exhibited (Figure 4a) violent oscillation after initial peak at higher strain rates (>0.01 s^{-1}) while at lower strain rates the flow reached a steady-state at larger strains. As the test temperature is increased beyond 420 °C, the flow curves showed (Figure 4b,c) initial peak followed by a steady state. The shapes of deformed specimens under different temperature and strain rate conditions are shown in Figure 5. It may be noted that the specimens have sheared under conditions where the flow curves exhibited oscillations, while at higher temperatures and lower strain rates, the deformation was homogenous leading to near circular shape.

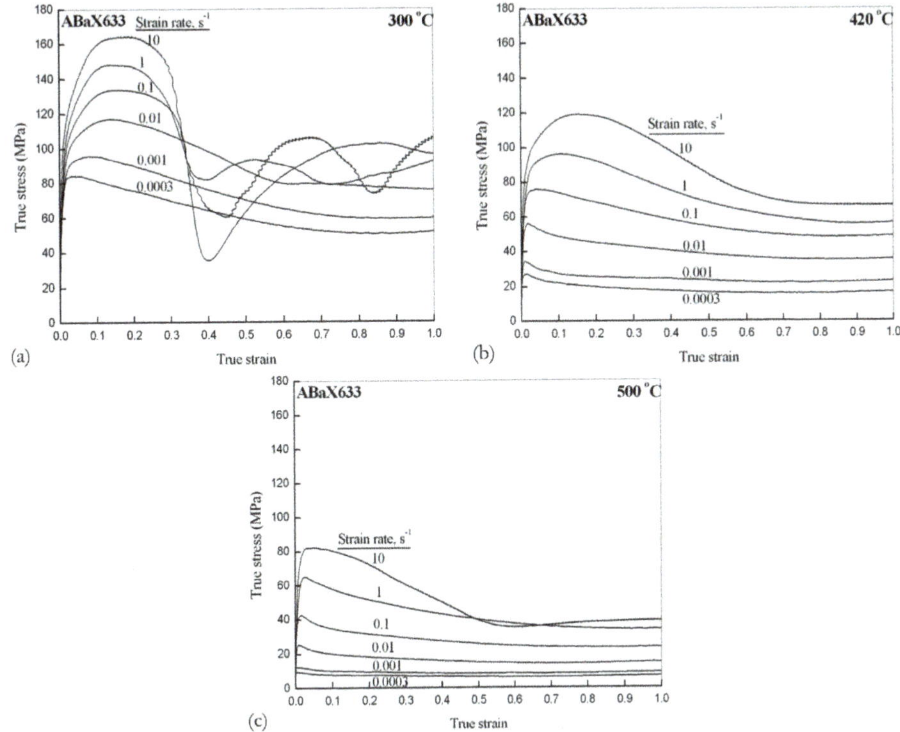

Figure 4. True stress-true strain curves obtained for ABaX633 alloy compressed at various strain rates and at temperatures of (**a**) 300 °C, (**b**) 420 °C, and (**c**) 500 °C.

3.4. Processing Map

In this approach, the material undergoing hot deformation is considered to be a non-linear dissipator of applied power [9,10]. The factor that partitions the applied power between deformation heat and microstructural changes is the strain rate sensitivity (*m*) of flow stress of the material. A measure of efficiency of power dissipation that is attributable to cause microstructural changes can be calculated by comparing with an ideal linear dissipator using the equation:

$$\eta = 2m/(m + 1) \tag{1}$$

The variation of η with temperature and strain rate at a chosen strain level can be represented as a three-dimensional "power dissipation map". However, it is more convenient to use as an iso-efficiency contour map so as to identify "domains" where the dissipation efficiencies reach peaks (mountains) and they are surrounded by troughs (valleys) that separate the domains or change-over from one domain to another. Conceptually, each domain represents a dominant mechanism through which microstructural change occurs, and this may be confirmed by the microstructural examination of the deformed specimens.

The material may also undergo non-uniform deformation under certain conditions of processing and leads to flow instability because of the irreversibility of plastic flow. This may lead to the development of adiabatic shear bands or flow localization within the material, which is undesirable in metal forming. The corresponding sets of conditions are termed as "regimes", in which the flow becomes unstable or non-uniform. On the basis of extremum principles of irreversible thermodynamics, as applied to large plastic flow [11], conditions that would cause such flow instabilities can be readily identified by estimating an instability parameter, ξ:

$$\xi(\dot{\varepsilon}) = \frac{\partial \ln[m/(m+1)]}{\partial \ln \dot{\varepsilon}} + m \leq 0 \tag{2}$$

The flow instability regimes are established, where ξ is negative in the selected range of temperature and strain rate. A "flow instability map" can thus be established and is normally presented as a contour map so as to delineate the "regimes". A processing map is the superimposition of flow instability map over the power dissipation map. With the help of such a composite map, one can easily identify suitable temperature-strain rate windows for hot working, in which microstructurally "safe" mechanisms like dynamic recrystallization (DRX) occur and flow instability regimes are avoided.

The kinetic rate equations are helpful in finding out the activation energy of rate controlling deformation mechanisms. Jonas et al. [12] have proposed an equation for hot deformation that relates flow stress (σ) to strain rate ($\dot{\varepsilon}$) and temperature (T):

$$\dot{\varepsilon} = A\sigma^n \exp[-Q/RT] \tag{3}$$

where A is a constant, n is a stress exponent, Q is the activation energy, and R is the gas constant. This rate equation can be used to estimate the activation parameters n and Q for each domain of the processing map so as to determine the corresponding rate-controlling mechanism.

The processing map developed at a strain of 0.5 is shown in Figure 6. At this strain and beyond, the material flow is of steady-state type. The features exhibited by maps obtained at other strains are not largely different from those seen in Figure 6. The map exhibited two domains at temperature and strain rate ranges:

(1). 380 to 475 °C and 0.0003 to 0.003 s^{-1} with a peak efficiency of 43% occurring at 430 °C and 0.0003 s^{-1} (Domain 1), and

(2). 480 to 500 °C and 0.003 to 0.5 s^{-1} with a peak efficiency of 39% occurring at 500 °C and 0.01 s^{-1} (Domain 2).

The map also exhibits two regimes of instability, one at lower temperatures and higher strain rates, and the other at higher temperatures (>470 °C) and higher strain rates (>0.5 s^{-1}). The microstructural manifestations of the flow instability are discussed later.

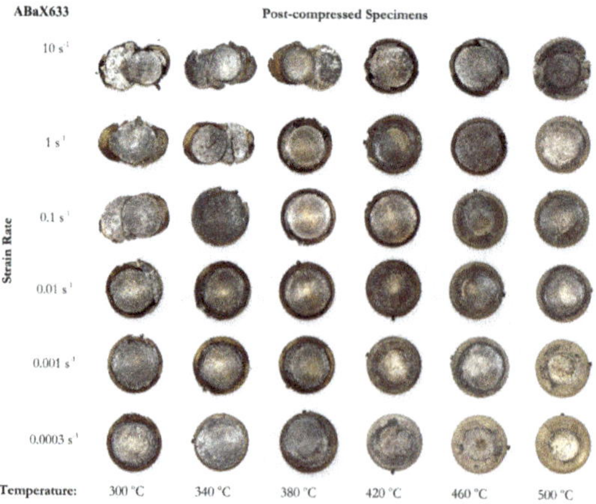

Figure 5. Top View of the ABaX633 alloy specimens compressed at different temperatures and strain rates.

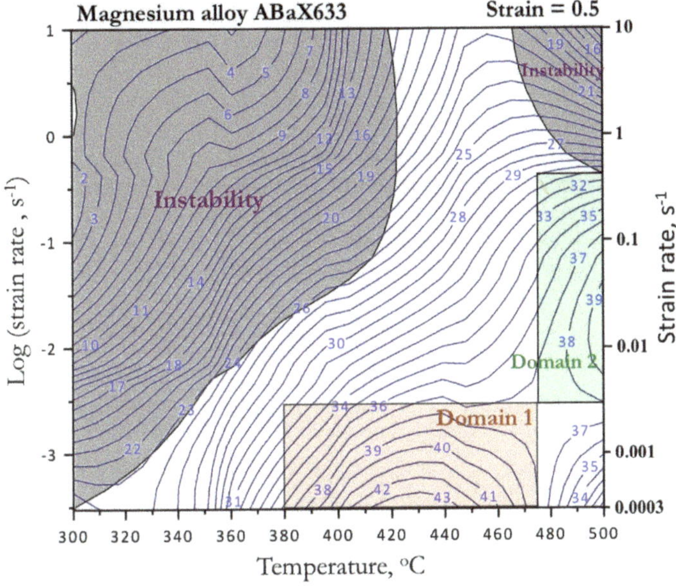

Figure 6. Processing map for the ABaX633 alloy obtained at a true strain of 0.5. The numbers associated with the contours represent the power dissipation efficiency values in percent.

The microstructure recorded on specimen deformed under peak efficiency conditions (420 °C/0.0003 s^{-1}) is shown in Figure 7a, which exhibits dynamic recrystallization (DRX). The initial as-cast microstructure is transformed into a wrought form due to this process. Similar microstructure is obtained at a higher strain rate (0.001 s^{-1}) within the domain, which is shown in Figure 7b. Under the temperature conditions of the domain (380–475 °C), basal and prismatic slip processes will be favored. Due to the low stacking fault energy for basal slip [13], the recovery mechanism that is associated with these slip processes and to nucleate DRX will be dislocation climb process, which is controlled by lattice diffusion.

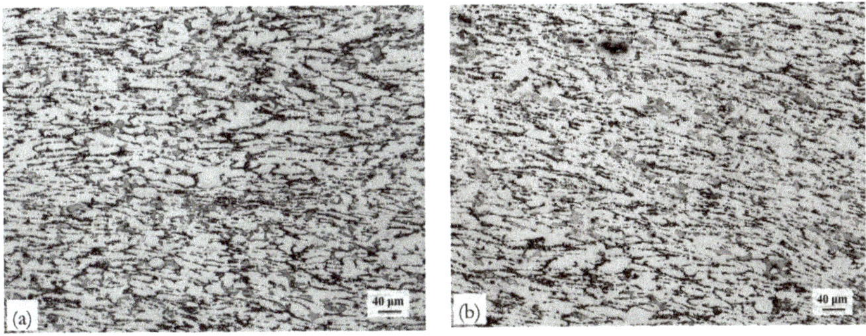

Figure 7. Optical microstructures of the ABaX633 alloy specimens deformed at (**a**) 420 °C/0.0003 s^{-1}, and (**b**) 420°C/0.001 s^{-1} exhibiting DRX (Domain 1). The compression axis is vertical.

Activation analysis of the temperature and strain rate dependency of flow stress using kinetic rate equation (Equation (3)) has been done with the help of data corresponding to Domain 1. The plot of the variation of flow stress with strain rate on a natural logarithmic scale is shown in Figure 8a for different test temperatures. The stress exponent evaluated from the linear relationship is 4.32. Arrhenius plot showing the variation of flow stress (normalized with shear modulus of pure magnesium (16.5 GPa) with an inverse of absolute temperature is shown in Figure 8b, from which apparent activation energy of 205 kJ/mol has been estimated. This value is higher than that for lattice self-diffusion (135 kJ/mol) in magnesium [14]. This apparent high value may be attributed to the contribution of back-stress to dislocation mobility that is caused by the presence of large volume fraction of intermetallic particles present in the as-cast microstructure. Taking this into account and discounting for the particle effect on the apparent activation energy, it may be concluded that lattice-diffusion is a plausible cause for recovery of dislocations on basal and prismatic slip systems. The slower strain rates at which Domain 1 occurs gives further support to diffusion-controlled mechanism for the nucleation of DRX.

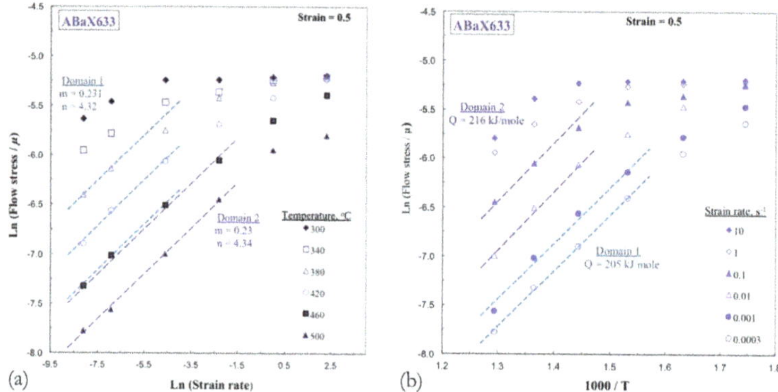

Figure 8. For ABaX633 specimens at a strain of 0.5: (**a**) Variation of the flow stress with strain rate; (**b**) Arrhenius plots showing the variation of normalized flow stress with inverse of absolute temperature.

The microstructure that is recorded under peak efficiency conditions (500 °C /0.01 s^{-1}) of Domain 2 is shown in Figure 9a, which exhibits recrystallized grain structure replacing the as-cast initial one. This suggests that DRX has occurred in this domain also. The microstructure obtained at a higher strain rate (0.1 s^{-1}) is shown in Figure 9b, which also confirms the occurrence of DRX. The temperature range for this domain is higher (>480 °C), which favors the activation of pyramidal slip systems, especially the second order pyramidal slip {11$\bar{2}$2} <11$\bar{2}$3>. Due to the large availability of intersecting slip planes in this system and a higher stacking fault energy [15], the likely recovery mechanism will involve cross-slip of screw dislocations. The values of stress exponent and apparent activation energy evaluated from the flow stress data in this domain, as plotted in Figure 8, are 4.34 and 216 kJ/mol, respectively. The presence of large back stress has an obvious contribution to the apparent activation energy.

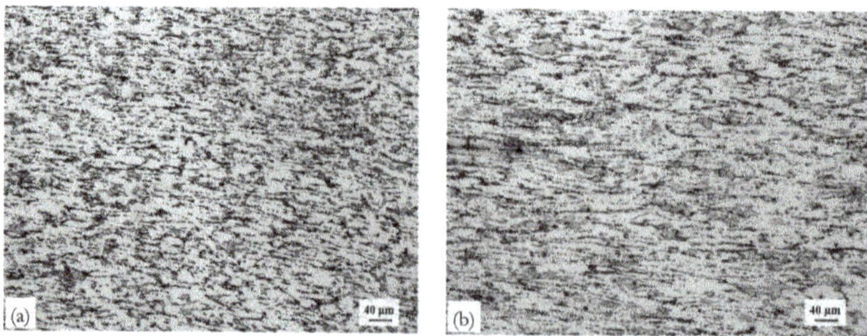

Figure 9. Optical microstructures (the compression axis is vertical) of the ABaX633 alloy deformed at (**a**) 500 °C /0.01 s^{-1}, and (**b**) 500 °C/0.1 s^{-1} (Domain 2).

As far as hot working is concerned, the above results suggest that bulk hot forming may be done in the second domain, the optimum conditions being 500 °C and 0.01 s^{-1}, since the workability will be maximum in this domain. On the other hand, finish forming may be done at slow speeds and lower temperatures, corresponding to Domain 1, the optimum conditions being 420 °C and 0.0003 s^{-1}. The grain size in the product will be finer that will result in better mechanical properties.

3.5. Instability Manifestation

The microstructures of ABaX633 specimens deformed in the lower and higher temperature instability regimes are shown in Figure 10a,b, respectively. At lower temperatures and higher strain rates (300 °C/10 s^{-1}), the microstructure exhibited adiabatic shear band. Shear cracking has occurred along the band due to the high intensity of flow localization. It is in this regime, that the stress-strain curves (Figure 4a) showed oscillations caused by the formation of localized shear and welding/re-welding of the cracks along the band due to compression. The instabilities at lower strain rates and higher temperatures manifested in the form of flow localization of lower intensity. However, no cracking was observed within the localized instability bands.

In the second regime of instability that occurred at higher temperatures and strain rates, the manifestation is that of intercrystalline cracking. The microstructure of the specimen corresponding to the conditions 500 °C and 10 s^{-1} is shown in Figure 10b, which confirms intercrystalline cracking occurring in localized regions. Under these conditions, the material exhibits low workability, as can be seen from the specimen deformed at 500 °C/10 s^{-1} in Figure 5.

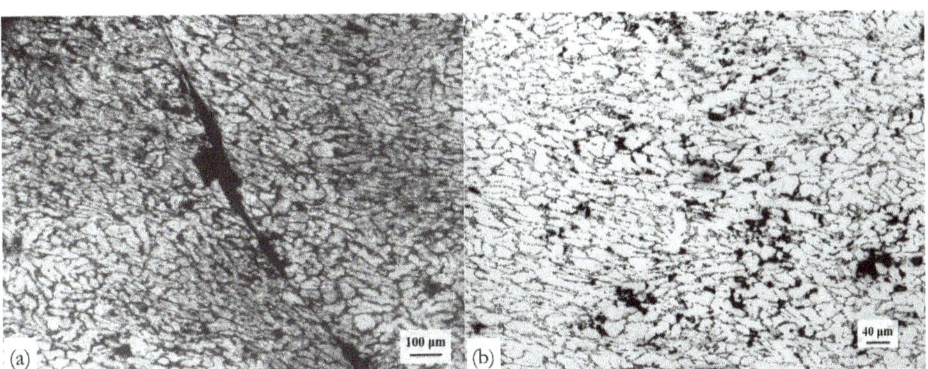

Figure 10. Optical microstructures of the ABaX633 alloy deformed corresponding to instability regimes at (**a**) lower temperatures (300 °C/10 s^{-1}), and (**b**) higher temperatures (500 °C/ 10 s^{-1}) in the processing map. The compression axis is vertical.

4. Validation of Processing Map with Forging

The results obtained from processing maps on the workability domains and the instability regimes are validated by forging a cylindrical preform into a rib-web (cup) shaped component, as shown in Figure 11. The chamfering on the bottom end of preform is to facilitate its accurate positioning on the bottom die. The forging was conducted using a MTS 810 servo-hydraulic machine (MTS, Minneapolis, MN, USA) using a pair of dies, as shown in the figure. An important feature of this semi-open die forging is that most of the material can flow outwards without any constraints. That is, the outer periphery of the deforming preform experiences tensile stresses during the entire forging process except for the small web portion that formed at the bottom of the component that experiences compressive stresses. Details of the overall experimental set up and procedure was given in an earlier publication [11]. The experiments were conducted at different temperatures in the range of 300 to

500 °C and at speeds of 0.01, 0.1, 1.0, and 10 mm/s^{-1}. The specimen, the die assembly, and the loading members were heated to the test temperature by surrounding them with a split furnace.

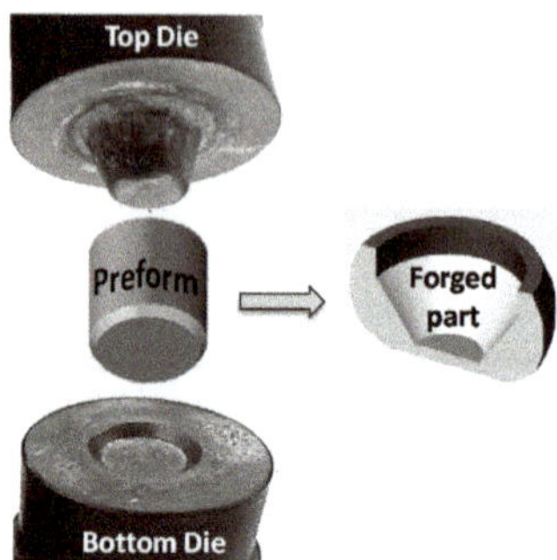

Figure 11. Sketch showing the configuration of the dies used for forging the workpiece (preform) to produce a rib-web (cup) shape component.

It is necessary to estimate the strain rate variations within the forging envelope, as well as the effective strain rate in order to locate the exact processing coordinates in the processing map. For this purpose, process simulation was done using analytical modeling of the forging process. The finite element software DEFORM 2D (axisymmetric version used for isothermal conditions, Version 11.0.1, Scientific Forming Technologies Corporation, Columbus, OH, USA) equipped with a pre-processor to input material data and object definition was used for simulation [16,17]. A post-processor was used for depicting the deformed geometry, state-of-stress, velocity vectors, strain, and strain rate. Process simulations were conducted at the temperature range of 300 to 500 °C at speeds of 0.01 to 10 mm·s^{-1}, until the stroke reached 11 in 0.1 mm increments. As an example, the effective strain distribution in the forged component at the end of the stroke is shown in Figure 12 for forging corresponding to 420 °C/0.01 mm·s^{-1} (Domain 1). From the simulations, the minimum and maximum effective strains range between 0.10 and 3.62, as indicated at the respective locations as $\triangle$ and $\square$, respectively. The average strain rates corresponding to the forging speeds of 0.01, 0.1, 1.0, and 10 mm·s^{-1} are approximately 0.001, 0.01, 0.1, and 1.0 s^{-1}, respectively. The flow patterns are not very different for other forging conditions.

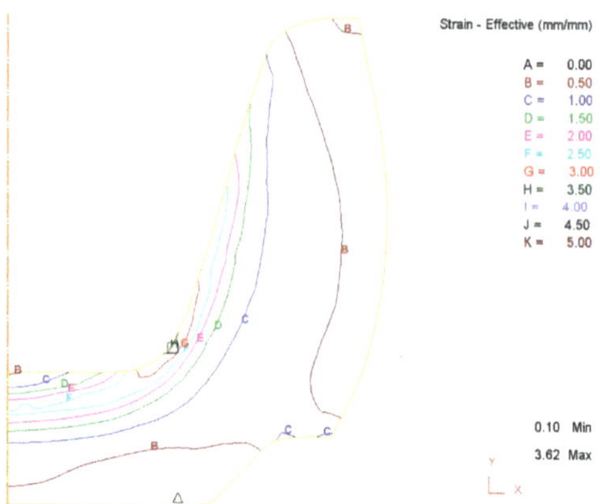

Figure 12. Strain contours obtained in finite element simulation at the end of stroke (11 mm) obtained at temperature and speed of 420 °C/0.01 mm·s^{-1} (Domain 1).

The microstructures recorded at various locations of the specimens forged at temperatures and speeds of 420 °C/0.01 mm·s^{-1} (Domain 1) and 500 °C/0.1 mm·s^{-1} (Domain 2) are shown in Figure 13a,b, respectively. These microstructures recorded in the bottom region of the cup, which has undergone maximum strain. However, the microstructure recorded in the outside rib region did not exhibit significant change when compared with the starting as-cast microstructure because the local strain is not large enough to cause DRX. The microstructures clearly showed the occurrence of DRX, and the grain size obtained in Domain 1 is finer than that in Domain 2.

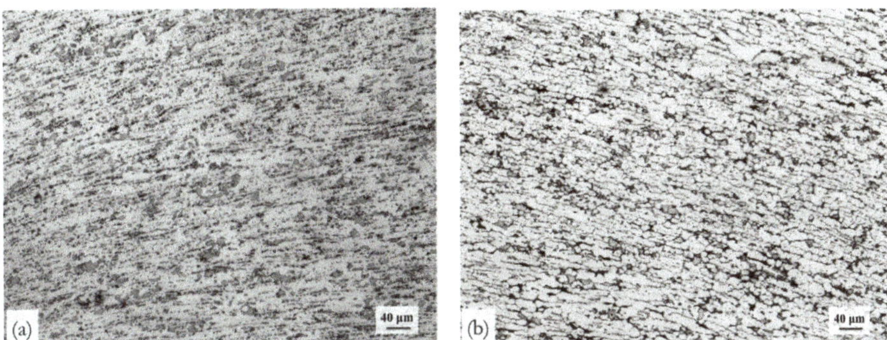

Figure 13. Optical microstructures of the ABaX633 alloy forged at (**a**) 420 °C/0.01 mm·s^{-1} (correspond to Domain 1 in the processing map) and (**b**) 500 °C/0.1 mm·s^{-1} (Domain 2). Forging axis is vertical.

The microstructures obtained in forged components conducted at conditions corresponding to the two flow instability regimes of the processing map (Figure 6), namely, 300 °C/10 mm·s^{-1} and 500 °C/10 mm·s^{-1}, are shown in Figure 14a,b. The flow localization is clearly observed in the bottom, radius, and inside-wall regions. In the lower temperature instability region, intense flow localization has occurred (Figure 14a), while in the higher temperature instability region, flow localization and intercrystalline cracking is observed (Figure 14b), both in line with the predictions of the processing map. In the outside wall region, the microstructure did not exhibit much change due to low local strain values. Thus, the microstructural features of the forgings match with those that were obtained on compression specimens and completely validate the predictions of the processing map both in workability domains, as well as in the flow instability regions.

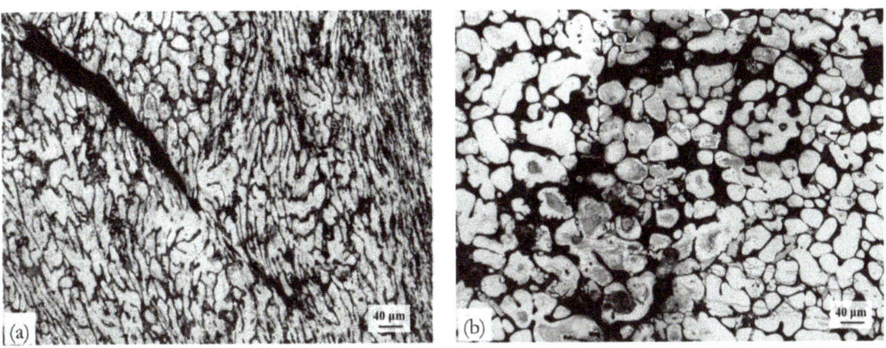

Figure 14. Optical microstructures of the ABaX633 alloy forged at conditions correspond to flow instability in the processing map: (**a**) 300 °C/10 mm·s^{-1} and (**b**) 500 °C/10 mm·s^{-1}. Forging axis is vertical.

5. Conclusions

Hot deformation characteristics of ABaX633 magnesium alloy have been studied in the temperature range of 300 to 500 °C, and a strain rate range of 0.0003 to 10 s^{-1} by developing a processing map. The results from the map have been validated by forging of a cup-shaped component under controlled conditions. The following conclusions are drawn from this investigation:

The as-cast microstructure of ABaX633 is fine grained and consisted of intermetallic phases (Al, Mg)$_2$Ca and Mg$_{21}$Al$_3$Ba$_2$ at the grain boundaries.

The processing map exhibits two workability domains in the temperature and strain rate ranges of: (1) 380–475 °C and 0.0003–0.003 s^{-1} with a peak efficiency of 43% occurring at 430 °C/0.0003 s^{-1}, and (2) 480–500 °C and 0.003–0.5 s^{-1} with a peak efficiency of 39% occurring at 500 °C and 0.01 s^{-1}. Optimum conditions for hot working correspond to those where peak efficiency occurs in these domains.

In both of the domains, dynamic recrystallization occurs. In Domain 1, basal slip + prismatic slip causes plastic flow with climb as the recovery process. In Domain 2, the flow mechanism involves second order pyramidal slip and recovery by cross-slip.

The apparent activation energy values estimated in the above two domains (204 and 216 kJ/mol) are higher than that for lattice self-diffusion, which is attributed to the large back-stress that is caused by the intermetallic particles.

Flow instability is exhibited at lower temperatures and higher strain rates, as well as at higher temperatures and higher strain rates, which is manifested in the form of adiabatic shear band formation and intercrystalline cracking, respectively.

The predictions of the map on the workability domains as well as the instability regimes are fully validated by the forging of a rib-web (cup) shape.

Acknowledgments: This work was supported by a General Research Fund (Project #11259116) from the Research Grants Council of the Hong Kong Special Administrative Region, China.

Author Contributions: Kamineni Pitcheswara Rao and Chalasani Dharmendra have performed the experimental work, analysis of the data, and writing the paper; Yellapregada Venkata Rama Krishna Prasad has contributed on the aspects related to the processing map, microstructure, and writing the paper; Norbert Hort and Hajo Dieringa have developed the alloy and creep tests and analysis.

Conflicts of Interest: The authors declare no conflict of interest.

References

1. Pekguleryuz, M.; Celikin, M. Creep resistance in magnesium alloys. *Int. Mater. Rev.* **2010**, *55*, 197–217. [CrossRef]

2. Dieringa, H.; Huang, Y.; Wittke, P.; Klein, M.; Walther, F.; Dikovits, M.; Poletti, C. Compression-creep response of magnesium alloy DieMag422 containing barium compared with the commercial creep-resistant alloys AE42 and MRI230D. *Mater. Sci. Eng. A* **2013**, *585*, 430–438. [CrossRef]

3. Dieringa, H.; Hort, N.; Kainer, K.U. Barium as alloying element for a creep resistant magnesium alloy. In Proceedings of the Magnesium: 8th International Conference on Magnesium Alloys and Their Applications, London, UK, 26–29 October 2009.

4. Dieringa, H.; Zander, D.; Gibson, M.A. Creep behaviour under compressive stresses of calcium and barium containing Mg-Al-based die casting alloys. *Mater. Sci. Forum* **2013**, *765*, 69–73. [CrossRef]

5. Rao, K.P.; Lam, S.W.; Hort, N.; Dieringa, H. High Temperature Deformation Behavior of a Newly Developed Mg Alloy Containing Al, Ba and Ca. In Proceedings of the 7th TSME International Conference on Mechanical Engineering, Chiang Mai, Thailand, 13–16 December 2016.

6. Prasad, Y.V.R.K.; Rao, K.P. Processing maps and rate controlling mechanisms of hot deformation of electrolytic tough pitch copper in the temperature range 300–950°C. *Mater. Sci. Eng. A* **2005**, *391*, 141–150. [CrossRef]

7. Rao, K.P.; Prasad, Y.V.R.K. Materials modeling and finite element simulation of isothermal forging of electrolytic copper. *Mater. Des.* **2011**, *32*, 1851–1858.

8. Rao, K.P.; Ip, H.Y.; Suresh, K.; Prasad, Y.V.R.K.; Wu, C.M.L.; Hort, N.; Kainer, K.U. Compressive strength and hot deformation mechanisms in as-cast Mg-4Al-2Ba-2Ca (ABaX422) alloy. *Philos. Mag.* **2013**, *93*, 4364–4377. [CrossRef]

9. Prasad, Y.V.R.K.; Rao, K.P.; Sasidhara, S. *Hot Working Guide: A Compendium of Processing Maps*; ASM International: Geauga, OH, USA, 2015.

10. Prasad, Y.V.R.K.; Seshacharyulu, T. Modelling of hot deformation for microstructural control. *Int. Mater. Rev.* **1998**, *43*, 243–258. [CrossRef]

11. Ziegler, H. *Progress in Solid Mechanics*; North-Holland Publishing Company: New York, NY, USA, 1963; Volume 4.

12. Jonas, J.J.; Sellars, C.M.; Tegart, W.J.M. Strength and structure under hot working conditions. *Metall. Rev.* **1969**, *14*, 1–24.

13. Sastry, D.H.; Prasad, Y.V.R.K.; Vasu, K.I. On the stacking fault energies of some close packed metals. *Scr. Metall.* **1969**, *3*, 927–930. [CrossRef]

14. Frost, H.J.; Ashby, M.F. *Deformation-Mechanism Maps*; Oxford University Press: Oxford, UK, 1989.

15. Morris, J.R.; Scharff, J.; Ho, K.M.; Turner, D.E.; Ye, Y.Y.; Yoo, M.H. Prediction of a {1122} hcp stacking fault using a modified generalized stacking-fault calculation. *Philos. Mag. A* **1997**, *76*, 1065–1077. [CrossRef]

16. Kobayashi, S.; Oh, S.-I.; Altan, T. *Metal Forming and the Finite-Element Method*; Oxford University Press: Oxford, UK, 1989.

17. Oh, S.-I. Finite element analysis of metal forming processes with arbitrarily shaped dies. *Int. J. Mech. Sci.* **1982**, *24*, 479–493.

Article

Fabrication of an Ultra-Fine Grained Pure Titanium with High Strength and Good Ductility via ECAP plus Cold Rolling

Haoran Wu [1], Jinghua Jiang [1,2,*], Huan Liu [1,3,*], Jiapeng Sun [1], Yanxia Gu [1], Ren Tang [1], Xincan Zhao [1] and Aibin Ma [1,2]

[1] College of Mechanics and Materials, Hohai University, Nanjing 211100, China; haoranwuted@163.com (H.W.); sunpengp@hhu.edu.cn (J.S.); guyanxiaj@126.com (Y.G.); 18262636237@163.com (R.T.); xincan_zhao@163.com (X.Z.); aibin-ma@hhu.edu.cn (A.M.)
[2] Suqian Institute, Hohai University, Suqian 223800, China
[3] Jiangsu Wujin Stainless Steel Pipe Group Company Limited, Changzhou 213111, China
* Correspondence: jinghua-jiang@hhu.edu.cn (J.J.); liuhuanseu@hhu.edu.cn (H.L.); Tel.: +86-025-8378-7239 (J.J. & H.L.)

Received: 6 November 2017; Accepted: 8 December 2017; Published: 14 December 2017

Abstract: Microstructure evolutions and mechanical properties of a commercially pure titanium (CP-Ti, grade 2) during multi-pass rotary-die equal-channel angular pressing (RD-ECAP) and cold rolling (CR) were systematically investigated in this work, to achieve comprehensive property for faster industrial applications. The obtained results showed that the grain size of CP-Ti decreased from 80 μm of as-received stage to 500 nm and 310 nm after four passes and eight passes of ECAP, respectively. Moreover, abundant dislocations were observed in ECAP samples. After subsequent cold rolling, the grain size of ECAPed CP-Ti was further refined to 120 nm and 90 nm, suggesting a good refining effect by combination of ECAP and CR. XRD (X-ray diffractometer) analysis and TEM (transmission electron microscope) observations indicated that the dislocation density increased remarkably after subsequent CR processing. Room temperature tensile tests showed that CP-Ti after ECAP + CR exhibited the best combination of strength and ductility, with ultimate tensile strength and fracture strain reaching 920 MPa and 20%. The high strength of this deformed CP-Ti originated mainly from refined grains and high density of dislocations, while the good ductility could be attributed to the improved homogeneity of UFG (ultra-fine grained) microstructure. Thus, a high strength and ductility ultra-fine grained CP-Ti was successfully prepared via ECAP plus CR.

Keywords: commercially pure titanium; rotary-die equal-channel angular pressing; cold rolling; ultra-fine grain; tensile property

---

1. Introduction

Titanium and its alloys have been widely used in many key fields, such as medical apparatus and instruments, ships, ocean industries, and so on. Among them Ti-6Al-4V alloy is the most commonly used due to its excellent mechanical property and corrosion resistance [1]. However, Ti-6Al-4V alloy is much more expensive than commercially pure titanium (CP-Ti), and the toxicity of the elements vanadium, aluminum and their corrosion products need to be concerned when using in biomedical fields [2]. Therefore, there is a strong need to develop a substitution for Ti-6Al-4V alloy. Grade 2 CP-Ti (TA2) with much lower cost, exhibits a similar corrosion resistance as titanium alloy, but with relatively low mechanical properties (especially its poor strength) [3]. Hence lots of efforts still need to be made to improve the strength of CP-Ti.

Grain refinement is an effective way to improve both strength and plasticity of polycrystalline metallic materials, and the improvement of strength follows a well-known Hall-Petch relationship approximately [4–6]. Therefore, the ultra-fine grained (UFG) metallic materials with an average grain size of 0.1–2 μm usually exhibit higher strength and ductility, compared with the coarse grained (CG) counterparts [7,8]. Severe plastic deformation (SPD), as a common technique to transform CG microstructure to UFG level, has been widely employed in fabricating bulk UFG aluminum and magnesium alloys without residual porosity and impurities [9–12]. In addition, recently various SPD technologies have been used to prepare high strength CP-Ti [13,14]. Terada et al. [13] employed accumulative roll-bonding (ARB) and successfully developed an UFG CP-Ti with grain size of 80–100 nm, which exhibited an ultimate tensile strength of 989 MPa and elongation of 10%. Stolyarov et al. [14] combined ECAP and cold rolling (CR) on CP-Ti, and prepared an UFG pure Ti (grain size lower than 170 nm) with the highest ultimate tensile strength of 1050 MPa, but with a poor elongation of 6%. It can be seen from the above studies that the SPD CP-Ti exhibited relatively high strength, but their elongations were generally lower than 10%.

Among various SPD techniques, equal channel angular pressing (ECAP) is one of the most promising techniques, which could prepare UFG material without changing the original shape of the materials [15–18]. Moreover, ECAP can easily produce large block material and displays great potential in industrial application [17]. Unfortunately, traditional ECAP processes with B_c route need to take the sample out of the die and rotate after each pass [14,19,20]. Furthermore, reheating is also needed before each new pass, which increases the complexity of processing and causes energy consumption, making it inefficient for large-scale industrial application. To solve this problem, Nishida et al. [21] developed a rotary-die ECAP (RD-ECAP), which tactfully avoids the limitation of the taking sample out and re-inserting procedure, consequently reducing the reheating procedure after each pass. As an ECAP process having a very wide application foreground, the ultimate value of grain refinement of RD-ECAP needs more attention. However, so far, little work has been reported on evolutions of microstructures and mechanical properties of CP-Ti during RD-ECAP, and without repeated annealing during each pass with B_c route, the microstructure evolutions and strengthening mechanism in higher passes of ECAP might be different.

Therefore, in this paper we systematically investigated the microstructure evolutions and mechanical properties of CP-Ti during multi-pass successive RD-ECAP processing. In addition, to refine the microstructure and improve its property more effectively, a cold rolling (CR) processing was further employed on the RD-ECAP CP-Ti. By means of this RD-ECAP + CR combined procedures, an UFG (almost nano-grained, grain size of 90 nm) CP-Ti with better comprehensive mechanical property than previous works was successfully developed, and its ultimate tensile strength and elongation reached 912 MPa and 21.3%, respectively. Moreover, we have also discussed the strengthening and toughing mechanism for this high strength and ductility CP-Ti, in order to provide sufficient guidance on preparing high-performance pure Ti by tuning the grain size and sub-microstructure.

2. Materials and Methods

The raw material used is commercially pure titanium (grade 2), and its composition is analyzed and listed in Table 1. The raw material was hot-forged rod with diameters of 200 mm and in annealed condition (TA2, Baoji Feiteng Metal Materials Co., Ltd., Baoji, China). The average grain size of the original material sample was around 80 μm.

The as-received rod was cut into cuboid billets with dimension of 19.5 mm × 19.5 mm × 45 mm for RD-ECAP. To remove the contaminations (oil, debris, etc.) on the surface, the samples were mechanical polished and ultrasonic cleaned in an acetone bath before RD-ECAP. Figure 1a shows the diagrammatic drawing of RD-ECAP, and the detailed information was described in reference [22,23]. The die was designed to have two perpendicular channels with equal cross-section. The channel angle is 90° and the curvature angle is zero. The die rests on the die holder, which have a removable bottom plate and a rigidly side plate, so the side plate can give some back pressure to the billet. There are four punches surrounding the cuboid sample, which can be denoted as bottom vertical punch, right horizontal punch, left horizontal punch and top vertical punch. When we start the ECAP process, the movement of the bottom vertical punch and right horizontal punch are confined by plate, and only left horizontal punch can move. Then the sample is pressed by the plunger and top vertical punch into the left horizontal channel and the first pass is finished. Rotating the die by 90° clockwise makes the die return to the initial state. Using this method, we can continuously make subsequent passes without taking the sample out each pass. All the punches and samples have been lubricated by molybdenum disulphide and graphite. Before the first pass and every four passes of ECAP, the sample was kept in the die and heated at 693 K for 10 min. Then the sample was continuously pressed for 4 and 8 ECAP passes at the processing speed of 0.5 mm·s^{-1}. The effective strain of per pass in RD-ECAP is about 1.15 [22].

Table 1. Analyzed composition of commercially pure titanium in this paper (wt %).

Element	C	Fe	O	N	H	Ti
Content	0.015	0.059	0.062	0.013	0.002	balence

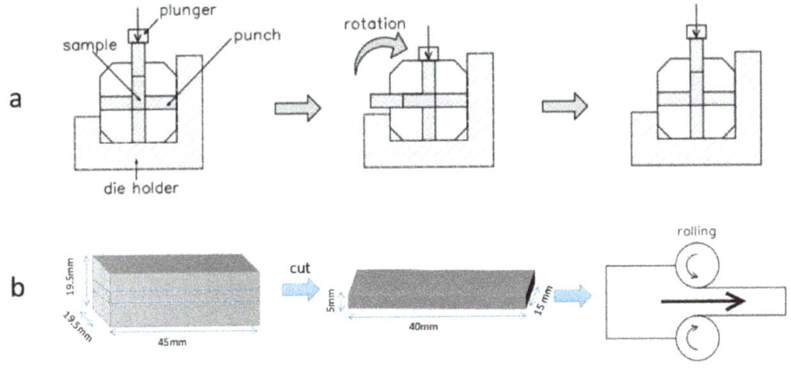

Figure 1. The diagrammatic drawing of (**a**) RD-ECAP (rotary-die equal-channel angular pressing) and (**b**) cold rolling.

In order to further refine grains, cold rolling (CR) was employed on ECAP samples. CR samples were cut from the central part of ECAP cuboid with slice thickness of 5 mm (shown in Figure 1b). These slices were also polished to avoid rough surface. Rolling of the slices were carried out on a 300 ton rolling mill at room temperature at a rolling speed of 0.1 m·s^{-1} with a total CR strain of 85%. After 8 cycles of rolling, the thickness of slices reached 0.2 mm, and no surface cracking was observed. The accumulative strains for samples after 4 passes ECAP + CR and 8 passes ECAP + CR are about 7.3 and 9.5, respectively.

Microstructure observations of as-received, ECAP and ECAP + CR samples were carried out by an optical microscope (OM, Olympus BHM, Tokyo, Japan), a scanning electron microscope (SEM, S-4800, Hitachi Ltd., Tokyo, Japan), and a transmission electron microscope (TEM, Tecnai G2 20, Field Electron and Ion Company, Hillsboro, OR, USA). Sample for OM observations were mechanically ground, polished, and chemically etched with 3 mL HF, 7 mL HNO_3 and 90 mL distilled water for 15 s. Tensile fracture samples for SEM observations were cleaned by ultrasonic cleaners with absolute ethyl alcohol for 10 min, and then dried by an electric hair dryer. TEM specimens were mechanically thinned to around 65 μm and then twin jet polished in a solution of 6% perchloric acid and 94% ethanol with the operating voltage of 200 KV. Moreover, phase identification was also conducted for all samples by X-ray diffractometer (XRD, Bruker D8 X-ray facility, Karlsruhe, Germany) using CuK$_\alpha$ irradiation (45 KV, 30 mA) at a scanning step of 0.5° and exposure time of 1 s from 30° to 80°. Tensile test was conducted by an electronic universal testing machine (TFW-50S, Shanghai Tuofeng

Instrument Technology Co., Ltd., Shanghai, China) at room temperature (298 K) with the displacement rate of 1 mm·min^{-1}. The tensile samples were machined to be dog bone stick shape with gauge length of 6 mm, gauge width and thickness of 2 mm. For each stage, at least three tensile samples were tested and the final curve which was closest to the average of the tests was selected.

3. Results and Discussions

3.1. Microstructure of As-Received and ECAP CP-Ti (Equal-Channel Angular Pressed Commercially Pure Titanium)

Figure 2 shows the XRD patterns of CP-Ti at all processing stages. For the as-received sample, all diffraction peaks correspond to α-Ti (hexagonal close-packed lattice structure). After multi-pass ECAP and CR processing, the locations of diffraction peaks remain unchanged, although the intensity of some peaks exhibits obvious change. Since there is no other phase detected by XRD for all samples, it can be confirmed that phase transformation was not activated during severe deformation in this work.

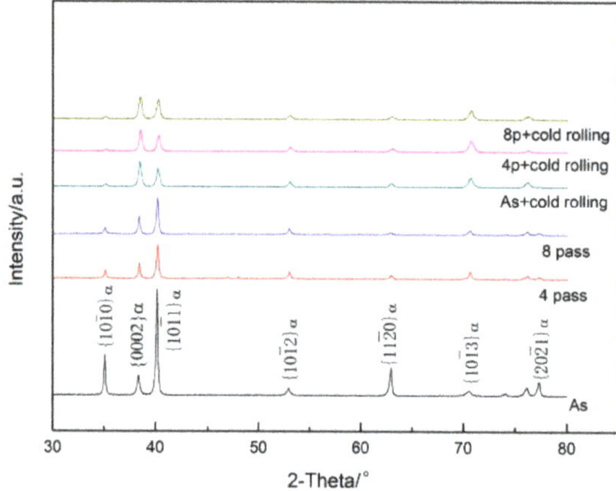

Figure 2. XRD (X-ray diffractometer) patterns of the CP-Ti (Commercially pure Titanium) at different processing stages.

Figure 3a shows the optical micrograph of the as-received CP-Ti. It is apparent that the as-received CP-Ti exhibits equiaxed grains with average grain size of 80 µm. Moreover, twins are also visible in some grains. Figure 3b,c displays the optical micrograph of ECAP CP-Ti after 4 passes and 8 passes, respectively. The grain size has been refined obviously, and abundant dislocation tangles can be observed. In addition, the twins which existed in as-received CP-Ti is barely observed after ECAP. To further explore the microstructure of ECAP CP-Ti and characterize their grain sizes, TEM observations were performed.

Figure 4a,b shows the TEM micrographs and corresponding SAED (selected area electron diffraction) patterns of ECAP samples after 4 passes and 8 passes, respectively. After 4 passes of RD-ECAP, it can be seen from Figure 4a that the grain size is refined to about 500 nm, and several grains are marked by white arrows in the figure. With increasing the number of ECAP passes to 8, the average grain size is decreased to 310 nm approximately, as is shown by arrows in Figure 4b. It should be noted that the decrease in grain size is significant compared with as-received CP-Ti. Moreover, it can be seen that the diffraction rings were intensive from corresponding SAED patterns, which also indicates that the grain size has been well refined. Moreover, Figure 4c shows that the density of dislocations is very high in 4p-ECAP CP-Ti, and distinct dislocation tangles can be easily observed. However, seen from Figure 4b, dislocation cells and sub-grain boundaries can be easily observed in 8p-ECAP CP-Ti, and the density of dislocations is decreased compared with 4p-ECAP sample.

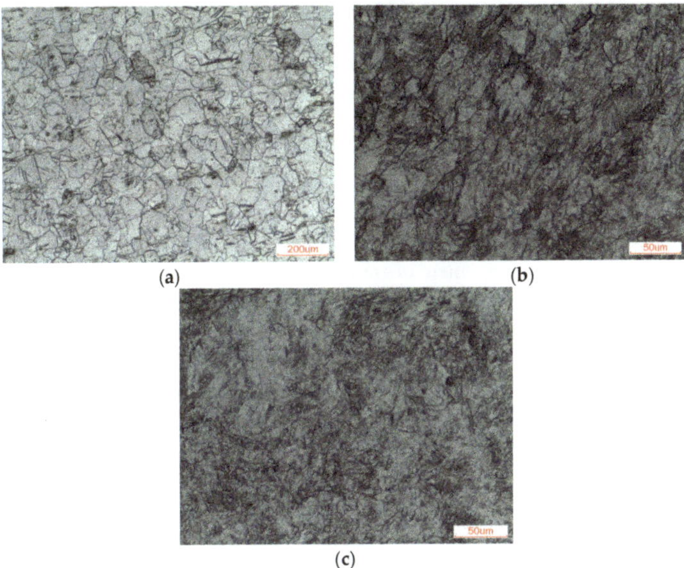

Figure 3. Optical micrographs of CP-Ti at different processing stages: (**a**) as-received; (**b**) after 4 passes ECAP; (**c**) after 8 passes ECAP.

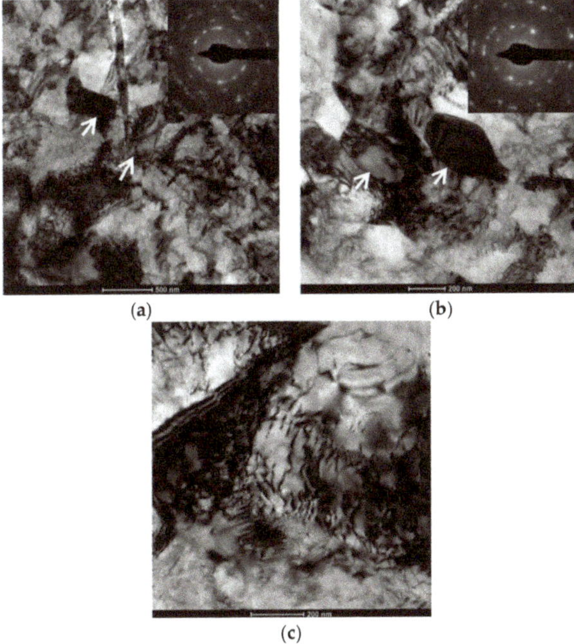

Figure 4. TEM (Transmission electron microscope) micrographs and corresponding SAED (Selected area electron diffraction) patterns of ECAP CP-Ti: (**a**) 4 passes; (**b**) 8 passes; (**c**) dislocations in 4 passes ECAP sample.

Continuous dynamic recrystallization (CDRX) could occur during high temperature deformation for metals with high stacking fault [16]. CDRX can promote the formation of sub-grains, and therefore refine the grain size. Many scholars had found that after eight passes ECAP, the grain size became larger than six passes or four passes ECAP sample [16,24]. Luo et al. [24] investigated the grain size evolution of CP-Ti during ECAP process with B_c route at 723 K, and found that the grain size was refined normally from two passes to four passes ECAP. However, when ECAP number was increased to eight passes, grains became coarse. The reason for this phenomenon could be attributed to the growing of DRX grains, as in their research CP-Ti had been subjected to repeated static annealing at the intervals of contiguous ECAP passes. In addition, Roodposhti et al. [16] approved the view that processing temperature could influence the refinement of grain size during the ECAP process. Higher temperature will promote the rate of recovery, and consequently dislocation annihilation becomes extensive which will lead to the decrease of dislocation density and increase of grain sizes [25,26].

In the case of this study, it is clear that with increasing the number of RD-ECAP passes, the grain size decreases continuously. The average grain size is about 500 nm after four passes RD-ECAP, which is almost 160 times finer than the original sample (80 μm). In addition, after the latter four passes ECAP, the grain size of eight pass ECAP (310 nm) is less than twice smaller compared with four-pass sample. However, it should be noted that the decrease in grain size after the latter four passes RD-ECAP is less obvious than initial four passes. This is because the refinement effect of ECAP is limited when initial microstructure is in sub-micron scale [27]. Previous study [28] showed that small grains may exhibit a low stability when metals are heavily deformed, and additional straining may occur as ECAP does not further reduce the grain size because of the intrinsic instability of nano-sized (below 100 nm) and submicrometre-sized (between 100 and 1000 nm) grains. There has been a dynamic balance of grain refinement between structure refinement and recovery (and coarsening) at ambient temperature [27]. Therefore, the grain refinement effect of ECAP exhibits a limit, and with a further increase of processing numbers, the grain size and dislocation density will achieve ultimate values. In order to further decrease the grain size, cold rolling (CR) was employed on ECAP CP-Ti in this work. CR is one of the most convenient ways to further process ECAP titanium, and recovery can hardly take place during rolling at room temperature [14,29].

3.2. Microstructure of ECAP + CR (Cold Rolling) CP-Ti

Figure 5 shows the optical microstructures of as-received and ECAP CP-Ti after cold rolling, and an obvious refining effect is observed after CR. It can be seen that CR process introduces a more homogeneous microstructure than the ECAP process, and grain size is refined obviously. Figure 6a shows the TEM micrograph of as-received CP-Ti after only CR process; it can be seen that the grain size is refined to about 3 μm, and some twins can be observed. However, it is hard to distinguish whether these twins were newly formed during CR or remained from the original microstructure. Figure 6b shows the TEM image of 4p-ECAP + CR sample; it is apparent that the grain size was further refined to 120 nm. In addition, high-density dislocations can be observed within finer grains, and the corresponding diffraction ring is more intensive and fluent than that of the ECAP sample, suggesting that there are more high-angle boundaries in ECAP + CR sample than four passes ECAP sample. As for 8p-ECAP + CR sample, it can be seen from Figure 6c that the grain size is decreased to 90 nm, almost forming a nano-grained (NG) CP-Ti. However, compared with 4p-ECAP + CR sample, the density of dislocations decreases.

During the CR process, accumulative strain continuously increases, as well as dislocations. As long as a dislocation cell is newly formed, it can absorb the high density dislocations inside the firstborn dislocation cell, and then transfer to the cell wall, thus a whole dislocation cell comes into being [16]. When these dislocation cells turn into high angle boundaries, the microstructure could be further refined. Based on the above considerations, it can be concluded that with the combination of multi-pass RD-ECAP and CR, a homogeneous UFG CP-Ti (even NG for 8p-ECAP + CR sample) is obtained, suggesting that the combined ECAP + CR technique is an effective method for preparation of UFG/NG CP-Ti.

Figure 5. Optical micrographs of cold rolled CP-Ti: (**a**) AS (As-received) + CR; (**b**) 4p-ECAP + CR; (**c**) 8p-ECAP + CR.

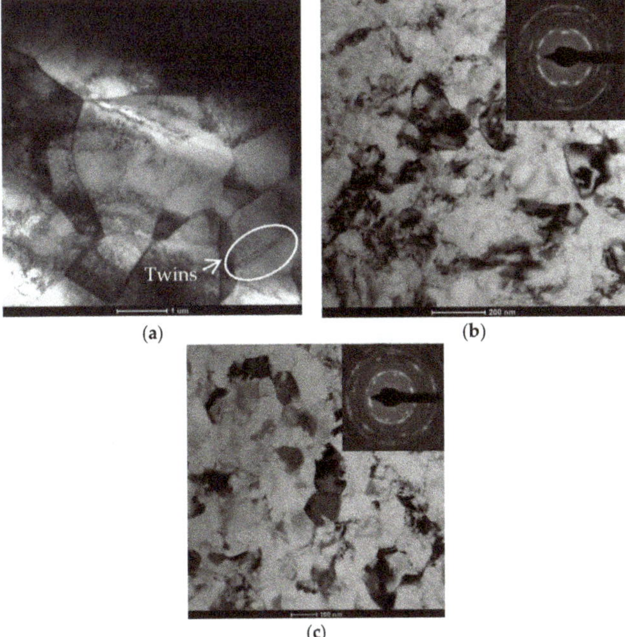

Figure 6. TEM micrographs and corresponding SAED patterns of cold rolled samples: (**a**) AS (As-received) + CR; (**b**) 4p-ECAP + CR; (**c**) 8p-ECAP + CR.

3.3. Mechanical Properties of CP-Ti at Different Processing Stages

Figure 7a shows the room temperature tensile curves of CP-Ti at different processing stages. Moreover, the variations of ultimate tensile strength (UTS), tensile yield strength (TYS), and strain to failure with different processing stages are also summarized in Figure 7b. The as-received sample exhibits a low UTS of 427 MPa and a high strain to failure of 45.3%. After four passes of ECAP, there is a boost of strength, which increased to 681 MPa, but the fracture strain decreased to 33.1%. As for 8p-ECAP sample, its UTS and fracture strain have decreased by 9% and 16%, respectively, when compared with that of 4p-ECAP sample. When as-received and ECAP CP-Ti were cold rolled, their strengths are improved remarkably, but their fracture strains are reduced. The as-received + CR sample shows high UTS of 799 MPa, which is almost twice that of the as-received sample, and is even higher than both ECAP samples, suggesting that cold rolling exhibits a more obvious strengthening effect than ECAP. However, its ductility is greatly decreased compared with ECAP metals. When ECAP metals are cold rolled, a more obvious improvement of strength is observed. The 4p-ECAP + CR and 8p-ECAP + CR metals display high strength with UTS of 923 and 912 MPa, and moderate ductility with fracture strain of 18.2% and 21.3%, respectively. Therefore, it can be concluded that the strengthening effect of combined ECAP + CR method is more effective than single CR or ECAP pressing. In addition, an ultra-fine grained (nano-grained) CP-Ti with both high strength and good ductility is successfully fabricated via the combination of ECAP and cold rolling.

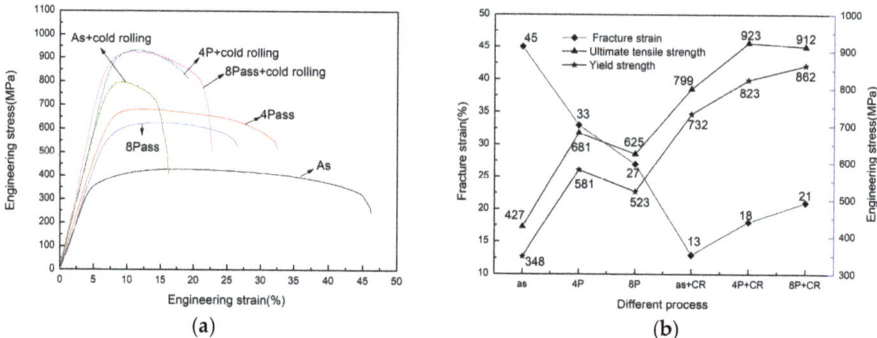

Figure 7. Mechanical properties of CP-Ti: (**a**) tensile curves of CP-Ti at different processing stages; (**b**) variation of ultimate tensile strength, tensile yield strength and fracture strain with different processing stages.

Moreover, Figure 8 shows the SEM micrographs of fracture appearance of CP-Ti after tensile tests. It is obvious that the dimples on the fracture surface of the as-received sample are bigger and deeper than other processed samples, which also demonstrates that the plasticity of the as-received sample is the best of all processing stages. As for ECAP metals, unequal-sized ovoid cavity appear, suggesting the existence of ductile fracture [30]. Seen from Figure 8d–f, the fracture surfaces of cold rolled samples exhibit more uniform characterizations with dense and shallow dimples. The decrease in dimple depth often implies the deterioration of ductility, which corresponds to the lower strain to failure in Figure 7.

Different technologies have already been employed to process high strength CP-Ti [4]. Figure 9 shows the comparison of mechanical properties of CP-Ti (grade 2) in this work and that fabricated via other different technologies in reference [8,11,27,29,31,32]. Stolyarov et al. [27,29] combined ECAP with CR and cold extrusion (CE) and successfully fabricated two high strength CP-Ti with high UTS of 1050 MPa. However, their elongations decreased obviously, which were 6% and 8%, respectively. Moreover, Stolyarov et al. [11] also employed two SPD methods together on processing of CP-Ti, and by this ECAP + HPT (high pressure torsion) method, CP-Ti gained a better elongation of 25% but with a lower strength of 730 MPa. It can be seen that in these studies, high strength and high ductility were not attained simultaneously for CP-Ti. In addition, other SPD methods were also employed for processing of CP-Ti. Fattah-Alhosseini et al. [32] used accumulative roll bonded to process CP-Ti and a UTS of 989 MPa and elongation of 10% were obtained. Kim et al. [8] introduced high-radio differential speed rolling (HRDSR) on CP-Ti processing, and prepared CP-Ti with UTS of 895 MPa and elongation of 11%. Although the ductility was improved in these works, it still needs further improvement. In this work, when the RD-ECAP + CR combined method was carried out, although the ultimate tensile strength is a bit lower than some work, both high strength (above 900 MPa) and good ductility (above 20%) is attained for CP-Ti. This exhibits a better comprehensive mechanical property than previous works and the typical annealed Ti-6Al-4V (TC4) alloys (with UTS of 950 MPa and elongation of 14%) [33].

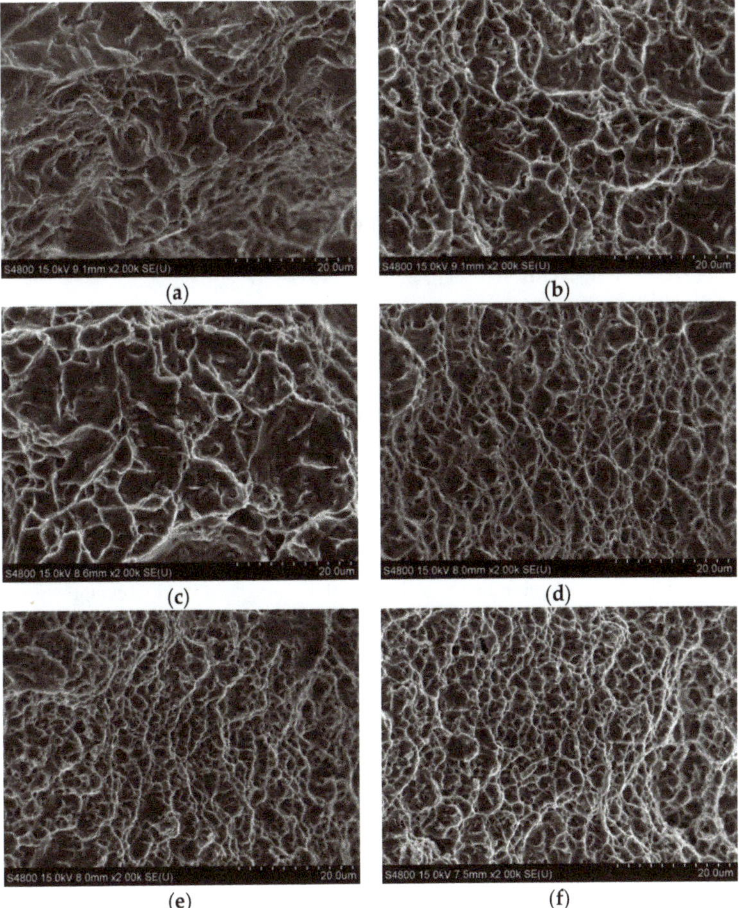

Figure 8. SEM micrographs of the tensile fracture of CP-Ti: (**a**) as-received; (**b**) 4p-ECAP; (**c**) 8p-ECAP; (**d**) as-received + CR; (**e**) 4p-ECAP + CR; (**f**) 8p-ECAP + CR.

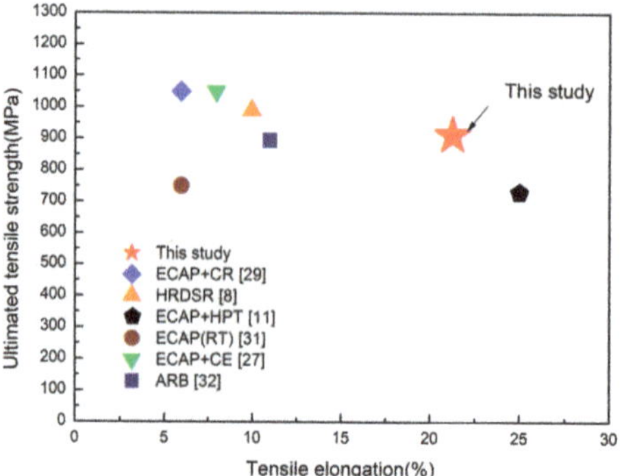

Figure 9. Comparison of mechanical properties of CP-Ti processed by different technologies in references [8,11,27,29,31,32] and in this work (HRDSR: high-radio differential speed rolling; HPT: high pressure torsion; RT: room temperature; CE: cold extrusion; ARB: accumulative roll-bonding).

3.4. Correlation between Microstructures and Mechanical Properties of CP-Ti

Grain size plays an important role in improving the strength of polycrystalline metallic materials [3,34]. Figure 10 shows the comparison of calculated Hall-Petch relationship for CP-Ti and the data obtained in this work. The Hall-Petch relationship is given by:

$$\sigma_{0.2} = \sigma_0 + Kd^{-1/2} \tag{1}$$

where $\sigma_{0.2}$ is the yield stress, σ_0 is a materials constant for the starting stress for dislocation movement (or the resistance of the lattice to dislocation motion), K is the strengthening coefficient (a constant specific to each material), and d is the average grain diameter. The coefficients σ_0 and K can be referred to reference [35] as σ_0 = 182 MPa and K = 0.36 MPa·m$^{1/2}$. It is generally accepted that Hall-Petch relationship is valid in submicrometer grains. When the grain size refined to around 100 nm, it has inverse Hall-Petch relationship, which has been reported by Tang et al. [36]. As for this work, it can be seen from Figure 10 that the strength of CP-Ti generally follows the well-known Hall-Petch relationship, even when the grain size has been refined to about 120 nm and 90 nm. However, it should be noted from Figure 10 that although the grain size for 8p-ECAP is finer than 4p-ECAP, the yield strength of 8p-ECAP is a little lower than 4p-ECAP. The same phenomenon is also observed for cold rolled ECAP samples. These abnormal phenomena imply that grain-boundary strengthening is not the only strengthening mechanism for CP-Ti obtained in this work.

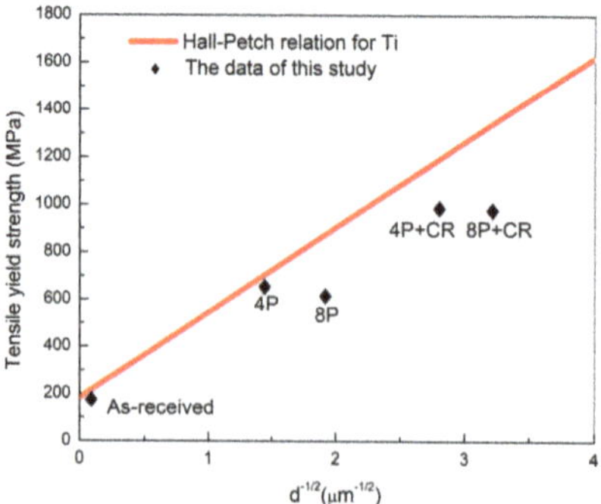

Figure 10. The Hall-Petch relation for Ti and the data of this study.

Combining the TEM micrographs and corresponding SAED patterns in Figures 4 and 6, it can be seen that the density of dislocations in 4p-ECAP sample is a little higher than 8p-ECAP, and it is the same situation in ECAP + CR samples. Therefore, the high-density dislocations may also play an important role in the strengthening of UFG CP-Ti.

To quantitatively analyze the evolution of dislocation densities with different processing stages, we evaluated the dislocation density by XRD according to the Williamson-Hall method [37,38]. Figure 11 shows the full width at half maximum (FWHM) of three representative peaks from XRD patterns of CP-Ti with different processing stages. It can be seen that after cold rolling, the FWHM exhibits a remarkable increase compared with that of ECAP CP-Ti. Previous studies [28] demonstrated that the increase of peak width is associated with grain refinement and higher dislocation density. Based on the Williamson-Hall method [39,40], the following formula can be deduced:

$$(\Delta K)^2 \cong \left(\frac{0.9}{D}\right)^2 + \left(\frac{\pi M^2 b^2}{2}\right)\rho K^2 \overline{C} + O\left(K^4 \overline{C}^2\right) \tag{2}$$

where $\Delta K = 2\cos\theta(\Delta\theta)/\lambda$, $K = 2\sin\theta/\lambda$, θ is the diffraction angle, λ is the X-ray wavelength, D is average particle size, ρ is the average dislocation density, b is the modulus of the Burgers vector of dislocations, $\overline{C}$ is the average contrast factor of dislocations, and M is a constant depending on the effective outer cut-off radius of dislocation [39–41].

We have calculated the dislocation density from the above formula and the results are shown in Figure 12. It can be seen that the calculated dislocation density of 4p-ECAP alloy is a little higher than that of 8p-ECAP CP-Ti, which corresponds to TEM results. However, the dislocation density of ECAP CP-Ti is relatively low as the ECAP processing was conducted at a high temperature in this work. After cold rolling, the dislocation density increases remarkably regardless of the initial state before CR. Although 8p-ECAP + CR CP-Ti exhibits a little higher dislocation density than 4p-ECAP + CR CP-Ti, they are in the same magnitude order when considering the error limits. In addition, since the average grain size in 8p-ECAP + CR CP-Ti is smaller than 4p-ECAP + CR, and most dislocations are inclined to aggregate at grain boundaries, the dislocation density within α-Ti grains must be lower for 8p-ECAP + CR Ti than 4p-ECAP + CR Ti. The above obtained results are in good agreement with TEM observations in Figures 4 and 6, which further demonstrates that high-density dislocations contribute to strengthening of this UFG CP-Ti.

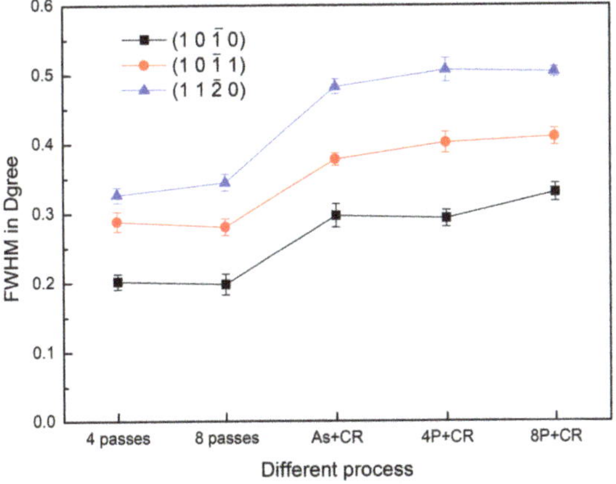

Figure 11. FWHM (full width at half maximum) for three characteristic peak of CP-Ti at different processing stage.

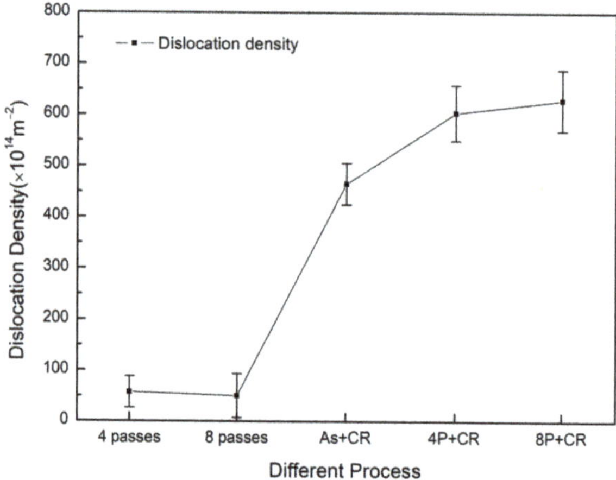

Figure 12. Evolutions of dislocation density in CP-Ti with different processing stages.

As for the good ductility of this UFG CP-Ti, it should be attributed to the increased fraction of high angle grain boundaries and the improved homogeneity of microstructures, which could increase the activation of grain boundary sliding and delay the propagation of cracks when they were nucleated during tensile [15,42].

Recently, it has been reported that $\alpha \to \omega$ and $\beta \to \omega$ phase transformations occurred for Ti-Fe alloys under high pressure deformation, and the mechanical properties might be improved after the high pressure phase transition [43,44]. Seen from Figure 9, two points are outside the strength-plasticity banana curve. One was prepared via combination of ECAP and HPT [11] and the other one is our work prepared via combined ECAP + CR. Both of these two studies are conducted under high pressure. However, during the XRD and TEM observations in our work and in reference [45], no phase transformation is confirmed, which might result from the relative lower Fe content in this CP-Ti, or from the undetec metastable phase. In our further study, we will focus on phase transformations of CP-Ti or Ti-Fe alloys during the combined ECAP + CR processing.

Based on the above considerations, it can be concluded that the improved strength of UFG CP-Ti is attributed to the refined grain size and high density of dislocations, while the good ductility might result from the refined and more homogeneous microstructure after multi-pass RD-ECAP and CR.

4. Conclusions

In this work, the microstructure evolution and mechanical properties of CP-Ti (Grade 2) during multi-pass RD-ECAP and subsequent CR were systematically investigated, and a high-strength and good-toughness UFG CP-Ti was successfully fabricated. The main conclusions are summarized as follows:

(1)　The grain size of CP-Ti is effectively refined after the combination of RD-ECAP and CR. It decreases from 80 μm of as-received state to around 500 nm and 310 nm after 4 passes and 8 passes RD-ECAP, and further decreases to about 120 nm and 90 nm after subsequent CR processing.

(2)　XRD analysis shows that no phase transformation happens during ECAP + CR processing, and α-Ti is the only phase in these SPD CP-Ti. Moreover, calculated results by Williamson-Hall analysis of XRD and TEM observations demonstrate that the dislocation density increases remarkably after combination of ECAP and CR.

(3)　Room temperature tensile tests showed that CP-Ti after ECAP + CR exhibited the best combination of strength and ductility, with ultimate tensile strength and fracture strain reached 920 MPa and

20%, respectively. The high strength of this UFG CP-Ti originated mainly from refined grains and high density of dislocations, while the good ductility could be attributed to the increased activation of grain boundary sliding and improved homogeneity of UFG microstructure.

Acknowledgments: This work was supported by the National Natural Science Foundation of China (Grant No. 51774109), the Natural Science Foundation of Jiangsu Province of China (Grant Nos. BK20160869 and BK20160867), the Key Research and Development Project of Jiangsu Province of China (Grant No. BE2017148), the Fundamental Research Funds for the Central Universities (Grant No. HHU2016B10314), and the Postgraduate Research & Practice Innovation Program of Jiangsu Province (Grant Nos. 2017B779X14 and 2017B780X14).

Author Contributions: Haoran Wu, Jinghua Jiang and Aibin Ma conceived and designed the experiments; Haoran Wu, Ren Tang and Yanxia Gu performed the experiments; Haoran Wu, Xincan Zhao and Jiapeng Sun analyzed the data; Haoran Wu and Huan Liu wrote the paper. All authors have discussed the results, read and approved the final manuscript.

Conflicts of Interest: The authors declare no conflict of interest.

References

1. Leyens, C.; Peters, M. Titanium and titanium alloys: Fundamentals and applications. In *Titanium and Titanium Alloys*; John Wiley & Sons: Hoboken, NJ, USA, 2003; Volume 401, pp. 23–67.
2. Myers, J.R.; Bomberger, H.B.; Froes, F.H. Corrosion behavior and use of titanium and its alloys. *JOM* **1984**, *36*, 50–60. [CrossRef]
3. Sanosh, K.P.; Balakrishnan, A.; Francis, L.; Kim, T.N. Vickers and knoop micro-hardness behavior of coarse- and ultrafine-grained titanium. *J. Mater. Sci. Technol.* **2010**, *26*, 904–907. [CrossRef]
4. Valiev, R.Z.; Langdon, T.G. Principles of equal-channel angular pressing as a processing tool for grain refinement. *Prog. Mater. Sci.* **2006**, *51*, 881–981. [CrossRef]
5. Wang, C.T.; Gao, N.; Gee, M.G.; Wood, R.J.K.; Langdon, T.G. Effect of grain size on the micro-tribological behavior of pure titanium processed by high-pressure torsion. *Wear* **2012**, *280–281*, 28–35. [CrossRef]
6. Wang, X.J.; Xu, D.K.; Wu, R.Z.; Chen, X.B.; Peng, Q.M.; Jin, L.; Xin, Y.C.; Zhang, Z.Q.; Liu, Y.; Chen, X.H.; et al. What is going on in magnesium alloys? *J. Mater. Sci. Technol.* **2017**, in press. [CrossRef]
7. Liu, H.; Chen, Z.J.; Yan, K.; Yan, J.L.; Bai, J.; Jiang, J.H.; Ma, A.B. Effect of multi-pass equal channel angular pressing on the microstructure and mechanical properties of a heterogeneous $Mg_{88}Y_8Zn_4$ alloy. *J. Mater. Sci. Technol.* **2016**, *32*, 1274–1281. [CrossRef]
8. Kim, H.S.; Yoo, S.J.; Jin, W.A.; Kim, D.H.; Kim, W.J. Ultrafine grained titanium sheets with high strength and high corrosion resistance. *Mater. Sci. Eng. A* **2011**, *528*, 8479–8485. [CrossRef]
9. Valiev, R.Z.; Korznikov, A.V.; Mulyukov, R.R. Structure and properties of ultrafine-grained materials produced by severe plastic deformation. *Mater. Sci. Eng. A* **1993**, *168*, 141–148. [CrossRef]
10. Valiev, R.Z.; Estrin, Y.; Horita, Z.; Langdon, T.G.; Zechetbauer, M.J.; Zhu, Y.T. Producing bulk ultrafine-grain materials by severe plastic deformation. *JOM* **2006**, *58*, 33–39. [CrossRef]
11. Stolyarov, V.V.; Zhu, Y.T.; Lowe, T.C.; Lslamgaliev, R.K.; Valiev, R.Z. A two step SPD processing of ultrafine-grained titanium. *Nanostruct. Mater.* **1999**, *11*, 947–954. [CrossRef]
12. Liu, H.; Ju, J.; Lu, F.M.; Yan, J.L.; Bai, J.; Jiang, J.H.; Ma, A.B. Dynamic precipitation behavior and mechanical property of an $Mg_{94}Y_4Zn_2$ alloy prepared by multi-pass successive equal channel angular pressing. *Mater. Sci. Eng. A* **2017**, *682*, 255–259. [CrossRef]
13. Terada, D.; Inoue, S.; Tsuji, N. Microstructure and mechanical properties of commercial purity titanium severely deformed by ARB process. *J. Mater. Sci.* **2007**, *42*, 1673–1681. [CrossRef]
14. Stolyarov, V.V.; Zhu, Y.T.; Alexandrov, I.V.; Lowe, T.C.; Valiev, R.Z. Grain refinement and properties of pure Ti processed by warm ECAP and cold rolling. *Mater. Sci. Eng. A* **2003**, *343*, 43–50. [CrossRef]
15. He, X.M.; Zhu, S.S.; Zhang, C.H. Analysis of the grain refinement mechanism for commercial pure titanium by ECAP and SMAT. *Adv. Mater. Res.* **2014**, *937*, 162–167. [CrossRef]
16. Roodposhti, P.S.; Farahbakhsh, N.; Sarkar, A.; Murty, K.L. Microstructural approach to equal channel angular processing of commercially pure titanium—A review. *Trans. Nonferr. Met. Soc. China* **2015**, *25*, 1353–1366. [CrossRef]
17. Zhao, X.; Fu, W.; Yang, X.; Langdon, T.G. Microstructure and properties of pure titanium processed by equal-channel angular pressing at room temperature. *Scr. Mater.* **2008**, *59*, 542–545. [CrossRef]

18. Meredith, C.S.; Khan, A.S. The microstructural evolution and thermo-mechanical behavior of UFG Ti processed via equal channel angular pressing. *J. Mater. Process. Technol.* **2015**, *219*, 257–270. [CrossRef]

19. Zhao, X.; Yang, X.; Liu, X.; Wang, X.; Langdon, T.G. The processing of pure titanium through multiple passes of ECAP at room temperature. *Mater. Sci. Eng. A* **2010**, *527*, 6335–6339. [CrossRef]

20. Kim, I.; Kim, J.; Shin, D.H.; Lee, C.S.; Hwang, S.K. Effects of equal channel angular pressing temperature on deformation structures of pure ti. *Mater. Sci. Eng. A* **2003**, *342*, 302–310. [CrossRef]

21. Nishida, Y.; Arima, H.; Kim, J.C.; Ando, T. Rotary-die equal-channel angular pressing of an Al-7 mass% Si-0.35 mass% Mg alloy. *Scr. Mater.* **2001**, *45*, 261–266. [CrossRef]

22. Liu, H.; Ju, J.; Bai, J.; Sun, J.P.; Song, D.; Yan, J.L.; Jiang, J.H.; Ma, A.B. Preparation, microstructure evolutions, and mechanical property of an ultra-fine grained Mg-10Gd-4Y-1.5Zn-0.5Zr alloy. *Metals* **2017**, *7*, 398. [CrossRef]

23. Gu, Y.X.; Ma, A.B.; Jiang, J.H.; Yuan, Y.C.; Li, H.Y. Deformation structure and mechanical properties of pure titanium produced by rotary-die equal-channel angular pressing. *Metals* **2017**, *7*, 297. [CrossRef]

24. Luo, P.; Mcdonald, D.T.; Zhu, S.M.; Palanisamy, S.; Dargusch, M.S.; Xia, K. Analysis of microstructure and strengthening in pure titanium recycled from machining chips by equal channel angular pressing using electron backscatter diffraction. *Mater. Sci. Eng. A* **2012**, *538*, 252–258. [CrossRef]

25. Yang, X.; Zhao, X.; Fu, W. Deformed microstructures and mechanical properties of CP-Ti processed by multi-pass ECAP at room temperature. *Rare Metall. Mater. Eng.* **2009**, *38*, 955–957.

26. Kim, I.; Kim, J.; Shin, D.H.; Park, K.T. Effects of grain size and pressing speed on the deformation mode of commercially pure Ti during equal channel angular pressing. *Metall. Mater. Trans. A* **2003**, *34*, 1555–1558. [CrossRef]

27. Stolyarov, V.V.; Zhu, Y.T.; Lowe, T.C.; Valiev, R.Z. Microstructure and properties of pure Ti processed by ECAP and cold extrusion. *Mater. Sci. Eng. A* **2001**, *303*, 82–89. [CrossRef]

28. Lu, K. Stabilizing nanostructures in metals using grain and twin boundary architectures. *Nat. Rev. Mater.* **2016**, *1*, 16019. [CrossRef]

29. Moradgholi, J.; Monshi, A.; Farmanesh, K. An investigation in to the mechanical properties of CP Ti/TiO$_2$ nanocomposite manufactured by the accumulative roll bonding (ARB) process. *Ceram. Int.* **2017**, *43*, 201–207. [CrossRef]

30. Nie, D.; Lu, Z.; Zhang, K. Grain size effect of commercial pure titanium foils on mechanical properties, fracture behaviors and constitutive models. *J. Mater. Eng. Perform.* **2017**, *26*, 1283–1292.

31. Zhang, Y.; Figueiredo, R.B.; Alhajeri, S.N.; Wang, J.T.; Gao, N.; Langdon, T.G. Structure and mechanical properties of commercial purity titanium processed by ECAP at room temperature. *Mater. Sci. Eng. A* **2011**, *528*, 7708–7714. [CrossRef]

32. Fattah-Alhosseini, A.; Keshavarz, M.K.; Mazaheri, Y.; Ansari, A.R.; Karimi, M. Strengthening mechanisms of nano-grained commercial pure titanium processed by accumulative roll bonding. *Mater. Sci. Eng. A* **2017**, *693*, 164–169. [CrossRef]

33. Purcek, G.; Yapici, G.G.; Karaman, I.; Maier, H.J. Effect of commercial purity levels on the mechanical properties of ultrafine-grained titanium. *Mater. Sci. Eng. A* **2011**, *528*, 2303–2308. [CrossRef]

34. Zhao, Z.; Chen, J.; Guo, S.; Tan, H.; Lin, X.; Huang, W.D. Influence of α/β interface phase on the tensile properties of laser cladding deposited Ti-6Al-4V titanium alloy. *J. Mater. Sci. Technol.* **2017**, *33*, 675–681. [CrossRef]

35. Luo, P.; Mcdonald, D.T.; Xu, W.; Palanisamy, S.; Dargusch, M.S.; Xia, K. A modified Hall-Petch relationship in ultrafine-grained titanium recycled from chips by equal channel angular pressing. *Scr. Mater.* **2012**, *66*, 785–788. [CrossRef]

36. Tang, Y.; Bringa, E.M.; Meyers, M.A. Inverse Hall-Petch relationship in nanocrystalline tantalum. *Mater. Sci. Eng. A* **2013**, *580*, 414–426. [CrossRef]

37. Woo, W.; Balogh, L.; Ungár, T.; Choo, H.; Feng, Z. Grain structure and dislocation density measurements in a friction-stir welded aluminum alloy using X-ray peak profile analysis. *Mater. Sci. Eng. A* **2008**, *498*, 308–313. [CrossRef]

38. Ungar, T.; Borbely, A. The effect of dislocation contrast on X-ray line broadening: A new approach to line profile analysis. *Appl. Phys. Lett.* **1996**, *69*, 3173–3175. [CrossRef]

39. Williamson, G.K.; Hall, W.H. X-ray line broadening from field aluminium and wolfram. *Acta Metall.* **1953**, *1*, 22–31. [CrossRef]

40. Williamson, G.K.; Smallman, R.E., III. Dislocation densities in some annealed and cold worked metals from measurement on the X-ray debye-scherrer spectrum. *Philos. Mag.* **1956**, *1*, 34–46. [CrossRef]

41. Schafler, E.; Zehetbauer, M.; Ungàr, T. Measurement of screw and edge dislocation density by means of X-ray Bragg profile analysis. *Mater. Sci. Eng. A* **2001**, *319*, 220–223. [CrossRef]

42. Gunderov, D.V.; Polyakov, A.V.; Semenova, I.P.; Raab, G.I.; Churakova, A.A.; Gimaltdinova, E.I.; Sabirov, I.; Segurado, J.; Sitdikov, V.D.; Alexandrov, I.V.; et al. Evolution of microstructure, macrotexture and mechanical properties of commercially pure Ti during ECAP-conform processing and drawing. *Mater. Sci. Eng. A* **2013**, *562*, 128–136. [CrossRef]

43. Kilmametov, A.R.; Ivanisenko, Y.; Mazilkin, A.A.; Straumal, B.B.; Gornakova, A.S.; Fabrichnaya, O.B.; Kriegel, M.J.; Rafaja, D.; Hahn, H. The $\alpha \rightarrow \omega$ and $\beta \rightarrow \omega$ phase transformations in Ti-Fe alloys under high-pressure torsion. *Acta Mater.* **2018**, *144*, 337–351. [CrossRef]

44. Kilmametov, A.; Ivanisenko, Y.; Straumal, B.; Mazilkin, A.A.; Gornakova, A.S.; Kriegel, M.J.; Fabrichnaya, O.B.; Rafaja, D.; Hahn, H. Transformations of α' martensite in Ti-Fe alloys under high pressure torsion. *Scr. Mater.* **2017**, *136*, 46–49. [CrossRef]

45. Kim, H.S.; Kim, W.J. Annealing effects on the corrosion resistance of ultrafine-grained pure titanium. *Corros. Sci.* **2014**, *89*, 331–337. [CrossRef]

Article

Characteristics of Resistance Spot Welded Ti6Al4V Titanium Alloy Sheets

Xinge Zhang [1], Jiangshuai Zhang [1], Fei Chen [1], Zhaojun Yang [1] and Jialong He [1,2,*]

[1] School of Mechanical Science and Engineering, Jilin University, Changchun 130025, China; zhangxinge@jlu.edu.cn (X.Z.); 18343126188@163.com (J.Z.); chenfeicn@jlu.edu.cn (F.C.); yangzj@jlu.edu.cn (Z.Y.)

[2] College of Computer Science and Technology, Jilin University, Changchun 130012, China

* Correspondence: hjl.star@163.com; Tel.: +86-0431-8509-5839

Received: 4 September 2017; Accepted: 3 October 2017; Published: 12 October 2017

Abstract: Ti6Al4V titanium alloy is applied extensively in the aviation, aerospace, jet engine, and marine industries owing to its strength-to-weight ratio, excellent high-temperature properties and corrosion resistance. In order to extend the application range, investigations on welding characteristics of Ti6Al4V alloy using more welding methods are required. In the present study, Ti6Al4V alloy sheets were joined using resistance spot welding, and the weld nugget formation, mechanical properties (including tensile strength and hardness), and microstructure features of the resistance spot-welded joints were analyzed and evaluated. The visible indentations on the weld nugget surfaces caused by the electrode force and the surface expulsion were severe due to the high welding current. The weld nugget width at the sheets' faying surface was mainly affected by the welding current and welding time, and the welded joint height at weld nugget center was chiefly associated with electrode force. The maximum tensile load of welded joint was up to 14.3 kN in the pullout failure mode. The hardness of the weld nugget was the highest because of the coarse acicular α' structure, and the hardness of the heat-affected zone increased in comparison to the base metal due to the transformation of the β phase to some fine acicular α' phase.

Keywords: Ti6Al4V titanium alloy; resistance spot welding; mechanical properties; microstructure

---

1. Introduction

Ti6Al4V titanium alloy is widely used in the aerospace, marine, pressure vessel, and chemical industries, as well as for surgical implant, etc., owing to its unique properties, such as a high strength-to-weight ratio, excellent high-temperature properties and corrosion resistance; furthermore, the usage of Ti6Al4V alloy accounts for over 50% of the total titanium alloy tonnage, globally [1–5]. Many welding methods, such as tungsten inert gas welding (TIG) [6,7], laser welding [8], electron beam welding [9], plasma arc welding [10], line friction welding [11–13], and friction stir welding [14,15] have been employed to weld titanium alloy in previous investigations.

Based on the existing literature, the TIG welding process is always used to weld butt joints of Ti6Al4V titanium alloy, but the weld structure is coarse, inducing high welding residual stress and distortion. The laser welding and electron beam welding processes are considered to be better replacements for the TIG welding process due to their advantages of high welding speed and slight welding distortion [6,7]. But the laser welding and electron beam welding processes are more suitable to weld butt joints than lap joints due to the high depth-to-width ratio penetration ability. Even more importantly, the equipment investment and operating cost are high [6–9]. The plasma arc welding process is a high-energy beam welding technology, and is utilized to weld the butt joint of thicker titanium alloy plate, which is similar to the laser welding process [10]. Line friction welding has been used to join titanium alloy bulk plates or parts because of its own process principles and features [11–13]. The brazing process is easy to implement for joining titanium alloy, but this process must be carried out under vacuum conditions, and a long welding time is normally required [14,15]. In the last three decades, friction stir welding has emerged as a new welding technology, and was adapted for welding thin sheets, not only for butt joints but also lap joints. However, owing to the high melting point of titanium alloy and the high contact forces between metal sheets and the friction tool, the choice of friction tool material is difficult and limited [16–18].

Resistance spot welding (RSW) is widely used for lap joint assembly of two metal sheets, especially in the automotive industry, mainly owing to its high production efficiency, low operation cost and high degree of automation [19]. During the resistance spot welding process, the metal sheets are assembled as lap joints and the lap joint is compressed by two water-cooled electrodes. Then, the welding current is applied to metal sheets through the electrodes provided by the welding power. Due to the large contact resistance at the faying surface of the metal sheets, the Joule heat generates mainly at the faying surface, and then the metal sheets melt in a short time [20]. Finally, when the welding current supply is turned off, the melted metal solidifies, resulting in the joining of the metal sheets. Resistance spot welding has been employed for welding steel, aluminum alloys, magnesium alloys, nickel alloys, copper alloys, pure titanium, and dissimilar metals [21–25]. However, to the best of the authors' knowledge, there are few reports on resistance spot welding of Ti6Al4V titanium alloy sheets in the previous literature. In the present study, assembling the Ti6Al4V titanium alloy sheets into lap joints was performed by resistance spot welding, and then the welding characteristics, including weld nugget formation and the effects of the welding parameters on it, were investigated. Meanwhile, the mechanical properties, including tensile strength, hardness, and microstructure features, of the typical spot-welded joint were evaluated and analyzed in detail.

2. Materials and Methods

In this study, the base metal is Ti6Al4V titanium alloy sheet and the sheet thickness is 1 mm. The chemical composition of the base metal is listed in Table 1. The schematic diagram of the resistance spot welding process is illustrated in Figure 1a, and the overlap joint configuration and dimensions of the specimen are shown in Figure 1b. Before welding, the specimens were cleaned to remove surface contamination and oxides. The resistance spot welding of the overlap joints was carried out using a DN-100 resistance spot welding machine (Denyo Kunshan Inc., Kunshan, China), and the tip diameter of Cu-Cr alloy electrodes was 6 mm. The resistance spot welding parameters include the welding current, welding time, and electrode force, and the experimental conditions are shown in Table 2. The unit of welding time is the cycle (1 cycle = 0.02 s), and the welding current is continuous and constant in each cycle. A schematic diagram of the applied of welding current is shown in Figure 2.

Table 1. Chemical composition of Ti6Al4V titanium alloy (wt %).

Component	Al	V	Fe	Si	O	C	N	Ti
wt %	6.15	3.96	0.3	0.15	0.2	0.02	0.01	Balance

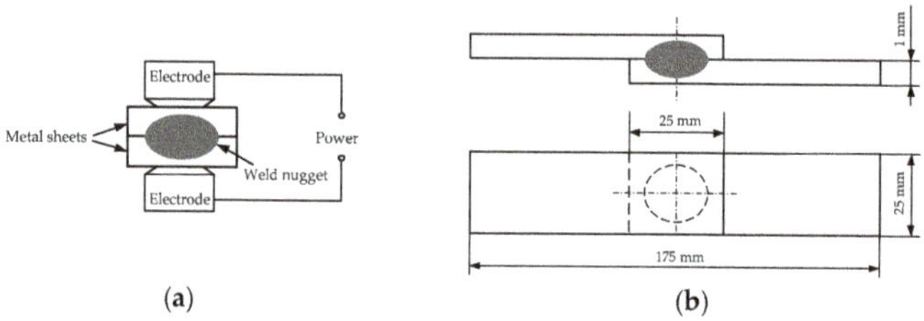

Figure 1. (**a**) Schematic diagram of resistance spot welding process; and (**b**) configuration and dimensions of specimen.

Table 2. Experimental conditions.

Condition	Welding Current (kA)	Welding Time (Cycle [1])	Electrode Force (kN)
Set 1	7, 8, 9, 10, 11	12	3
Set 2	9	4, 8, 12, 16, 20	3
Set 3	9	12	2, 3, 4, 5

[1] 1 cycle = 0.02 s.

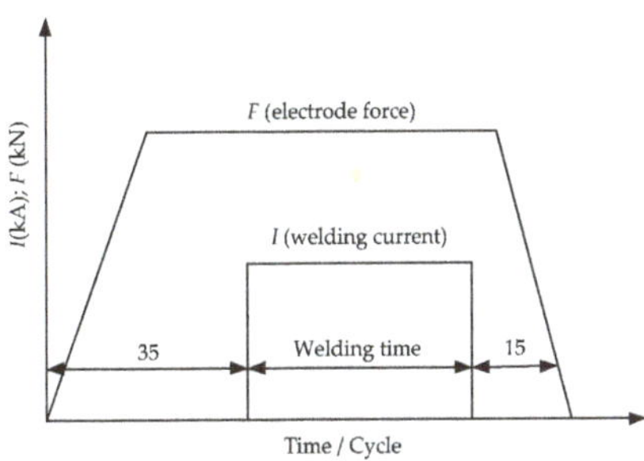

Figure 2. Schematic diagram of the applied of welding current.

After welding, the welded joints were sectioned along the spot weld nugget center, followed by mounting, polishing, and etching for microstructure observation. The etch solution consisted of 1 vol % HCl, 1.5 vol % HF, 2.5 vol % HNO_3, and 95 vol % H_2O. The microstructure was examined using an Eclipse E200 optical microscope (Nikon Instruments (Shanghai) Co., Ltd., Shanghai, China) and S-4700 scanning electron microscopy (SEM; Hitachi High-Technologies in China, Suzhou, China). The hardness measurement was performed on the cross-sections of spot welded joints using a HV-30 hardness machine (Shanghai Tuming Instrument Co., Ltd., Shanghai, China) with a 5 kgf test load. The tensile shear test was carried out by means of Instron 5500R material tensile machine (Instron, Division of Illinois Tool Works Inc., Norwood, MA, USA). The fracture surface morphology of spot welded joints was observed using SEM.

3. Results

3.1. Weld Nugget Formation

The typical appearances of resistance spot welded Ti6Al4V titanium alloy joints are shown in Figure 3. Visible indentations could be seen on the weld nugget surfaces for all welded joints, which is a local plastic deformation on the sheets' surface owing to the high temperature and applied electrode force at the faying surface between electrode and sheet. During the resistance spot welding process, the expulsion on the weld nugget surface was an unfavorable phenomenon, and is mainly associated with the welding current. Although there was no expulsion on the weld nugget surface when the welding current was no more than 9 kA, a slight expulsion appeared along the edge of the weld nugget while the welding current was increased to 10 kA, as shown in Figure 3b. As seen in Figure 3c, when the welding current was increased to 11 kA, the expulsion became very severe. It is well known that severe expulsion may decrease the surface quality of the spot weld and cause adhesion between the electrode and specimen surfaces. Therefore, a low-level welding current should be chosen to suppress the expulsion on the weld nugget surface, thus improving the spot weld quality and electrode life.

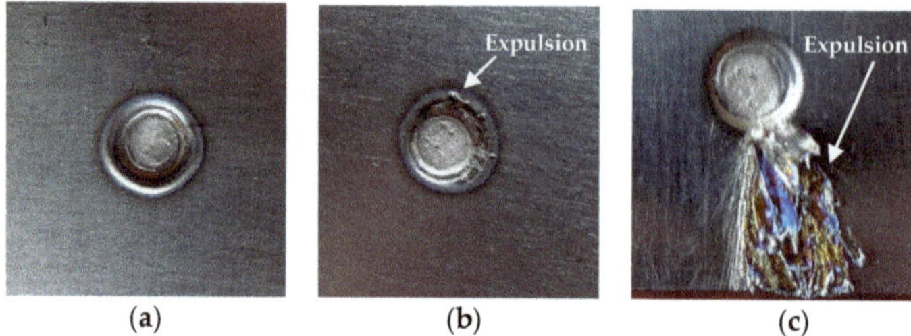

(a)   (b)   (c)

Figure 3. Appearances of resistance spot welded joints with different welding currents: (a) 9 kA; (b) 10 kA; and (c) 12 kA.

Figure 4 illustrates the typical cross-section of spot welded joint with 9 kA welding current, 12 cycles welding time, and 3 kN electrode force. The joint of the resistance spot welded Ti6Al4V titanium alloy presents characteristics such as a large weld nugget width at the faying surface and full penetration in the weld nugget center. In this study, because the thickness of the Ti6Al4V titanium alloy sheet was only 1 mm, when the high-level welding heat input (high welding current or welding time) was used to obtain enough weld nugget width at the faying surface, this resulted in full melting from the faying surfaces outward to the electrodes. However, the two types of materials (metal sheet: Ti6Al4V alloy; electrode: Cu-Cr alloy) exhibited large differences in physical and chemical properties, which made the adhesion between them difficult, and thus no significant contamination of or damage to the electrodes occurred. A spot-welded joint consists of the weld nugget, the heat affected zone (HAZ), and the base metal. A distinct boundary exists between three different regions, and the HAZ is located between the weld nugget and the base metal. The indentations on the top surface and bottom surface of the welded joint are distinct, which causes the thickness of weld zone to decrease, especially in the weld center. In general, the spot weld morphology significantly affects the welded joint strength, and then two geometric dimensions are extracted as the evaluation index; namely, the weld nugget width (W) at the faying surface, and the welded joint height (H) in the spot weld center, as shown in Figure 4.

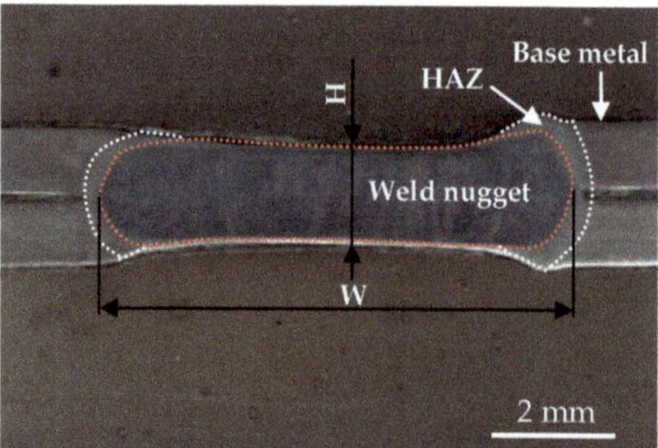

Figure 4. Typical cross-section of the spot welded joint.

Figure 5 shows the effect of the single welding parameter on the weld nugget width (W) at the faying surface and the welded joint height (H) in the spot weld center.

In resistance spot welding, the welding current flows from the electrode through the metal sheets' faying surface, with electrical resistance heat primarily being generated at the faying surface. Then, the metal sheets around the faying surface are heated and melted, which finally results in the formation of the weld nugget. The weld nugget morphology is mainly controlled by this heat generation. The heat generation can be determined as:

$$Q = I^2 R t \tag{1}$$

where Q is the heat generation in joules, I is the welding current in amperes, R is the contact resistance in ohms, and t is the welding time in seconds [19]. As seen in Figure 5a, according to the Equation (1), when the welding current was increased from 7 kA to 10 kA, the heat generation increased correspondingly, which caused the weld nugget width (W) to increase from 6.3 mm to 7.9 mm; but, when the welding current was increased to 11 kA, the weld nugget width (W) tended to decrease, owing to the severe expulsion shown in Figure 3c. Since the higher welding current results in larger heat generation and a longer time above the melting temperature of the sheet material, the indentation is deeper. On the other hand, a higher welding current leads to severe expulsion. Thus, the welded joint height (H) decreased with the increase of the welding current, as shown in Figure 5a.

As seen in Figure 5b, when the welding time was increased from 4 cycles to 12 cycles, the weld nugget width (W) evidently increased due to higher heat generation, in accordance with Equation (1); however, if the welding time continues to increase, the weld nugget width (W) has no obvious change, owing to the basic balance between the heat input rate and heat output rate. The longer the welding time was, the longer the electrode force applied, which resulted in deeper indentation, following which the welded joint height (H) slightly decreased with the increase of welding time.

As seen in Figure 5c, when the electrode force is increased from 2 kN to 5 kN, the weld nugget width (W) and welded joint height (H) all increase at first, and then obviously reduce. When the electrode force was 2 kN, the two metal sheets were not compressed tightly by the electrode force, leading to a high current density passing through the weld nugget zone, and then resulting in severe expulsion at the faying surface between the two metal sheets, as shown in Figure 6. The severe expulsion at the faying surface was of no benefit to the growth of the weld nugget. While the electrode force increased to 3 kN, there was no evident expulsion; therefore, the weld nugget width (W) and welded joint height (H) had all increased, compared with of the values for 2 kN electrode force. If the electrode force continues to increase, two metal sheets are compressed tightly, and the current density passing through the weld nugget zone decreases, causing the weld nugget width (W) to sharply reduce. Meanwhile, the larger the electrode force, the deeper the indentation, as the welded joint height (H) decreased, as seen in Figure 5c.

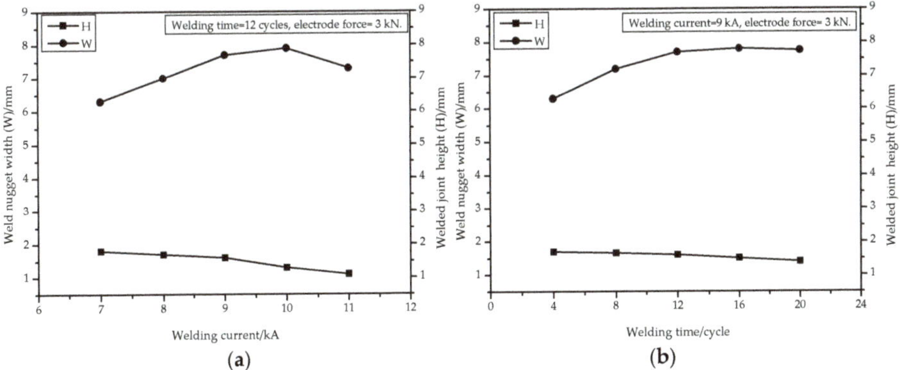

(a)   (b)

Figure 5. *Cont.*

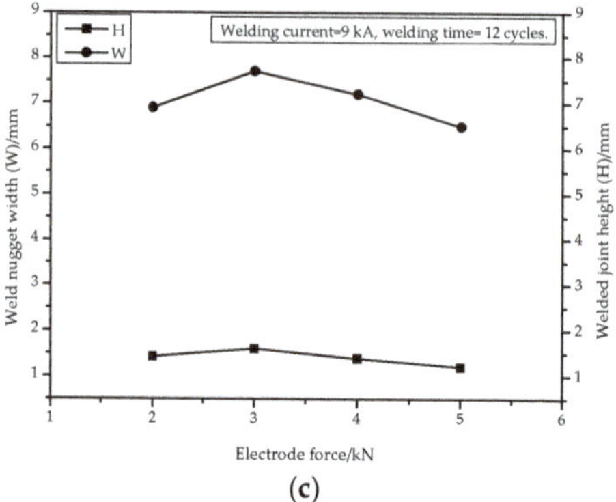

(**c**)

Figure 5. Effects of welding parameters on the spot weld morphology (W, H): (**a**) Welding current; (**b**) welding time; and (**c**) electrode force.

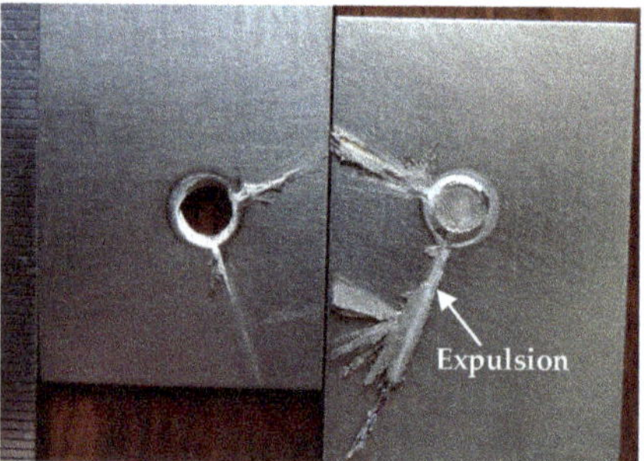

Figure 6. Expulsion at the faying surface between two metal sheets.

3.2. Tensile Strength

The tensile shear test results demonstrated that the welding parameters had a direct effect on the tensile strength of tensile shear test specimens. The relationships between the weld parameters and tensile shear load are given in Figure 7. In this investigation, the maximum tensile load of the specimen was up to 14.3 kN with the favorable welding parameters: 9 kA welding current, 12 cycles welding time, and 3 kN electrode force.

The tensile loads of specimens increase fast from 7 kA up to 9 kA welding current, where the maximum tensile load occurs, as shown in Figure 7a. The higher welding current caused severe expulsion and deep indentation, so the tensile load of the specimen decreased. As shown in Figure 7b, with the increase in welding time, the tensile loads of specimens increased at first, and then obviously tended to a constant level, which is consistent with the influence of welding time on weld nugget width (W). As seen in Figure 7c, when the electrode force is increased from 2 kN to 3 kN, the tensile loads of the specimens increase, owing to the increase of the weld nugget width (W). However, when the electrode force continues to increase, the tensile loads of the specimens sharply decrease due to the deeper indentation, resulting in the lower welded joint height (H).

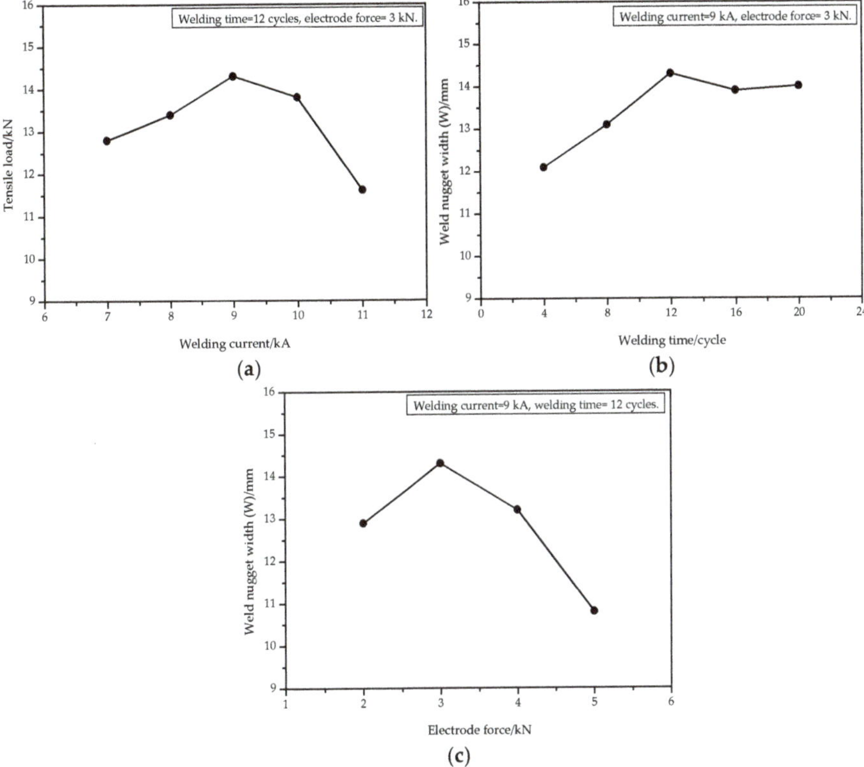

Figure 7. Relations between welding parameters and tensile shear load: (**a**) Welding current; (**b**) welding time; and (**c**) electrode force.

During the tensile shear test, the partial interfacial failure (PIF) mode and pullout failure (PF) mode were observed, as shown in Figures 8 and 9. For the partial interfacial failure mode, the tensile testing crack initiated at the two metal sheets faying surface owing to the stress concentration resulting from the notch at the periphery of the weld nugget shown in Figure 10a, and then propagated along the weld nugget circumference due to the electrode indentation resulting in the lower welded joint height (H). The fracture surface morphology of the tear weld nugget was observed at higher magnification with SEM. The upper part of the fracture surface displayed a brittle fracture feature, as shown in Figure 8b, and the lower part of fracture surface displayed a ductile fracture feature, as shown in Figure 8c. For the pullout failure mode, the tensile testing crack also originated at the two metal sheets' faying surface, but the crack propagated along the vertical thickness direction, and then the tensile testing specimen broke in the normal fracture mode, as shown in Figure 9a. The fracture surface shows dimple features, as shown in Figure 9b. In this investigation, the transition from the PIF mode to the PF mode was determined by the electrode indentation owing to enough weld nugget width (W) for all the welded joints, and if the welded joint height (H) was used to replace the electrode indentation as an evaluation index, the critical value of the welded joint height (H) was 1.55 mm.

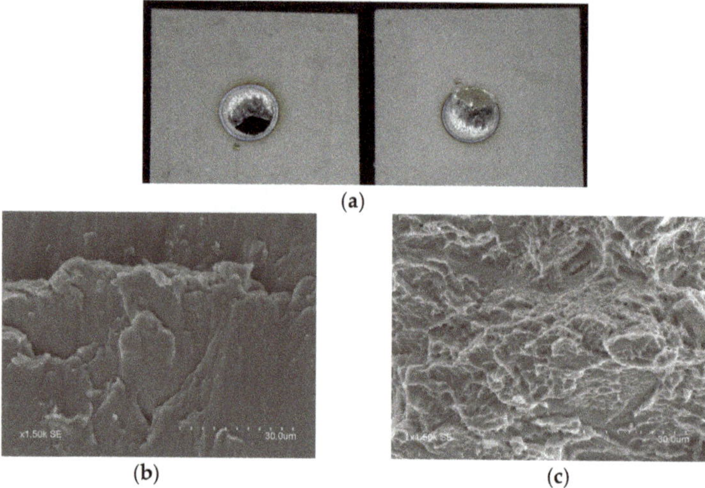

Figure 8. Partial interfacial failure mode: (**a**) Fracture appearance; (**b**) SEM (scanning electron microscopy) observation of the upper part of fracture surface; (**c**) SEM observation of the lower part of fracture surface.

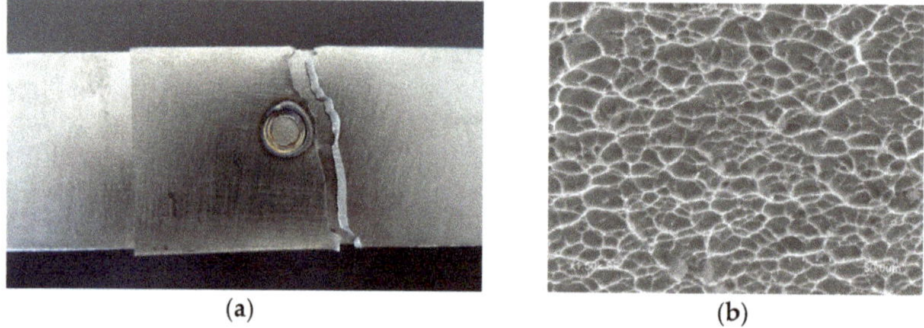

Figure 9. Pullout failure mode: (**a**) Fracture appearance; and (**b**) SEM observation of the fracture surface.

3.3. Microstructure

Figure 10 indicates the typical microstructure features of a resistance spot welded Ti6Al4V titanium alloy joint obtained using 9 kA welding current, 12 cycles welding time, and 3 kN electrode force. The welded joint consists of three main zones—weld nugget, HAZ and base metal—and there is usually a clear delineation between the three zones, owing to the different microstructure feature. As seen in Figure 10a, there is an obvious notch in the edge of the weld nugget at the two metal sheets' faying surface resulting from the insufficient heat input in this region, which is always the initiation of a crack under load conditions. The microstructure of the base metal consists of a small percentage of β phase distributed at the elongated α phase grain boundary, which is the typical anneal structure for the α + β titanium alloy, as shown in Figure 10b. The microstructure in the HAZ is a mixture of primary α, primary β and transformation of β phase to some fine acicular α′ phase, as shown in Figure 10c. It could be inferred that the HAZ maximum temperature excursion was below the temperature of the β phase transit at 980–1000 °C and the cool rate was relatively rapid in this region due to its being close to the base metal. The grain structure of the weld nugget was coarsened compared with that of the base metal, owing to the large heat input and the slower cool rate in the weld nugget. In the weld nugget zone, the β phase structure entirely transformed to a coarse acicular martensite α′ phase, as shown in Figure 10d.

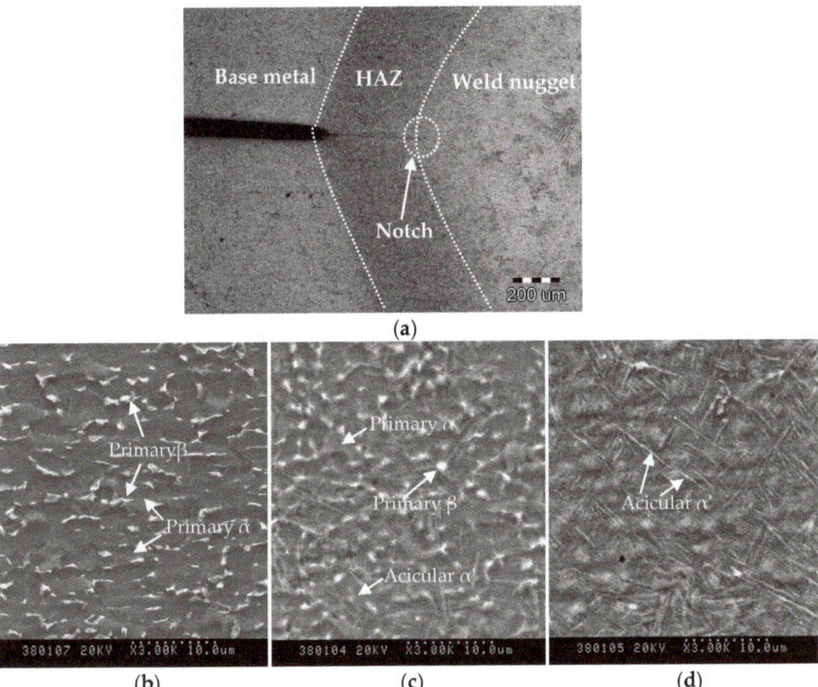

Figure 10. Microstructure of: (**a**) Three zones; (**b**) base metal; (**c**) HAZ (heat affected zone); and (**d**) weld nugget zone.

3.4. Hardness Distribution

Hardness measurements were carried out on the cross-sections of the typical resistance spot welded joint obtained using the optimal welding parameters using 9 kA welding current, 12 cycles welding time and 3 kN electrode force. The measurements were performed along two directions, as illustrated in Figure 11a,b. The measurement results are indicated in Figure 11c,d, respectively. The average hardness of the base metal is 300 HV. The hardness of HAZ increases to about 318 HV. The average hardness of the weld nugget is the highest in the welded joint because of the acicular martensite microstructure in this region, shown in Figure 10d, and the peak hardness of the weld nugget reaches about 358 HV. Along the vertical measurement location, the hardness varies slightly, except for two location points close to the top and bottom surfaces, owing to the nearly full penetration, as shown in Figure 11d. Therefore, during the tensile test, because the weld nugget width at the faying surface was wide enough and the indentation was slight, a tensile testing crack initiated at the two metal sheets' faying surface and propagated along the vertical thickness direction in the HAZ owing to the lower hardness compared with weld nugget zone.

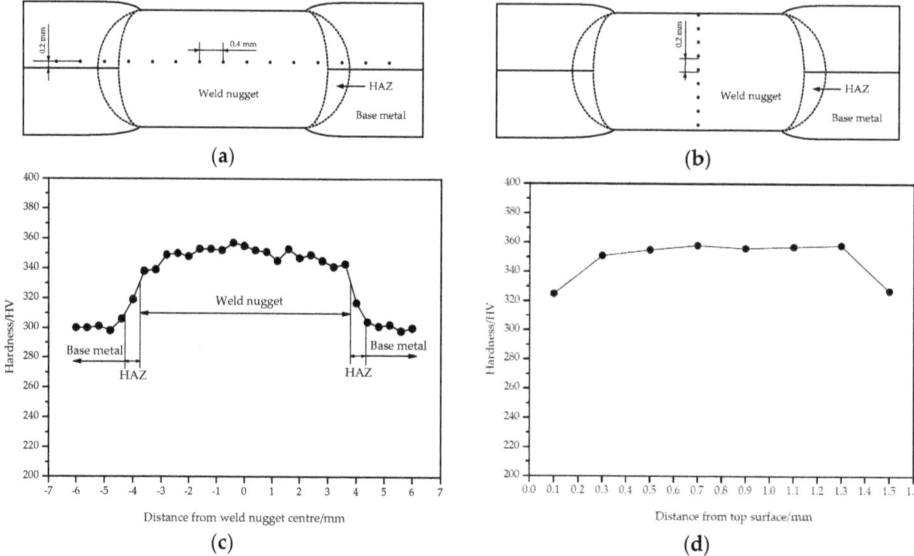

Figure 11. (**a**) Horizontal measurement location; (**b**) vertical measurement location; (**c**) horizontal hardness distribution; and (**d**) vertical hardness distribution.

4. Conclusions

Ti6Al4V titanium alloy sheets were well joined using resistance spot welding, and the welding characteristics were investigated in detail. The conclusions of this study are presented as follows:

(1) The visible indentations resulting from a local plastic deformation on the sheets' surface were caused by the electrode force at the faying surface between electrode and sheet. The expulsion on the weld nugget surface was mainly associated with the welding current. Although there was no expulsion on the weld nugget surface when the welding current was no more than 9 kA, with the welding current raised to 11 kA, the expulsion became very severe.

(2) With an increase in welding current, the weld nugget width (W) increased, and the welded joint height (H) at the weld nugget center decreased. When the welding time was increased from 4 cycles to 20 cycles, the weld nugget width (W) at first increased, and then changed less obviously. When the electrode force was increased from 2 kN to 5 kN, the weld nugget width (W) and welded joint height (H) all increased at first, and then evidently decreased.

(3) The maximum tensile load of the welded joint was up to 14.3 kN. Two types of failure mode were observed: partial interfacial failure and pullout failure. In this investigation, the transition from PIF mode to PF mode was determined by the electrode indentation owing to enough weld nugget width (W) for all welded joints. The critical value of the welded joint height (H) was 1.55 mm.

(4) The welded joint mainly consisted of three zones—weld nugget, HAZ and base metal—and there was an obvious notch in the edge of the weld nugget at the two metal sheets' faying surface resulting from the insufficient heat input in this region. In the weld nugget zone, it could be seen that the β phase structure entirely transformed to a coarse acicular martensite α' phase; therefore, the hardness of the weld nugget was the highest. The microstructure in HAZ was a mixture of primary α, primary β, and a transformation of the β phase to fine acicular α' phase, resulting in the increase of hardness in the HAZ compared with the base metal.

Acknowledgments: The work was financially supported by the Jilin Scientific and Technological Development Program (20160520055JH).

Author Contributions: Xinge Zhang designed the experiments, performed the experiments, analyzed the data and wrote the paper. Jiangshuai Zhang and Fei Chen performed the microstructure observation and hardness measurement experiments. Zhaojun Yang and Jialong He contributed to the discussion and interpretation of the experiment results.

Conflicts of Interest: The authors declare no conflict of interest.

References

1. Lütjering, G.; Williams, J.C. *Titanium*, 2nd ed.; Springer: Berlin/Heideberg, Germany, 2007; pp. 449–450.

2. Cao, X.; Jahazi, M. Effect of welding speed on butt joint quality of Ti6Al4V alloy welded using a high-power Nd:YAG laser. *Opt. Lasers Eng.* **2009**, *47*, 1231–1241. [CrossRef]

3. Köse, C.; Karaca, E. Robotic Nd:YAG Fiber laser welding of Ti-6Al-4V alloy. *Metals* **2017**, *7*, 221. [CrossRef]

4. Attanasio, A.; Gelfi, M.; Pola, A.; Ceretti, E.; Giardini, C. Influence of material microstructures in micromilling of Ti6Al4V alloy. *Materials* **2013**, *6*, 4268–4283. [CrossRef] [PubMed]

5. Akman, E.; Demir, A.; Canel, T.; Sınmazçelik, T. Laser welding of Ti6Al4V titanium alloys. *J. Mater. Process. Technol.* **2009**, *209*, 3705–3713. [CrossRef]

6. Gao, X.L.; Zhang, L.J.; Liu, J.; Zhang, J.X. Comparison of tensile damage evolution in Ti6A14V joints between laser beam welding and gas tungsten arc welding. *J. Mater. Eng. Perform.* **2014**, *23*, 4316–4327. [CrossRef]

7. Qi, Y.L.; Deng, J.; Hong, Q.; Zeng, L.Y. Electron beam welding, laser beam welding, gas tungsten arc welding of titanium sheet. *Mater. Sci. Eng. A* **2000**, *280*, 177–181.

8. Caiazzo, F.; Curcio, F.; Daurelio, G.; Memola Capece Minutolo, F. Ti6Al4V sheets lap and butt joints carried out by CO_2 laser: Mechanical and morphological characterization. *J. Mater. Process. Technol.* **2004**, *149*, 546–552. [CrossRef]

9. Prasad Rao, K.; Angamuthu, K.; Bala Srinivasan, P. Fracture toughness of electron beam welded Ti6Al4V. *J. Mater. Process. Technol.* **2008**, *199*, 185–192.

10. Chen, J.C.; Pan, C.X. Welding of Ti-6Al-4V alloy using dynamically controlled plasma arc welding process. *Trans. Nonferr. Met. Soc. China* **2011**, *21*, 1506–1512. [CrossRef]

11. Wen, G.D.; Ma, T.J.; Li, W.Y.; Li, J.L.; Guo, H.Z.; Chen, D.L. Cyclic deformation behavior of linear friction welded Ti6Al4V joints. *Mater. Sci. Eng. A* **2014**, *597*, 408–414. [CrossRef]

12. Buffa, G.; Campanella, D.; Cammalleri, M.; Ducato, A.; Astarita, A.; Squillace, A.; Esposito, S.; Fratini, L. Experimental and numerical analysis of microstructure evolution during linear friction welding of Ti6Al4V. *Procedia Manuf.* **2015**, *1*, 429–441. [CrossRef]

13. Maio, L.; Liberini, M.; Campanella, D.; Astarita, A.; Esposito, S.; Boccardi, S.; Meola, C. Infrared thermography for monitoring heat generation in a linear friction welding process of Ti6Al4V alloy. *Infrared Phys. Technol.* **2017**, *81*, 325–338. [CrossRef]

14. Jiang, C.Y.; Wu, M.F.; Yu, C.; Liang, C. Morphology and strength of TC4/TC4 joint with 72Ag-28Cu filler metal. *Rare Met. Mater. Eng.* **2003**, *32*, 295–297.

15. Wu, Z.; Mei, J.; Voice, W.; Beech, S.; Wu, X. Microstructure and properties of diffusion bonded Ti–6Al–4V parts using brazing-assisted hot isostatic pressing. *Mater. Sci. Eng. A* **2011**, *528*, 7388–7394. [CrossRef]

16. Zhang, Y.; Sato, Y.S.; Kokawa, H.; Seung Park, S.H.C.; Hirano, S. Microstructural characteristics and mechanical properties of Ti–6Al–4V friction stir welds. *Mater. Sci. Eng. A* **2008**, *485*, 448–455. [CrossRef]

17. Mironov, S.; Zhang, Y.; Sato, Y.S.; Kokawa, H. Development of grain structure in β-phase field during friction stir welding of Ti–6Al–4V alloy. *Scr. Mater.* **2008**, *59*, 27–30. [CrossRef]

18. Buffa, G.; Ducato, A.; Fratini, L. FEM based prediction of phase transformations during friction stir welding of Ti6Al4V titanium alloy. *Mater. Sci. Eng. A* **2013**, *581*, 56–65. [CrossRef]

19. Ertek Emre, H.; Kaçar, R. Resistance spot weldability of galvanize coated and uncoated TRIP steels. *Metals* **2016**, *6*, 299. [CrossRef]

20. Aslanlar, S.; Ogur, A.; Ozsarac, U.; Ilhan, E. Welding time effect on mechanical properties of automotive sheets in electrical resistance spot welding. *Mater. Des.* **2008**, *29*, 1427–1431. [CrossRef]

21. Kazdal Zeytin, H.; Ertek Emre, H.; Kaçar, R. Properties of resistance spot-welded TWIP steels. *Metals* **2017**, *7*, 14. [CrossRef]

22. Wu, S.N.; Ghaffari, B.; Hetrick, E.; Li, M.; Liu, Q.; Jia, Z.H. Thermo-mechanically affected zone in AA6111 resistance spot welds. *J. Mater. Process. Technol.* **2017**, *249*, 463–470. [CrossRef]

23. Kahraman, N. The influence of welding parameters on the joint strength of resistance spot-welded titanium sheets. *Mater. Des.* **2007**, *28*, 420–427. [CrossRef]
24. Shi, H.X.; Qiu, R.F.; Zhu, J.H.; Zhang, K.K.; Yu, H.; Ding, G.J. Effects of welding parameters on the characteristics of magnesium alloy joint welded by resistance spot welding with cover plates. *Mater. Des.* **2010**, *31*, 4853–4857. [CrossRef]
25. Chen, N.N.; Wang, H.P.; Carlson, B.E.; Sigler, D.R.; Wang, M. Fracture mechanisms of Al/steel resistance spot welds in lap shear test. *J. Mater. Process. Technol.* **2017**, *243*, 347–354. [CrossRef]

MDPI

St. Alban-Anlage 66

4052 Basel

Switzerland

Tel. +41 61 683 77 34

Fax +41 61 302 89 18

www.mdpi.com

Metals Editorial Office

E-mail: metals@mdpi.com

www.mdpi.com/journal/metals